Statics and Dynamics with Background Mathematics

This book uniquely covers both statics and dynamics, together with a section on background mathematics, providing the student with everything needed to complete typical first year undergraduate courses in these areas. Students often find statics and dynamics difficult subjects, since the skills needed to visualize problems and handle the mathematics can be tricky to master. Roberts' friendly approach makes life easier for both student and tutor, tackling concepts from first principles with many examples, exercises and helpful diagrams. The inclusion of a revision section on introductory mathematics is a huge bonus, allowing students to catch up on the prerequisite mathematics needed to work through both courses.

Dr Adrian Roberts was for 25 years a full Professor in the Faculty of Engineering at Queen's University, Belfast, including three years as Dean of the Faculty. He is now retired.

D1421761

Statics and Dynamics
with Background Mathematics

A. P. Roberts

CAMBRIDGE
UNIVERSITY PRESS

PUBLISHED BY THE PRESS SYNDICATE OF THE UNIVERSITY OF CAMBRIDGE
The Pitt Building, Trumpington Street, Cambridge, United Kingdom

CAMBRIDGE UNIVERSITY PRESS
The Edinburgh Building, Cambridge CB2 2RU, UK
40 West 20th Street, New York, NY 10011-4211, USA
477 Williamstown Road, Port Melbourne, VIC 3207, Australia
Ruiz de Alarcón 13, 28014 Madrid, Spain
Dock House, The Waterfront, Cape Town 8001, South Africa

http://www.cambridge.org

First published 2003

Printed in the United Kingdom at the University Press, Cambridge

Typefaces Times 10.5/14 pt and Helvetica   *System* LaTeX 2_ε [TB]

A catalogue record for this book is available from the British Library

Library of Congress Cataloguing in Publication data

Roberts, A. P.
Statics and dynamics with background mathematics / A. P. Roberts.
        p.   cm.
Includes bibliographical references and index.
ISBN 0 521 81766 8 – ISBN 0 521 52087 8 (pbk.)
1. Statics.   2. Dynamics.   I. Title.
QA821 .R64   2003
531′.12 – dc21   2002073607

ISBN 0 521 81766 8 hardback
ISBN 0 521 52087 8 paperback

To my wife Mary

Contents

Part II Dynamics   139

Preface

On their first encounter with statics and dynamics, students often have difficulty with three particular aspects of the subject. These are (1) visualization of the physical properties involved, (2) expression of the latter in mathematical terms and (3) manipulation of the mathematics in order to solve the various problems that arise. The aim of this book is to teach the basics of statics and dynamics in a way which will help them to overcome these difficulties. It will be a valuable supplement for A-level mathematics in schools and further education colleges and for first year university courses in mathematics and engineering.

The book starts with an in-depth discussion of the concept and description of forces. The concepts are usually taken for granted in textbooks on statics and dynamics, leaving students with rather hazy ideas regarding the physical nature of forces. The quantitative description leading onto conditions for equilibrium involves the use of vectors, coordinate geometry and trigonometry, all of which are dealt with in Part IV: Background Mathematics.

Part I of this book is entitled Statics, but much of the material in the chapters on forces, moments, centre of gravity and friction is required in Part II: Dynamics. Hydrostatics is considered briefly in the chapter on distributed forces. There are also chapters on trusses, beams, non-coplanar forces and couples, and virtual work.

Part II: Dynamics is basically concerned with the two-dimensional motion of rigid bodies. It is ordered into five chapters: kinematics of a point; kinetics of a particle; plane motion of a rigid body; impulse and momentum; and, finally, work, power and energy.

Parts I and II are liberally illustrated with helpful diagrams.

Besides having worked examples in the text, exercises are set in each section with worked solutions given at the end of each chapter. Part III is devoted entirely to supplementary problems. Reference to appropriate problems in Part III is given throughout the text and the answers (unworked) are listed as an appendix at the end of the book.

Finally, Part IV contains all the background mathematics which might be required in the study of the main text. This is mainly for quick reference but it is given in sufficient detail for a first-time study when a topic has not been encountered previously. It is also very useful for refreshing one's memory.

Acknowledgements I was introduced to the mathematics of elementary mechanics in my last years at Kirkham Grammar School by an excellent teacher called Mr Barton. To him goes the credit (or the blame) for the two-diagram approach which I have used in solving many of the problems in dynamics. I am grateful to Queen's University, Belfast for the opportunity to teach the subject and for the invaluable feedback from my students. My thanks also go to the University for the facilities given me as a Professor Emeritus while producing the book. I am indebted to various members of staff but in particular to Stuart Ferguson for his help in using LaTeX to produce the text and for sorting out numerous computing problems for me. Thank you also to my son-in-law Mike Winstanley for help in using the drawing package ViSiO. Last but not least, I wish to thank Eric Willner and his colleagues at Cambridge University Press for the final production of the book.

Part I

Statics

1 Forces

1.1 Force

In the study of statics we are concerned with two fundamental quantities: length or distance, which requires no explanation, and force. The quantity length can be seen with the eye but with force, the only thing that is ever seen is its effect. We can see a spring being stretched or a rubber ball being squashed but what is seen is only the effect of a force being applied and not the force itself. With a rigid body there is no distortion due to the force and in statics it does not move either. Hence, there is no visual indication of forces being applied.

We detect a force being applied to our human body by our sense of touch or feel. Again, it is not the force itself but its effect which is felt – we feel the movement of our stomachs when we go over a humpback bridge in a fast car; we feel that the soles of our feet are squashed slightly when we stand.

We have now encountered one of the fundamental conceptual difficulties in the study of mechanics. Force cannot be seen or measured directly but must always be imagined. Generally the existence of some force requires little imagination but to imagine all the different forces which exist in a given situation may not be too easy. Furthermore, in order to perform any analysis, the forces must be defined precisely in mathematical terms.

For the moment we shall content ourselves with a qualitative definition of force. 'A force is that quantity which tries to move the object on which it acts.' This qualitative definition will suffice for statical problems in which the object does not move but we shall have to give it further consideration when we study the subject of dynamics. If the object does not move, the force must be opposed and balanced by another force. If we push with our hand against a wall, we know that we are exerting a force; we also know that the wall would be pushed over if it were not so strong. By saying that the wall is strong we mean that the wall itself can produce a force to balance the one applied by us.

EXERCISE 1

Note down a few different forces and state whether, and if so how, they might be observed.

1.2   Forces of contact

Before giving a precise mathematical description of force, we shall discuss two general categories. We shall start with the type which is more easily imagined; this is that due to contact between one object and another.

In the example of pushing against a wall with one's hand, the wall and hand are in contact, a force is exerted by the hand on the wall and this is opposed by another force from the wall to the hand. In the same way, when we are standing on the ground we can feel the force of the ground on our feet in opposition to the force due to our weight transmitted through our feet to the ground. Sometimes we think of a force being a pull but if we analyse the situation, the force of contact from one object to another is still a push. For instance, suppose a rope is tied around an object so that the latter may be pulled along. When this happens, the force from the rope which moves the object is a push on the rear of the object.

Another form of contact force is that which occurs when a moving object strikes another one. Any player of ball games will be familiar with this type of force. It only acts for a short time and is called an impulsive force. It is given special consideration in dynamics but it also occurs in statics in the following sense. When the surface of an object is in contact with a gas, the gas exerts a pressure, that is a force spread over the surface. The pressure is caused by the individual particles of the gas bouncing against the surface and exerting impulsive forces. The magnitude of each force is so small but the frequency of occurrence is so high that the effect is that of a force continuously distributed over the whole surface.

Forces of contact need not be exerted normal to the surface of contact. It is also possible to exert what is called a tangential or frictional component of force. In this case the force is applied obliquely to the surface; we can think of part being applied normally, i.e. perpendicular to the surface, and part tangentially. The maximum proportion of the tangential part which may be applied depends upon the nature of the surfaces in contact. Ice skaters know how small the tangential component can be and manufacturers of motor car tyres know how high.

A fact which must be emphasized concerning contact forces is that the forces each way are always equal and opposite, i.e. action and reaction are equal and opposite. When you push against the wall with your hand, the force from your hand on the wall is equal and opposite to the force from the wall against your hand. The rule is true for any pair of contact forces.

EXERCISE 2

Note down some of the contact forces which you have experienced or which have been applied to objects with which you have been concerned during the day. For each contact force, note the equal and opposite force which opposed it.

1.3   Mysterious forces

It is not too difficult to imagine the contact forces already described from our everyday experience but what is it that prevents a solid object from bending, squashing or just falling apart under the action of such forces? Mysterious forces of attraction act between the separate molecules of the material binding them together in a particular way and resisting outside forces which try to disturb the pattern. These *intermolecular forces* constitute the strength of the material. Although we shall not be concerned with it here, knowledge of the strength of materials is of great importance to engineers when designing buildings, machinery, etc.

Another mysterious force which will concern us deeply is the *force of gravity*. The magnitude of the force of gravity acting on a particular object depends on the size and physical nature of the object. In our study this force will remain constant and it will always act vertically downwards, this being referred to as the *weight* of the object. This is sufficient for most earthbound problems but when studying artificial satellites and space-craft it is necessary to consider the full properties of gravity.

Gravity is a force of attraction between any two bodies. It needs no material for its transmission nor is it impeded or changed in any way by material placed in between the bodies in question. The magnitude of the force was given mathematical form by Sir Isaac Newton and published in his *Philosophiae Naturalis Principia Mathematica* in 1687. The force is proportional to the product of the masses of the two bodies divided by the square of the distance between them. We shall say more about mass when studying dynamics but it is a constant property of any body. The law of gravitation, i.e. the inverse square law, was deduced by correlating it with the elliptical motion of planets about the sun as focus. Newton proved that such motion would be produced by the inverse square law of attraction to the sun acting on each planet.

Given that a body generates an attractive force proportional to its mass, it is reasonable that an inverse square law with respect to distance should apply. The force acts in towards the body from all directions around. However, the force acts over a larger area as the distance from the body increases. Since the effort is spread out over a larger area we can expect the strength at any particular point in the area to decrease accordingly. Thus we expect the magnitude of the force to be inversely proportional to the area of the sphere with the body at the centre and the point at which the force acts being on the surface of the sphere. The area is proportional to the square of the radius of the sphere; hence the inverse square law follows.

Magnetic and electrostatic forces are also important mysterious forces. However, they will not be dicussed here since we shall not be concerned with them in this text.

EXERCISE 3

Find the altitude at which the weight of a body is one per cent less than its weight at sea level. (Assume that the radius of the earth from sea level is 6370 km.)

EXERCISE 4

Find the percentage reduction in weight when the body is lifted from sea level to a height of 3 km.

1.4 Quantitative definition of force

In statics, force is that quantity which tries to move the object on which it acts. The magnitude of a force is the measure of its strength. It is then necessary to define basic units of measurement.

In lifting different objects we are very familiar with the concept of weight, which is the downward gravitational force on an object. It is tempting to use the weight of a particular object as the unit of force. However, weight varies with altitude (see Exercises 3 and 4) and also with latitude. To avoid this, a dynamical unit of force has been adopted.

The basic SI unit (Système International d'Unités) is the newton (symbol N). It is the force which would give a mass of one kilogramme (1 kg) an acceleration of one metre per second per second (1 m/s^2 or 1 m s^{-2}). The kilogramme is the mass of a particular piece of platinum–iridium. Of course, once a standard has been set, other masses can easily be evaluated by comparing relative weights. Incidentally, the mass of 1 kg is approximately the mass of one cubic decimetre of distilled water at the temperature (3.98°C) at which its density is maximum.

If you are more familiar with the pound-force (lbf) as the unit of force, then
1 lbf = 4.449 N or 1 N = 0.2248 lbf.

In quantifying a force, not only must its magnitude be given but also its direction of application, i.e. the direction in which it tries to move the object on which it acts. Having both magnitude and direction, force is a *vector* quantity. Sometimes it is convenient to represent a force graphically by an arrow (see Figure 1.1) which points in a direction corresponding to the direction of the force and has a length proportional to the magnitude of the force.

EXERCISE 5

Consider an aeroplane (see Figure 1.2) flying along at constant speed and height. Since there is no acceleration, forces should balance out in the same way that they do in statics. Draw vectors which might correspond to (a) the weight of the aeroplane, (b) the thrust from its engines and (c) the force from the surrounding air on the aeroplane which is a combination of lift and drag (lift/drag).

Figure 1.1. Force vector.

Figure 1.2. Simple sketch of an aeroplane.

1.5   Point of application

In studying forces acting on a rigid body, it is necessary to know the points of the body to which the forces are applied. For instance, consider a horizontal force applied to a stone which is resting on horizontal ground. If the force is strong enough the stone will move, but whether it moves by toppling or slipping depends on where the force is applied.

Forces rarely act at a single point of a body. Usually the force is spread out over a surface or volume. If the stone mentioned above is pushed with your hand, then the force from your hand is spread out over the surface of contact between your hand and the stone. The force from the ground which is acting on the stone is spread out over the surface of contact with the ground. The gravitational force acting on the stone is spread out over the whole volume of the stone. In order to perform the analysis in minute detail it would be necessary to consider each small force acting on each small element of area and on each small element of volume. However, since we are only considering rigid bodies, we are not concerned with internal stress. Thus we can replace many small forces by one large force. In our example, the small forces from the small elements of area of contact of your hand are represented by a single large force acting on the stone. Similarly, we have a single large force acting from the ground. Also, for the small gravitational forces acting on all the small elements of volume of the stone, we have instead a single force equal to the weight of the stone acting at a point in the stone which is called the *centre of gravity*.

The derivation of the points of action of these equivalent resultant forces will be discussed later. For the time being we shall assume that the representation is valid so that we can study the example of the stone as though there were only three forces acting on it, one from your hand, one from the ground and one from gravity.

EXERCISE 6

Continue Exercise 5 by drawing in the three force vectors on a rough sketch of the aeroplane.

1.6   Line of action

In the answer to Exercise 6, it appears that the three resultant forces of weight, thrust and lift/drag all act at the same point. Of course this may not be so but it does not matter provided the lines of action of the three resultant forces intersect at one point. For instance, this point need not coincide with the centre of gravity but it must be in the same vertical line as the centre of gravity.

Thus, with a rigid body the effect of a force is the same for any point of application along its line of action. This property is referred to as the *principle of transmissibility*. If two non-parallel but coplanar forces act on a body, it is convenient to imagine them to be acting at the point of intersection of their lines of action.

EXERCISE 7

Suppose that a smooth sphere is held on an inclined plane by a string which is fastened to a point on the surface of the sphere at one end and to a point on the plane at the other end. Sketch the side view and draw in the force vectors at the points of intersection of their lines of action.

EXERCISE 8

Do the same as in Exercise 7 for a ladder leaning against a wall, assuming that the lines of action of the three forces (weight and reactions from wall and ground) are concurrent.

Problems 1 and 2.

1.7 Equilibrium of two forces

A force tries to move its point of application and it will move it unless there is an equal and opposite counterbalancing force. When you push a wall with your hand, the wall will move unless it is strong enough to produce an equal and opposite force on your hand. If you are holding a dog with a lead, you will only remain stationary if you pull on the lead with the same amount of force as that exerted by the dog. By considering such physical examples we can see that for two forces to balance each other, they must be equal in magnitude and opposite in direction.

Yet another property is also required for the balance to exist. Suppose we have a large wheel mounted on a vertical axle. If one person pushes the wheel tangentially along the rim on one side and another person pushes on the other side, the wheel will start to move if the two pushes are equal in magnitude and opposite in direction. In fact two forces only balance each other if not only are they equal in magnitude and opposite in direction but also have the same line of action. When you are holding the dog, the line of the lead is the line of action of both the force from your hand and the force from the dog.

When the three conditions hold, we say that the two forces are in *equilibrium*. If a rigid body is acted on by only two such forces, the body will not move and we say that the body is in equilibrium. When a stone rests in equilibrium on the ground, the resultant contact force from the ground is equal, opposite and collinear to the resultant gravitational force acting on the stone.

EXERCISE 9

Suppose that a rigid straight rod rests on its side on a smooth horizontal surface. Let two horizontal forces of equal magnitude be applied to the rod simultaneously, one at either end. Consider what will happen to the rod immediately after the forces have been applied for a few different situations regarding the directions in which the separate forces are applied. Show that there will be only two possible situations in which the rod will remain in equilibrium.

1.8   Parallelogram of forces (vector addition)

If two non-parallel forces $\mathbf{F}_1$ and $\mathbf{F}_2$ act at a point A, they have a combined effect equivalent to a single force $\mathbf{R}$ acting at A. The single force $\mathbf{R}$ is called the *resultant* and it may be found as follows. Let $\mathbf{F}_1$ and $\mathbf{F}_2$ be represented in magnitude and direction by two sides of a parallelogram meeting at A. Then $\mathbf{R}$ is represented in magnitude and direction by the diagonal of the parallelogram from A, as shown in Figure 1.3. This is an empirical result referred to as the parallelogram law.

The parallelogram law may be illustrated by the following experiment. Take three different known weights of magnitudes W_1, W_2 and W_3, and attach W_1 and W_2 to either end of a length of string. Drape the string over two smooth pegs set a distance apart at about the same height. Then attach W_3 with a small piece of string to a point A of the other string between the two pegs. Finally, allow W_3 to drop gently and possibly move sideways until an equilibrium position is established (see Figure 1.4).

Now measure the angles to the horizontal made by the sections of string between the two pegs and A. Make an accurate drawing of the strings which meet at A and mark off distances proportional to W_1 and W_2 as shown in Figure 1.5. Since the pegs are smooth,

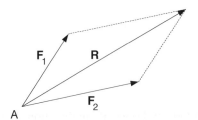

Figure 1.3. Parallelogram of forces.

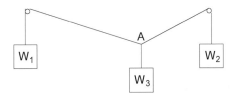

Figure 1.4. String over two smooth pegs.

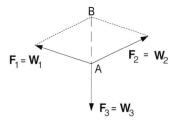

Figure 1.5. Three forces acting at A.

Figure 1.6. Vector addition.

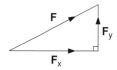

Figure 1.7. Cartesian components $\mathbf{F}_x$ and $\mathbf{F}_y$ of vector $\mathbf{F}$.

the tensions in the string of magnitudes F_1 and F_2 must be equal to the weights W_1 and W_2, respectively. Complete the parallelogram on the sides F_1 and F_2 and let B be the corner opposite A.

Since the point A is in equilibrium, the resultant of F_1 and F_2 should be equal, opposite and collinear to F_3 which is the tension in the string supporting W_3 with $F_3 = W_3$. If the parallelogram law holds, then AB should be collinear with the line corresponding to the vertical string and the length AB should correspond to the weight W_3.

The parallelogram law also applies to the *vector sum* of two vectors. Hence, the resultant of two forces acting at a point is their vector sum. Thus, if we use boldface letters to indicate vector quantities, the resultant $\mathbf{R}$ of two forces $\mathbf{F}_1$ and $\mathbf{F}_2$ acting at a point may be written as $\mathbf{R} = \mathbf{F}_1 + \mathbf{F}_2$.

Also, since opposite sides of a parallelogram are equal, $\mathbf{R}$ may be found by drawing $\mathbf{F}_2$ onto the end of $\mathbf{F}_1$ and joining the start of $\mathbf{F}_1$ to the end of $\mathbf{F}_2$ as shown in Figure 1.6.

A force vector $\mathbf{F}$ may also be written in terms of its *Cartesian components* $\mathbf{F} = \mathbf{F}_x + \mathbf{F}_y$ as shown in Figure 1.7.

Similarly, if we want the resultant $\mathbf{R}$ of two forces $\mathbf{F}_1$ and $\mathbf{F}_2$ acting at a point, then

$$\mathbf{R} = \mathbf{F}_1 + \mathbf{F}_2 = \mathbf{F}_{1x} + \mathbf{F}_{1y} + \mathbf{F}_{2x} + \mathbf{F}_{2y} = (\mathbf{F}_{1x} + \mathbf{F}_{2x}) + (\mathbf{F}_{1y} + \mathbf{F}_{2y}) = \mathbf{R}_x + \mathbf{R}_y.$$

In other words, the x-component of $\mathbf{R}$ is the sum of the x-components of $\mathbf{F}_1$ and $\mathbf{F}_2$ and the y-component of $\mathbf{R}$ is the sum of the y-components of $\mathbf{F}_1$ and $\mathbf{F}_2$. This

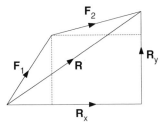

Figure 1.8. Addition of Cartesian components in vector addition.

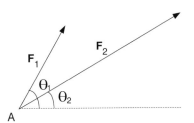

Figure 1.9. Three elastic bands used to demonstrate the parallelogram law.

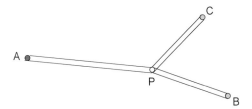

Figure 1.10. Two forces $\mathbf{F}_1$ and $\mathbf{F}_2$ acting at a point A.

can be seen diagramatically by drawing in the Cartesian components as illustrated in Figure 1.8.

EXERCISE 10

Use a piece of cotton thread to tie together three identical elastic bands. Having measured the unstretched length of the bands, peg them out as indicated in Figure 1.9, so that each band is in a stretched state but not beyond the elastic limit. The points A, B and C represent the fixed positions of the pegs but the point P takes up its equilibrium position pulled in three directions by the tensions in the bands. Use the fact that tension in each band is proportional to extension in order to verify the parallelogram law for the resultant of two forces acting at a point.

EXERCISE 11

Calculate the magnitude and direction of the resultant $\mathbf{R}$ of the two forces $\mathbf{F}_1$ and $\mathbf{F}_2$ acting at the point A given the magnitudes $F_1 = 1\,\text{N}$, $F_2 = 2\,\text{N}$ and directions $\theta_1 = 60°$, $\theta_2 = 30°$ (see Figure 1.10).

Problems 3 and 4.

1.9 Resultant of three coplanar forces acting at a point

Consider the three forces $\mathbf{F}_1$, $\mathbf{F}_2$ and $\mathbf{F}_3$ shown in Figure 1.11. The resultant $\mathbf{R}_1$ of $\mathbf{F}_1$ and $\mathbf{F}_2$ can be found by drawing the vector $\mathbf{F}_2$ on the end of $\mathbf{F}_1$ and joining the start of $\mathbf{F}_1$ to the end of $\mathbf{F}_2$. Then the final resultant $\mathbf{R}$ of $\mathbf{R}_1$ and $\mathbf{F}_3$, i.e. of $\mathbf{F}_1$, $\mathbf{F}_2$ and $\mathbf{F}_3$, is found by drawing the vector $\mathbf{F}_3$ on the end of $\mathbf{R}_1$ and joining the start of $\mathbf{R}_1$ to the end of $\mathbf{F}_3$. Having done this, we see that the intermediate step of inserting $\mathbf{R}_1$ may be omitted. Hence, the construction shown in Figure 1.11 is replaced by that of Figure 1.12. The procedure is simply to join the vectors $\mathbf{F}_1$, $\mathbf{F}_2$ and $\mathbf{F}_3$ end-on-end; then the resultant $\mathbf{R}$ corresponds to the vector joining the start of $\mathbf{F}_1$ to the end of $\mathbf{F}_3$.

The resultant vector $\mathbf{R}$ corresponds to the vector addition

$$\mathbf{R} = \mathbf{F}_1 + \mathbf{F}_2 + \mathbf{F}_3.$$

In terms of Cartesian components:

$$R_x = F_{1x} + F_{2x} + F_{3x}$$

and

$$R_y = F_{1y} + F_{2y} + F_{3y}.$$

The three forces $\mathbf{F}_1$, $\mathbf{F}_2$ and $\mathbf{F}_3$ will be in equilibrium if their resultant $\mathbf{R}$ is zero. In this case, joining the vectors end-on-end, the end of $\mathbf{F}_3$ will coincide with the start of $\mathbf{F}_1$. Thus we have the triangle of forces, which states that three coplanar forces acting at a point are in equilibrium if their vectors joined end-on-end correspond to the sides of a triangle, as illustrated in Figure 1.13.

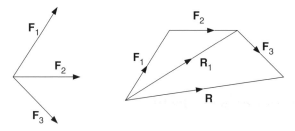

Figure 1.11. Constructing the resultant of three coplanar forces acting at a point.

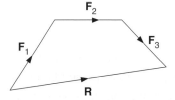

Figure 1.12. The final construction $\mathbf{R} = \mathbf{F}_1 + \mathbf{F}_2 + \mathbf{F}_3$.

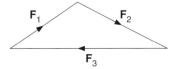

Figure 1.13. Triangle of forces.

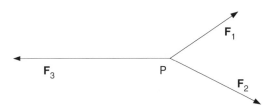

Figure 1.14. Three coplanar forces acting at a point.

Referring to Figures 1.13 and 1.14, we notice that the angle in the triangle between $\mathbf{F}_1$ and $\mathbf{F}_2$ is $180°$ minus the angle between the forces $\mathbf{F}_1$ and $\mathbf{F}_2$ acting at the point P, and similarly for the other angles. Now, the sine rule for a triangle states that the length of each side is proportional to the sine of the angle opposite. Since $\sin(180° - \theta) = \sin\theta$, we have Lamy's theorem which states that 'three coplanar forces acting at a point are in equilibrium if the magnitude of each force is proportional to the sine of the angle between the other two forces'.

EXERCISE 12

Let three forces $\mathbf{F}_1$, $\mathbf{F}_2$ and $\mathbf{F}_3$ have magnitudes in newtons of $\sqrt{6}$, $1 + \sqrt{3}$ and 2, respectively. If the angles made with the positive x-direction are $45°$ for $\mathbf{F}_1$, $180°$ for $\mathbf{F}_2$ and $-60°$ for $\mathbf{F}_3$, show that the three forces are in equilibrium by (a) calculating their resultant, (b) triangle of forces and (c) Lamy's theorem.

Problems 5 and 6.

1.10 Generalizations for forces acting at a point

Firstly, consider more than three coplanar forces acting at a point. The vector addition procedure can be continued. For instance, drawing the vectors end-on-end gives $\mathbf{R}_3$, say, for the resultant of the first three. Then, as shown in Figure 1.15, drawing the vector for $\mathbf{F}_4$ onto the end of $\mathbf{R}_3$, the resultant of $\mathbf{R}_3$ and $\mathbf{F}_4$ is given by the vector $\mathbf{R}_4$ joining the start of $\mathbf{R}_3$ to the end of $\mathbf{F}_4$. $\mathbf{R}_3$ can now be omitted.

This procedure is obviously valid for finding the resultant of any number of coplanar forces acting at a point. Furthermore, the forces will be in equilibrium if the final resultant is zero, i.e. when the end of the last vector coincides with the start of the first. Hence, we have the *polygon of forces*, which states that 'n coplanar forces acting

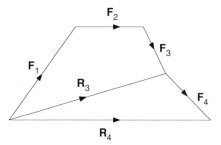

Figure 1.15. Constructing the resultant of four coplanar forces acting at a point.

at a point are in equilibrium if their vectors joined end-on-end complete an n-sided polygon'.

Although it is not convenient for two-dimensional drawing, the basic concept can be extended to finding the resultant of non-coplanar forces acting at a point. Any two of the forces are coplanar, so the resultant $\mathbf{R}_2$ of $\mathbf{F}_1$ and $\mathbf{F}_2$ is the vector sum $\mathbf{R}_2 = \mathbf{F}_1 + \mathbf{F}_2$. Then $\mathbf{R}_2$ and $\mathbf{F}_3$ must be coplanar with resultant $\mathbf{R}_3 = \mathbf{R}_2 + \mathbf{F}_3 = \mathbf{F}_1 + \mathbf{F}_2 + \mathbf{F}_3$. Thus, if there are n forces, their resultant is $\mathbf{R}_n = \mathbf{F}_1 + \mathbf{F}_2 + \cdots + \mathbf{F}_n$.

We have now moved from two-dimensional to three-dimensional space. Each vector has three *Cartesian components*, i.e. its x-, y- and z-components. The Cartesian components of $\mathbf{R}_n$ are :

$$R_{nx} = F_{1x} + F_{2x} + \cdots + F_{nx}$$
$$R_{ny} = F_{1y} + F_{2y} + \cdots + F_{ny}$$
$$R_{nz} = F_{1z} + F_{2z} + \cdots + F_{nz}$$

where x, y and z signify the corresponding component in each case.

As in the coplanar case, the forces will be in equilibrium if their resultant $\mathbf{R}_n$ is zero, i.e. $R_{nx} = R_{ny} = R_{nz} = 0$.

EXERCISE 13

Let four coplanar forces $\mathbf{F}_1, \mathbf{F}_2, \mathbf{F}_3$ and $\mathbf{F}_4$ acting at a point have magnitudes in newtons of 2, $\sqrt{6}$, 2 and $\sqrt{2}$, and directions relative to the positive x-axis of $150°, 45°, -60°$ and $-135°$, respectively. Show that the four forces are in equilibrium by (a) calculating their resultant and (b) using a polygon of forces.

EXERCISE 14

Find the resultant of three non-coplanar forces acting at a point, where each is given in newtons in terms of its Cartesian components: $\mathbf{F}_1 = (1, -2, -1)$, $\mathbf{F}_2 = (2, 1, -1)$ and $\mathbf{F}_3 = (-1, -1, 1)$. Besides giving the resultant in terms of its Cartesian components, find its magnitude and the angles which it makes with the x-, y- and z-axes.

Problems 7 and 8.

1.11 More exercises

EXERCISE 15

Figure 1.16 shows weights W_1 and W_2 attached to the ends of a string which passes over two smooth pegs A and B at the same height and 0.5 m apart. A third weight W_3 is suspended from a point C on the string between A and B.

(a) If $W_1 = 3\,\text{N}$, $W_2 = 4\,\text{N}$ and $W_3 = 5\,\text{N}$, find the horizontal and vertical distances a and d of C from A when the weights are in equilibrium.

(b) If $W_1 = 10\,\text{N}$, the angle between CB and the vertical is $30°$ and $\angle ACB = 90°$ when the weights are in equilibrium, find the weights W_2 and W_3.

EXERCISE 16

Two smooth spheres, each of weight W and radius r, are placed in the bottom of a vertical cylinder of radius $3r/2$. Find the magnitudes of the forces R_a, R_b, R_c and R_p which act on the spheres as indicated in Figure 1.17.

EXERCISE 17

Four identical smooth spheres, each of weight W and radius r, are placed in the bottom of a hollow vertical circular cylinder of inner radius $2r$. Two spheres rest on the bottom and the other two settle as low as possible above the bottom two. Find all the forces acting on the spheres. Because of the symmetry, only one of the top and one of the bottom spheres need be considered.

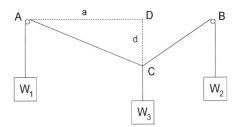

Figure 1.16. Three weights on a string.

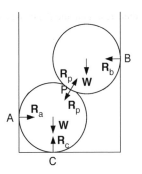

Figure 1.17. Two spheres in a hollow cylinder.

1.12 Answers to exercises

1.   There are infinitely many possible answers to Exercise 1. The following are a couple of examples.

     If you hang a wet and heavy piece of clothing on a clothes-line, the weight of the clothing exerts a downward force on the line, which is opposed by an increased tension in the line. The latter is observed in a downward sag of the line from its unloaded position.

     If you stand on weighing scales in your bathroom, your own weight exerts a downward force via your feet which is opposed by an upward force from the scales. The latter is observed both by the feeling of pressure on your feet and by the measurement of your weight as indicated by the scales.

2.   There will be many different answers to Exercise 2. Here are a couple of examples from my own experience.

     When I came out of the house this morning, I was carrying a brief-case. The weight of the case exerted a downward force via the handle on the fingers of my hand. The latter exerted an equal and opposite force on the handle of the brief-case.

     Later, as I was driving my car at a steady speed, I had to exert a constant force on the accelerator pedal to keep it in a certain position. This force was transmitted to the pedal through the sole of my shoe. At the same time, the pedal was exerting an equal and opposite force on the sole of my shoe.

3.   The distance in the inverse square law is measured from the centre of each body. Let W be the weight of a body and r be its distance from the centre of the earth. Also, let the Greek letter δ (delta) mean 'change in' so that δW is change in weight and δr is change in distance from the centre of the earth. Using k as a constant of proportionality, by the inverse square law, $W = k/r^2$.

     If the weight reduces by δW when the body is raised from sea level to an altitude of δr, then $W - \delta W = k/(r + \delta r)^2$. Dividing this by W gives:

$$1 - \frac{\delta W}{W} = 1 - 0.01 = 0.99 = \frac{r^2}{(r + \delta r)^2} = \frac{1}{(1 + \frac{\delta r}{r})^2}.$$

     Hence, $1 + \frac{\delta r}{r} = 1/\sqrt{0.99}$ and $\delta r = (\frac{1}{\sqrt{0.99}} - 1)r$.   If $r = 6370\,\text{km}$,   altitude $\delta r = 6370(\frac{1}{\sqrt{0.99}} - 1) = 32.09\,\text{km}$.

4.   It follows from Exercise 3 that $\frac{\delta W}{W} = 1 - \frac{1}{(1 + \frac{\delta r}{r})^2}$. With $\delta r = 3$ and $r = 6370$, $\frac{\delta W}{W} = 0.00094 = 0.094\%$.

5.   The weight of the aeroplane acts vertically downwards, so the force may be represented by an arrow pointing downwards (Figure 1.18a). The thrust of the engines is a force in the direction in which the aeroplane is travelling (Figure 1.18b). The force from the air on the aeroplane has two components: lift upwards and drag backwards, so the two together may be represented by an upward arrow sloping backwards (Figure 1.18c).

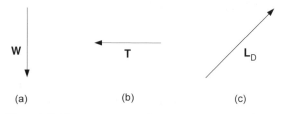

      (a)   (b)   (c)

Figure 1.18. Forces on an aeroplane.

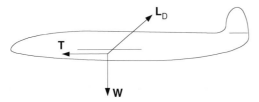

Figure 1.19. Forces shown acting on an aeroplane.

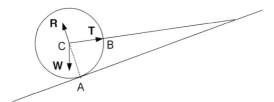

Figure 1.20. A smooth sphere held by a string on an inclined plane.

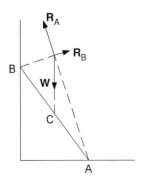

Figure 1.21. Forces acting on a ladder.

6. The weight vector **W** must act downwards through the centre of gravity somewhere in the middle of the aeroplane. The thrust vector **T** is level with the engines which are assumed in Figure 1.19 to be in pods under the wings. The lift/drag vector $\mathbf{L}_D$ must act through the point of intersection of **T** and **W** in order to avoid any turning effect.

7. The weight **W** acts through the centre of the sphere and so does the reaction **R** from the inclined plane, since the sphere is smooth and therefore the reaction is perpendicular to the plane. The tension **T** in the string must also apply a force acting through the centre of the sphere and hence may be represented as shown in Figure 1.20. Notice that, by the principle of transmissibility, we can draw in **R** and **T** as forces acting at the centre C even though the points of application are actually A and B, respectively.

8. The weight **W** acts vertically downwards through the centre C of the ladder. Assuming that there are frictional components of reaction, the lines of action of the corresponding total reactions $\mathbf{R}_A$ and $\mathbf{R}_B$ will be inclined in the manner shown in Figure 1.21. Notice, this time, that the forces act through a point outside the body, i.e. outside the ladder. We shall see later that this does not affect the usefulness of this procedure even though we can no longer appeal to the principle of transmissibility.

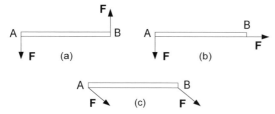

Figure 1.22. A rod acted on by two coplanar forces.

Figure 1.23. A rod in equilibrium under the action of two forces.

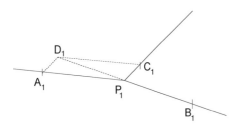

Figure 1.24. Testing the parallelogram law.

9.　　If the forces are equal and opposite and perpendicular to the rod (Figure 1.22a), the forces would start to rotate the rod. If the force at end A is as before, but that at end B is along its length (Figure 1.22b), the rod would start to both rotate and translate. If the forces both act in the same direction (Figure 1.22c), the rod would start to translate in that direction.

Obviously, there are many more possible examples but moving on to those which result in equilibrium, we remember that for this to exist, the two forces must not only be equal in magnitude but also opposite in direction and collinear. For the latter to be true, both forces must act along the length of the rod. Then to be opposite in direction as well, they must either both pull outwards (Figure 1.23a) or both push inwards (Figure 1.23b).

10.　　Draw three straight lines from a point P_1 in exactly the directions of the bands PA, PB and PC shown in Figure 1.9. Mark off distances from P_1 proportional to the band extensions and therefore to their tensions. Denote these distance marks A_1, B_1 and C_1, respectively, as shown in Figure 1.24. Complete the parallelogram on the sides P_1A_1 and P_1C_1, and denote the fourth corner D_1. If the parallelogram law for the resultant of two forces holds, the diagonal P_1D_1 should be collinear with and of equal length to P_1B_1.

Note that the parallelogram law may also be tested by completing a parallelogram on the sides P_1A_1 and P_1B_1 or on P_1B_1 and P_1C_1.

11.　　Referring to Figure 1.25, $F_{1x} = F_1 \cos\theta_1 = 1/2$. $F_{1y} = F_1 \sin\theta_1 = \sqrt{3}/2$. $F_{2x} = F_2 \cos\theta_2 = 2\sqrt{3}/2 = \sqrt{3}$. $F_{2y} = F_2 \sin\theta_2 = 2/2 = 1$.

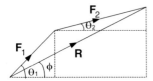

Figure 1.25. Resultant **R** of two forces **F**$_1$ and **F**$_2$ acting at a point.

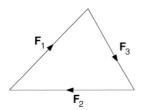

Figure 1.26. Triangle of forces.

Thus $R_x = F_{1x} + F_{2x} = \dfrac{1}{2} + \sqrt{3} = \dfrac{1 + 2\sqrt{3}}{2}$ and $R_y = F_{1y} + F_{2y} = \dfrac{\sqrt{3}}{2} + 1 = \dfrac{\sqrt{3} + 2}{2}$

$R^2 = R_x^2 + R_y^2 = 5 + 2\sqrt{3}, \quad R = 2.91\,\text{N}$

$\tan\phi = R_y/R_x = \dfrac{\sqrt{3} + 2}{1 + 2\sqrt{3}}, \quad \phi = 39.9°.$

12. (a) Calculate the Cartesian components of **F**$_1$, **F**$_2$ and **F**$_3$ as follows. $F_{1x} = \sqrt{6}\cos 45° = \sqrt{6}/\sqrt{2} = \sqrt{3}$, $F_{2x} = -(1 + \sqrt{3})$, $F_{3x} = 2\cos(-60°) = 1$, $F_{1y} = \sqrt{6}\sin 45° = \sqrt{6}/\sqrt{2} = \sqrt{3}$, $F_{2y} = 0$, $F_{3y} = 2\sin(-60°) = 2(-\sqrt{3}/2) = -\sqrt{3}$. Then the x- and y-components of the resultant are:

$R_x = F_{1x} + F_{2x} + F_{3x} = \sqrt{3} - (1 + \sqrt{3}) + 1 = 0$ and

$R_y = F_{1y} + F_{2y} + F_{3y} = \sqrt{3} + 0 - \sqrt{3} = 0.$

Hence, the resultant **R** = 0 and the forces **F**$_1$, **F**$_2$ and **F**$_3$ are in equilibrium.

(b) Draw the vectors corresponding to **F**$_1$, **F**$_2$ and **F**$_3$ end-on-end as shown in Figure 1.26. Since they form the sides of a triangle, the forces must be in equilibrium. Note that the order in which the vectors are joined does not matter provided that all point the same way around the triangle, i.e. all clockwise or all anti-clockwise.

(c) Referring to Figure 1.27: $\dfrac{F_1}{\sin 120°} = \dfrac{\sqrt{6}}{\sin 120°} = 2.828$, $\dfrac{F_2}{\sin 105°} = \dfrac{1+\sqrt{3}}{\sin 105°} = 2.828$ and $\dfrac{F_3}{\sin 135°} = \dfrac{2}{\sin 135°} = 2.828$. The forces are in equilibrium since the magnitude of each is proportional to the sine of the angle between the other two.

13. (a) $F_{1x} = 2\cos 150° = 2(-\sqrt{3}/2) = -\sqrt{3}$, $F_{2x} = \sqrt{6}\cos 45° = \sqrt{6}/\sqrt{2} = \sqrt{3}$, $F_{3x} = 2\cos(-60°) = 2/2 = 1$, $F_{4x} = \sqrt{2}\cos(-135°) = \sqrt{2}(-1/\sqrt{2}) = -1$. Therefore, $R_x = F_{1x} + F_{2x} + F_{3x} + F_{4x} = -\sqrt{3} + \sqrt{3} + 1 - 1 = 0.$

$F_{1y} = 2\sin 150° = 2/2 = 1$, $F_{2y} = \sqrt{6}\sin 45° = \sqrt{6}/\sqrt{2} = \sqrt{3}$, $F_{3y} = 2\sin(-60°) = 2(-\sqrt{3}/2) = -\sqrt{3}$, $F_{4y} = \sqrt{2}\sin(-135°) = \sqrt{2}(-1/\sqrt{2}) = -1$. Therefore, $R_y = F_{1y} + F_{2y} + F_{3y} + F_{4y} = 1 + \sqrt{3} - \sqrt{3} - 1 = 0.$

Since both R_x and R_y are zero, the resultant **R** is zero and therefore the four forces **F**$_1$, **F**$_2$, **F**$_3$ and **F**$_4$ are in equilibrium.

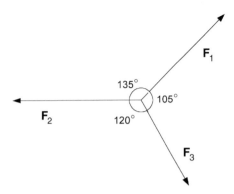

Figure 1.27. Three forces acting at a point.

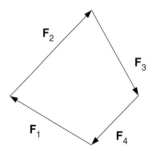

Figure 1.28. Tetragon of forces.

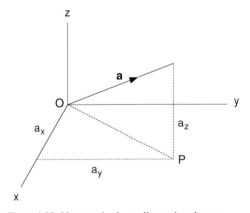

Figure 1.29. Vector **a** in three-dimensional space.

(b) Draw the vectors corresponding to F_1, F_2, F_3 and F_4 end-on-end as shown in Figure 1.28. Since they complete the sides of a tetragon (four-sided polygon), by the polygon of forces, the forces must be in equilibrium.

14.   Before answering the specific problem, let us consider the properties of a vector **a** as shown supposedly in three-dimensional space in Figure 1.29. This has been simplified by localizing the vector **a** at the

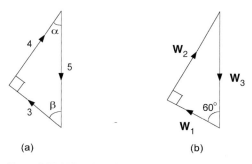

Figure 1.30. Triangles of tension forces acting at C.

origin of the three-dimensional Cartesian coordinate system. If P is the projection of the end of **a** onto the x, y plane, by Pythagoras' theorem, $a^2 = OP^2 + a_z^2 = a_x^2 + a_y^2 + a_z^2$. In other words, the square of the length of a vector equals the sum of the squares of its x-, y- and z-components. Also, $a_x = a\cos\theta_x$, $a_y = a\cos\theta_y$ and $a_z = a\cos\theta_z$, where θ_x, θ_y and θ_z are the angles which the vector **a** makes with the positive x-, y- and z-axes, respectively.

Returning to the specific problem, the resultant **R** of the three forces $\mathbf{F}_1$, $\mathbf{F}_2$ and $\mathbf{F}_3$ will have x-, y- and z-components as follows:

$$R_x = F_{1x} + F_{2x} + F_{3x} = 1 + 2 - 2 = 2$$
$$R_y = F_{1y} + F_{2y} + F_{3y} = -2 + 1 - 1 = -2$$
$$R_z = F_{1z} + F_{2z} + F_{3z} = -1 - 1 + 1 = -1.$$

Therefore, $\mathbf{R} = (2, -2, -1)$.
$$R^2 = R_x^2 + R_y^2 + R_z^2 = 4 + 4 + 1 = 9. \text{ Hence, } R = 3\,\text{N}.$$

$$R_x = R\cos\theta_x, \ \theta_x = \cos^{-1}(2/3) = 48.2°$$
$$R_y = R\cos\theta_y, \ \theta_y = \cos^{-1}(-2/3) = 131.8°$$
$$R_z = R\cos\theta_z, \ \theta_z = \cos^{-1}(-1/3) = 109.5°.$$

15.   (a) The three tensions acting at C are in equilibrium so they must obey the triangle of forces as shown in Figure 1.30a. W_3 is vertically downwards and since $3^2 + 4^2 = 5^2$, the angle between the W_1 and W_2 tensions is a right-angle. Comparing Figure 1.16 with Figure 1.30a, $\angle BAC = \alpha$ and $\angle ABC = \beta$. Therefore, the triangle of forces is similar to $\triangle ABC$ which in turn is similar to $\triangle ACD$. Since $AB = 0.5\,\text{m}$, $AC = 0.5(4/5) = 0.4\,\text{m}$. Then, $a = (4/5)AC = 0.32\,\text{m}$ and $d = (3/5)AC = 0.24\,\text{m}$.

(b) In this case, the triangle of forces is as shown in Figure 1.30b. Remembering that a 30° right-angled triangle has sides of length proportional to $1 : \sqrt{3} : 2$, we see that $W_3 = (2/1)W_1 = 20\,\text{N}$ and $W_2 = (\sqrt{3}/1)W_1 = 10\sqrt{3}\,\text{N}$.

16.   The diameter of the cylinder is $3r$ and the diameter of each sphere is $2r$. Thus the length of the line joining the centres of the spheres is $2r$ and the horizontal displacement between the centres is r, as shown in Figure 1.31a. Consequently, the line of centres makes an angle of 30° with the vertical, since $\sin 30° = 1/2$.

We can now draw the triangle of forces for the three forces acting on the upper sphere, as in Figure 1.31b. From this we see that $R_b = W\tan 30° = W/\sqrt{3}$. Also, $R_p = W\sec 30° = 2W/\sqrt{3}$.

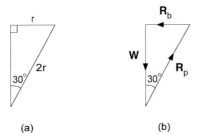

Figure 1.31. Triangle of forces acting on upper sphere.

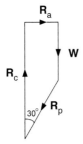

Figure 1.32. Tetragon of forces acting on lower sphere.

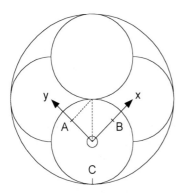

Figure 1.33. Top view of the four spheres in the cylinder.

Next we draw the polygon of forces for the four forces acting on the lower sphere (see Figure 1.32). From this, we see that $R_a = R_p \sin 30° = 2W/2\sqrt{3} = W/\sqrt{3}$. Also, $R_c = W + R_p \cos 30° = W + \frac{2}{\sqrt{3}} W \frac{\sqrt{3}}{2} = 2W$.

17.   Figure 1.33 shows the top view of the four spheres in the cylinder. Consider one of the top spheres and draw in x- and y-axes as shown with the origin at the centre of the sphere. The z-axis will be at right-angles vertically upwards. The points of contact with the bottom spheres will be on the lines of centres below the points A and B. The top view diagram (Figure 1.33) shows that $AO = BO = r/\sqrt{2}$.

If P is the point of contact below A, we see from Figure 1.34 of the APO triangle in the y, z plane that PO and therefore the direction of the reaction force at P is at 45° to the vertical, PO being the sphere radius r.

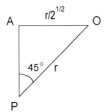

Figure 1.34. Triangle APO in the vertical y, z plane.

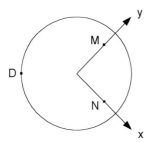

Figure 1.35. Horizontal x- and y-axes with origin at centre of left lower sphere.

We now deduce that the reaction force from P in the direction of O can be written in terms of its Cartesian components as $\mathbf{R}_p = (0, -R/\sqrt{2}, R/\sqrt{2})$, where R is its magnitude. We now have another reaction force at the point of contact Q below B given by $\mathbf{R}_q = (-R/\sqrt{2}, 0, R/\sqrt{2})$.

Since the top two spheres try to move down and out, we can assume that there will be no reaction force between the two. However, there will be one from the wall of the cylinder directed towards O which can be written as $\mathbf{R}_c = (R_c/\sqrt{2}, R_c/\sqrt{2}, 0)$, where R_c is the magnitude and its x- and y-components are at 45° to the direction of $\mathbf{R}_c$. Finally, the weight of the sphere can be written as the force $\mathbf{W} = (0, 0, -W)$.

Hence, the sphere is kept in equilibrium by the four forces $\mathbf{R}_p$, $\mathbf{R}_q$, $\mathbf{R}_c$ and $\mathbf{W}$ all acting through its centre O. For equilibrium, the sum of the x- components must be zero, the sum of the y-components must be zero and the sum of the z-components must be zero. Thus, $0 - R/\sqrt{2} + R_c/\sqrt{2} + 0 = 0$ and $-R/\sqrt{2} + 0 + R_c/\sqrt{2} + 0 = 0$, each of which implies that $R = R_c$, and $R/\sqrt{2} + R/\sqrt{2} + 0 - W = 0$, i.e. $R = W/\sqrt{2}$.

Now consider the bottom two spheres. The top two try to push them apart, so we can assume that there is no reactive force between the bottom two spheres. Again, because of symmetry, we only need to study one of the spheres. Let us take the one on the left and draw in x- and y-axes with origin at the centre as shown in Figure 1.35. The z-axis will again be vertically upwards. The point D is the point of contact with the cylinder and a reactive force will act on the sphere at D towards its centre. This force may be written as $\mathbf{R}_d = (R_d/\sqrt{2}, R_d/\sqrt{2}, 0)$. The points of contact with the upper spheres are above M and N in Figure 1.35 and the corresponding downward sloping forces acting on our bottom sphere can be written as $\mathbf{R}_m = (0, -R/\sqrt{2}, -R/\sqrt{2})$ and $\mathbf{R}_n = (-R/\sqrt{2}, 0, -R/\sqrt{2})$. We have already shown that $R = W/\sqrt{2}$, so $R/\sqrt{2} = W/2$.

Besides these three forces, we have the weight of the sphere $\mathbf{W} = (0, 0, -W)$ and an upward reaction force through the base of the sphere given by $\mathbf{R}_b = (0, 0, R_b)$.

There are thus five forces acting on the sphere through its centre. For equilibrium we can in turn equate to zero the sum of the x-components, the sum of the y-components and the sum of the

z-components. Hence, $R_d/\sqrt{2} + 0 - W/2 + 0 + 0 = 0$ and $R_d/\sqrt{2} - W/2 + 0 + 0 + 0 = 0$, each of which gives $R_d = W/\sqrt{2}$, and $0 - W/2 - W/2 - W + R_b = 0$, i.e. $R_b = 2W$.

To summarize the results: (1) the force between each top sphere and the cylinder is $W/\sqrt{2}$; (2) the forces at the points of contact between upper and lower spheres are each equal to $W/\sqrt{2}$; (3) the force between each bottom sphere and the cylinder is $W/\sqrt{2}$; (4) the force between each bottom sphere and the base of the cylinder is $2W$.

2 Moments

2.1 Moment of force

When forces are applied to a rigid body, they may have a translational or a rotational effect or both. When you push or pull on furniture to move it across a room, your force has a translational effect. When you push on a door to open it, your force has a rotational effect. Some doors have springs to keep them closed. If you push on the middle of such a door, it is much more difficult to open it than if you push it on the side furthest away from the hinge. Thus the force must have a greater turning effect the further it is away from the hinge. In fact, it can be shown by experiment that the turning effect of a force is directly proportional to the perpendicular distance of the force from the point about which the turning is to take place. As we might expect, the turning effect is also proportional to the magnitude of the force.

We can now define a measure of the turning effect of a force. It is called the *moment* of the force and is equal to the magnitude of the force multiplied by its perpendicular distance from the point about which the turning effect is being measured. A turning moment is also called a *torque* and since it is measured as force times distance, its SI unit of measurement is the newton metre (N m). To illustrate the measurement of moment by a diagram, suppose a force $\mathbf{F}$ is acting at a point P and that we want to find its moment about a point O. It is then necessary to extend its line of action as shown in Figure 2.1 back to Q, which is the point on the line nearest to O. Then the measurement of moment is the magnitude of $\mathbf{F}$ multiplied by the distance OQ, i.e. $F \cdot OQ$. Finally, to distinguish between a clockwise turning effect and an anti-clockwise one as in Figure 2.1, it is conventional to let clockwise be negative and anti-clockwise be positive.

Another way of denoting a moment is to use vectors. From Figure 2.1 we see that

$$F \cdot OQ = F \cdot OP \sin \alpha = |\mathbf{r} \times \mathbf{F}|$$

where the $\times$ indicates *vector product* and the two vertical lines $|\ |$ denote the *modulus* or magnitude of the enclosed vector. $\mathbf{r} \times \mathbf{F}$ is a vector with direction perpendicular to

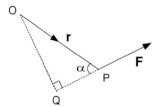

Figure 2.1. Moment of a force.

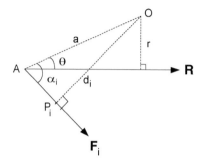

Figure 2.2. Theorem of Varignon.

both **r** and **F** and in this case pointing out of the paper. This is the *right-hand (screw) thread rule*. If the ridge in the screw were turned from the **r**-direction to the **F**-direction, the screw would screw upwards out of the paper. Notice that this turning is always in the same sense as the turning moment. Having agreed this principle, it is essential that the position vector **r** should precede the force vector **F** when writing the moment as the vector product **r** × **F**.

While in the introductory section on moments, it is convenient to prove the *theorem of Varignon* which states that if several coplanar forces act at a point, the moment of their resultant about another point in the plane is equal to the sum of the moments of the separate forces. In Figure 2.2, we show just one of several forces F_i. It is assumed that all act through the point A. The resultant force $\mathbf{R} = \sum_i \mathbf{F}_i$, i.e. the vector sum of the separate forces.

Referring to Figure 2.2: the sum of the moments of F_i about O

$$= \sum_i F_i d_i = \sum_i F_i a \sin \alpha_i = a \sum_i F_i \sin \alpha_i$$

$= a \cdot (\text{sum of the components of } \mathbf{F}_i \perp AO) = a \cdot (\text{component of } \mathbf{R} \perp AO) = aR \sin \theta =$
$Rr = $ the moment of **R** about O ($\perp$ means 'perpendicular to').

EXERCISE 1

If a force of 10 N acts at a point A as shown in Figure 2.3 and its line of action crosses the x-axis at P such that $OP = 0.5$ m, find the moment of the force about O.

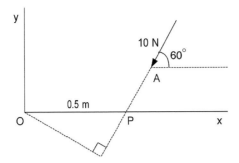

Figure 2.3. Moment about O.

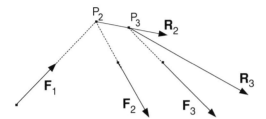

Figure 2.4. Resultant $\mathbf{R}_3$ of three non-concurrent coplanar forces $\mathbf{F}_1$, $\mathbf{F}_2$ and $\mathbf{F}_3$.

EXERCISE 2

Solve Exercise 1 again using Varignon's theorem, i.e. let the force act at P and split it into its Cartesian components.

Problems 9 and 10.

2.2   Three or more non-parallel non-concurrent coplanar forces

Consider coplanar forces acting on a rigid body such that the point of intersection of the lines of action of one pair of forces is different from that of any other pair.

Suppose there are n forces, labelled $\mathbf{F}_1$, $\mathbf{F}_2$, ..., $\mathbf{F}_n$, acting on a rigid body. (Figure 2.4 shows an example of three such forces.) If we examine $\mathbf{F}_1$ and $\mathbf{F}_2$, their lines of action will meet in a point P_2, say. We can therefore find a resultant force $\mathbf{R}_2$ which is the vector sum of $\mathbf{F}_1$ and $\mathbf{F}_2$ and has a line of action passing through P_2. Next we find the resultant $\mathbf{R}_3$ of $\mathbf{R}_2$ and $\mathbf{F}_3$, i.e. $\mathbf{R}_3 = \mathbf{R}_2 + \mathbf{F}_3 = \mathbf{F}_1 + \mathbf{F}_2 + \mathbf{F}_3$. The line of action of $\mathbf{R}_3$ passes through P_3 which is the point of intersection of the lines of action of $\mathbf{R}_2$ and $\mathbf{F}_3$.

This process may be continued until ultimately the final resultant of the n forces is found as $\mathbf{R}_n = \mathbf{R}_{n-1} + \mathbf{F}_n = \mathbf{F}_1 + \mathbf{F}_2 + \cdots + \mathbf{F}_n$, i.e. the vector sum of the original n forces. The line of action of $\mathbf{R}_n$ passes through the point of intersection of the lines of action of $\mathbf{R}_{n-1}$ and $\mathbf{F}_n$.

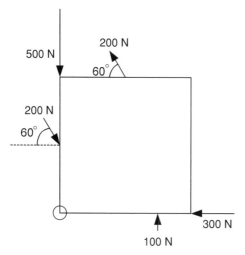

Figure 2.5. Forces acting on the rim of a square plate.

Notice that no drawing is necessary to find the magnitude and direction of the resultant since $\mathbf{R}_n$ is simply the vector sum of the separate forces $\mathbf{F}_i$ as i goes from 1 to n. Thus the x-component of $\mathbf{R}_n$ is the sum of the x-components of $\mathbf{F}_i$ and the y-components of $\mathbf{R}_n$ is the sum of the y-components of $\mathbf{F}_i$.

The next question is: can the line of action of the resultant $\mathbf{R}_n$ be found without performing the possibly tedious and inaccurate drawing procedure? The answer is 'yes' by using the concept of moment and the theorem of Varignon.

Consider moments about a given point O in the plane of action of the forces $\mathbf{F}_i$. By the theorem of Varignon, the moment of $\mathbf{R}_2$ equals the sum of the moments of $\mathbf{F}_1$ and $\mathbf{F}_2$. Then the moment of $\mathbf{R}_3$ equals the sum of the moments of $\mathbf{R}_2$ and $\mathbf{F}_3$ which is equal to the sum of the moments of $\mathbf{F}_1$, $\mathbf{F}_2$ and $\mathbf{F}_3$. The process ends with the moment of $\mathbf{R}_n$ equals the sum of the moments of $\mathbf{R}_{n-1}$ and $\mathbf{F}_n$ equals the sum of the moments of $\mathbf{F}_1$, $\mathbf{F}_2$, ..., $\mathbf{F}_{n-1}$ and $\mathbf{F}_n$. Therefore, the line of action of $\mathbf{R}_n$ must be such that its moment about O is equal to the sum of the moments about O of $\mathbf{F}_1$, $\mathbf{F}_2$, ..., $\mathbf{F}_n$.

EXERCISE 3

Five forces are applied to the rim of a square plate, the points of application being either a corner, mid-point or quarter-length point of a side as indicated in Figure 2.5. The numbers give the magnitudes of the forces in newtons. Find the magnitude, direction and line of action of the resultant force. (Remember that the moment of a force equals the sum of the moments of its Cartesian components.)

Problems 11 and 12.

2.3   Parallel forces

Let two parallel forces of magnitudes P and Q act in the same direction at points A and B, respectively. These are shown in Figure 2.6 where we have also introduced two

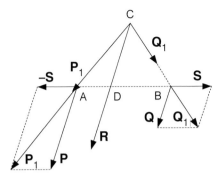

Figure 2.6. Constructing the resultant of two parallel forces acting in the same direction.

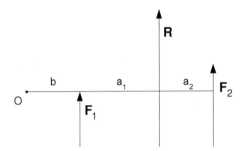

Figure 2.7. Positioning **R** by taking moments.

equal and opposite forces of magnitude S at A and B with the same line of action AB. Since they are equal, opposite and collinear, they have no overall effect but they allow us to replace the parallel forces **P** and **Q** by non-parallel forces $\mathbf{P}_1$ and $\mathbf{Q}_1$. The lines of action of the latter intersect at C. The effect of the original forces is the same as $\mathbf{P}_1$ and $\mathbf{Q}_1$ acting at C which in turn is the same as the resultant **R** acting at C. Since the equal and opposite forces of magnitude S cancel, $\mathbf{R} = \mathbf{P} + \mathbf{Q}$, which has magnitude $R = P + Q$ and direction parallel to **P** and **Q**.

Let the line of action of **R** cut AB at the point D. Then by comparing similar triangles: $\frac{AD}{CD} = \frac{S}{P}$ and $\frac{DB}{CD} = \frac{S}{Q}$. Dividing one equation by the other gives $\frac{AD}{DB} = \frac{Q}{P}$. Hence, **R** is nearer to the larger of **P** and **Q**, and the ratio of the distances of **R** from **P** and **Q** equals the ratio of the magnitudes Q and P, respectively.

Having established the latter rule, we now show how the same result would be obtained by taking moments about a point O in the plane of **P** and **Q**, and equating the moment of **R** to the sum of the moments of **P** and **Q**. This can be proved to be true by the introduction of equal and opposite **S** forces again but it is also physically logical that the turning effect of **R** should be the same as the total turning effect of **P** and **Q**.

As shown in Figure 2.7, we draw a line through O perpendicular to $\mathbf{F}_1$, $\mathbf{F}_2$ and **R**. Let b, a_1 and a_2 be the distances of O from $\mathbf{F}_1$, $\mathbf{F}_1$ from **R** and $\mathbf{F}_2$ from **R**, respectively.

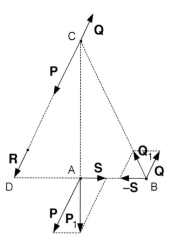

Figure 2.8. Constructing the resultant of two parallel forces acting in opposite directions.

Taking moments about O:

$$F_1 b + F_2(b + a_1 + a_2) = R(b + a_1) = (F_1 + F_2)(b + a_1)$$
$$= F_1 b + F_1 a_1 + F_2 b + F_2 a_1.$$

Cancelling like terms leaves us with

$$F_2 a_2 = F_1 a_1, \text{ i.e. } \frac{a_1}{a_2} = \frac{F_2}{F_1}.$$

Next we consider two parallel forces **P** and **Q** which act in opposite directions at the points A and B, respectively, of a rigid body. Again introduce equal and opposite forces of magnitude S at A and B and acting along the line AB as shown in Figure 2.8. This produces forces **P**$_1$ and **Q**$_1$ with lines of action which intersect at C. Considering them to be acting at C, the S-components cancel leaving a resultant **R** of magnitude $R = P - Q$ parallel to **P** and **Q** and acting through D. Assuming, as in this case, that $P > Q$, then **R** is on the opposite side of **P** from **Q**. By similar triangles:

$$\frac{DC}{DB} = \frac{Q}{S} \quad \text{and} \quad \frac{DC}{DA} = \frac{P}{S}.$$

Dividing one equation by the other gives:

$$\frac{DA}{DB} = \frac{Q}{P}.$$

Hence, **R** is on the outside of the larger of **P** and **Q**, and the ratio of the distances of **R** from **P** and **Q** equals the ratio of the magnitudes Q and P, respectively.

To obtain the same rule by taking moments, draw a line through O perpendicular to **F**$_1$, **F**$_2$ and **R** as shown in Figure 2.9. Let b, a_1 and a_2 be the distances of O from **F**$_2$, **F**$_1$ from **R** and **F**$_2$ from **R**, respectively. Taking moments about O:

$$F_1(b + a_2 - a_1) - F_2 b = R(a_2 + b) = (F_1 - F_2)(a_2 + b) = F_1(a_2 + b) - F_2(a_2 + b).$$

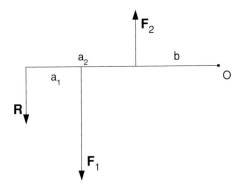

Figure 2.9. Finding the resultant **R** of two opposite parallel forces by taking moments.

Cancelling like terms leaves us with:

$$-F_1a_1 = -F_2a_2, \text{ i.e. } \frac{a_1}{a_2} = \frac{F_2}{F_1}.$$

EXERCISE 4

Two parallel forces of magnitudes 2 N and 3 N are 0.5 m apart and acting in the same direction. Find the magnitude, direction and line of action of the resultant force.

EXERCISE 5

Two parallel forces of magnitudes 1 N and 3 N are 0.4 m apart and acting in opposite directions. Find the magnitude, direction and line of action of the resultant force.

Problems 13 and 14.

2.4   Couples

We now turn our attention to the situation in which two forces acting on a rigid body are parallel and opposite in direction as in the previous exercise but this time the magnitudes of the two forces are equal. The trick of introducing equal and opposite forces of magnitude S (see Figure 2.10) does not help this time since the resulting forces (marked $\mathbf{F}_1$ in the diagram) are also equal, opposite and parallel.

Presumably, we can assume that the resultant force is still the vector sum of the original forces. Thus $\mathbf{R} = \mathbf{F} - \mathbf{F} = \mathbf{0}$ and in this case there is no resultant force. However, the two forces obviously have a turning effect, so let us investigate their moment.

Suppose the two forces have points of application A and B with $AB = a$ and also with AB perpendicular to the forces $\mathbf{F}$ and $-\mathbf{F}$ (see Figure 2.10). Then extend AB to C with $BC = b$ and take moments about C. Since a force can be moved to any point along its line of action, there is no restriction on the position of C other than that it be in the plane of $\mathbf{F}$ and $-\mathbf{F}$. The total moment of the two forces $\mathbf{F}$ and $-\mathbf{F}$ about C is:

$$M_c = F(a + b) - Fb = Fa.$$

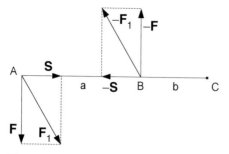

Figure 2.10. Equal, opposite and parallel forces.

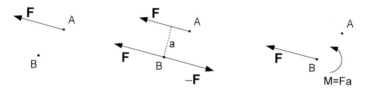

Figure 2.11. Equivalence of a force at A to a force at B and a couple.

Hence, the two forces **F** and −**F** have no resultant force but they have a turning moment equal to *Fa* which is the same when measured about any point C in the plane of **F** and −**F**. Such a pair of forces is referred to as a *couple* and its effect is completely characterized by its moment. Hence, any two couples with the same moment (which includes the same sense, i.e. either both anti-clockwise or both clockwise) are equivalent. Also any two couples with moments M_1 and M_2 applied to a rigid body are equivalent to a single couple with moment $M_1 + M_2$.

The effect of a force **F** with one line of action is the same as that of the force **F** with a different but parallel line of action together with a couple and vice versa. This is illustrated by the sequence of diagrams in Figure 2.11.

Starting with the force **F** at A, the situation is unaffected by the introduction of forces **F** and −**F** at B. However, the force **F** at A and −**F** at B constitute a couple of moment *Fa*, where *a* is the distance between the lines of action of **F** through A and −**F** through B. Hence, the force **F** at A is equivalent to the force **F** at B together with a couple of moment $M = Fa$.

EXERCISE 6

A hoisting drum, as sketched in Figure 2.12, has a diameter of 0.6 m and the cable tension applies a tangential force $F = 5$ kN. A drive motor applies a torque (couple) of moment $M = 1.5$ kN m. What is the magnitude, direction and location of the resultant force acting on the drum due to these two factors?

Problems 15 and 16.

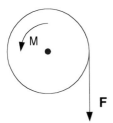

Figure 2.12. A hoisting drum.

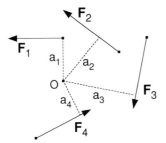

Figure 2.13. Four coplanar forces acting on a rigid body.

2.5 Equations of equilibrium of coplanar forces

A rigid body will be in statical equilibrium if the forces acting on it have no translational effect and no rotational effect. This means that the resultant force must be zero and there must be no resultant couple. This will be so if both of the following conditions apply.
1. The vector sum of the forces is zero.
2. The sum of the moments of the forces about any point is zero.
   For example, for the forces $\mathbf{F}_1$, $\mathbf{F}_2$, $\mathbf{F}_3$ and $\mathbf{F}_4$ corresponding to the vectors shown in Figure 2.13 to be in equilibrium, we must have both:

1. $\mathbf{F}_1 + \mathbf{F}_2 + \mathbf{F}_3 + \mathbf{F}_4 = \mathbf{0}$ and
2. $F_1 a_1 + F_2 a_2 - F_3 a_3 + F_4 a_4 = 0.$

Since the first equation is a vector equation, it would usually be split into two scalar equations by *resolving* into Cartesian components. Thus equation 1 above would be replaced by the two equations:

1_a. $F_{1x} + F_{2x} + F_{3x} + F_{4x} = 0$
1_b. $F_{1y} + F_{2y} + F_{3y} + F_{4y} = 0.$

Also, if it makes it any easier, equation 2 may be replaced by equating to zero the sum of the moments of both the x- and y-components of the forces. If $\mathbf{F}$ in Figure 2.14

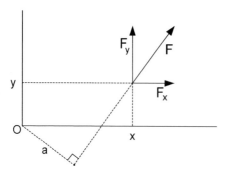

Figure 2.14. Moment of **F** equals the sum of moments of its components.

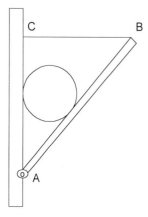

Figure 2.15. A smooth sphere wedged between a plank and a wall.

represents a force acting at a point with coordinates (x, y), its moment about the origin O is $Fa = F_y x - F_x y$, which is the sum of the moments of the y- and x-components of **F** about O.

Alternatively, three equations sufficient to ensure equilibrium may be obtained by taking moments about three non-collinear points. The sum of the moments of the forces about each of the points must be zero. It is not sufficient to take moments about two points since the forces may have a resultant force with line of action passing through both of those points. However, in that case, the force would have a moment about a third non-collinear point.

EXERCISE 7

A uniform smooth sphere of radius 0.2 m and weighing 500 N rests between a support AB and a vertical wall as shown in Figure 2.15. The support is a uniform rigid plank, of weight 200 N and length 1.3 m, hinged at A and kept at an angle of 40° to the vertical by a light horizontal cable CB attached to the wall at C.

Firstly, find the forces acting on the sphere from the wall and the plank. Secondly, find the tension in the cable CB and thirdly, find both the magnitude and the direction of the reactive force at the hinge A.

2.6 Applications

This section on 'applications' lists a series of physical problems which can be solved by using techniques already developed. Try to solve them yourself first but refer to the worked solutions given in Section 2.7 if you experience difficulties.

EXERCISE 8

A uniform solid cube of weight W rests on a horizontal surface but is attached to the surface with a smooth hinge along a bottom edge denoted by A in Figure 2.16. What is the minimum magnitude of the horizontal force **F**, shown in the diagram, which is necessary to topple the cube? Also, what will be the reaction through A when that minimum force **F** is applied?

EXERCISE 9

A light rigid beam AB is secured in a horizontal position at end A, as shown in Figure 2.17, and supports a weight **W** at B. If $W = 50\,\mathrm{N}$ and $AB = 2\,\mathrm{m}$, find the reaction (force and couple) at A.

EXERCISE 10

The same light beam AB, as in Exercise 9, has fixed supports at A and C, as shown in figure 2.18. With the same weight **W**, find the reactions at the supports A and C, given that the distance along the beam between A and C is 0.2 m.

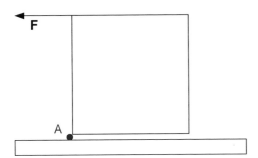

Figure 2.16. Toppling a block.

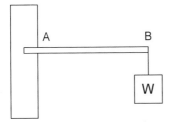

Figure 2.17. A cantilever beam projecting from a wall.

Figure 2.18. Cantilever beam with two supports at A and C.

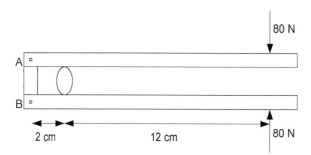

Figure 2.19. Nutcracker.

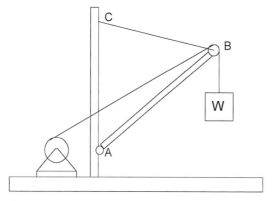

Figure 2.20. A boom hoist.

EXERCISE 11

A nutcracker is squeezed with a force of 80 N on each handle, as shown in Figure 2.19. Find the force transmitted to the nut and also the tension in the linkage AB.

EXERCISE 12

Figure 2.20 illustrates a boom hoist with hoisting drum. CB is a light cable which supports a light boom at B. The boom is smoothly hinged to the support at A. Neglecting the weight of the hoisting cable which passes over a smooth pulley at B, find the compressive force in the boom AB and the tension in the supporting cable CB if the weight **W** exerts a downward force of 4 kN and the angles made to the horizontal by CB, AB and the hoisting cable are 17°, 40° and 27°, respectively.

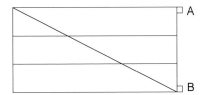

Figure 2.21. A gate with hinges at A and B.

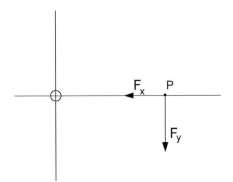

Figure 2.22. Force **F** split into x- and y-components.

EXERCISE 13

A gate of breadth 2 m, as shown in Figure 2.21, has two hinges A and B with A 1 m above B. The weight is supported entirely at B and A will pull away from its mounting if the horizontal pull there exceeds 1550 N. If the weight of the gate is 500 N, what is the maximum safe distance away from A for a person weighing 700 N to sit on the gate?

Problems 17, 18, 19 and 20.

2.7   **Answers to exercises**

1. Referring back to Figure 2.3, since the turning effect of the force is clockwise about O, the moment is given as a negative quantity. The magnitude of the moment is the magnitude of the force multiplied by the distance of the line of action of the force from O. Hence, we can write the moment as:

$$M_\mathrm{o} = -F \cdot OP \sin 60° = -10 \times 0.5 \frac{\sqrt{3}}{2} = -\frac{5\sqrt{3}}{2}\,\mathrm{N\,m}.$$

2. The force can be taken to act at any point along its line of action, so let us suppose that it acts at P on the x-axis as shown in Figure 2.3. Then, by the theorem of Varignon, its moment about O is equal to the sum of the moments of its x- and y-components, shown as F_x and F_y in Figure 2.22. However, the x-component F_x acts through O and thus has zero moment about O. Therefore, the moment is:

$$M_\mathrm{o} = -F_y \cdot OP = -10 \cos 30° \times 0.5 = -\frac{5\sqrt{3}}{2}\,\mathrm{N\,m}.$$

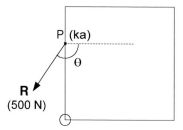

Figure 2.23. Resultant force acting on a square plate.

3. Referring back to Figure 2.5 showing forces acting on a square plate, let corner O be the origin of Cartesian coordinates with the x-axis being the side along which the 300 N force acts and the y-axis the side along which the 500 N force acts, both forces acting in negative directions.

    Since the resultant force is the vector sum of the separate forces, the x-component of the resultant will be the sum of the x-components of the separate ones, i.e.

    $$R_x = 200\cos 60° - 200\cos 60° - 300 = -300.$$

    Similarly, for the y-components, i.e.

    $$R_y = -500 - 200\sin 60° + 200\sin 60° + 100 = -400.$$

    Hence, the magnitude of the resultant is

    $$R = \sqrt{R_x^2 + R_y^2} = 500\,\text{N}.$$

    The direction of **R** is such that both the x- and y-components are negative with angle θ to the positive x-axis such that

    $$\tan\theta = \frac{R_y}{R_x} = \frac{4}{3}, \text{ i.e. } \theta = -127°.$$

    To find the line of action of the resultant, let the length of each side of the square plate be $4a$ and take moments about O. Then the moment of the resultant is

    $$M_o = -200\cos 60° \times 2a + 200\cos 60° \times 4a + 200\sin 60° \times 2a + 100 \times 3a$$
    $$= -200a + 400a + 200\sqrt{3}a + 300a = 846 \cdot 4a.$$

    Suppose that, as shown in Figure 2.23, the resultant acts at a point P on the side corresponding to the y-axis such that the y-coordinate of P is ka. Since the moment about O of the y-components of **R** acting at P will be zero, it follows that $M_o = -R_x \times ka = 300ka$. Hence, $k = \frac{846.4}{300} = 2.82$ and P is $\frac{2.82}{4} = 0.705$ of the way along the y-axis side from O.

4. The resultant **R** acts as shown in Figure 2.24 with magnitude $R = 2 + 3 = 5\,\text{N}$. The distance of **R** from the original forces is determined by $\frac{a}{b} = \frac{3}{2}$ and $a + b = 0.5$

    Therefore, $\frac{3}{2}b + b = \frac{5}{2}b = 0.5, \ b = 0.2\,\text{m}.$
    Then, $a = 0.5 - b = 0.5 - 0.2 = 0.3\,\text{m}.$

5. The resultant **R** acts as shown in Figure 2.25, i.e. with direction of the larger force and on the opposite side of the larger force from the smaller one. The magnitude $R = 3 - 1 = 2\,\text{N}$. The line of action of **R** is determined by:

    $$\frac{0.4 + a}{a} = \frac{3}{1}, \text{ i.e. } 0.4 + a = 3a, \ 2a = 0.4, \ a = 0.2\,\text{m}.$$

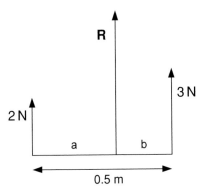

Figure 2.24. Resultant of parallel forces acting in the same direction.

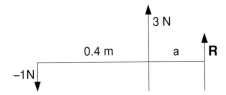

Figure 2.25. Resultant of parallel forces with opposite directions.

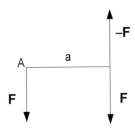

Figure 2.26. Finding the resultant force on a hoisting drum.

6. The force $\mathbf{F}$ and couple of moment M can be represented as shown in Figure 2.26. We replace the couple by another force $-\mathbf{F}$ on the rim of the drum and a force $\mathbf{F}$ at A, a distance a to the left of $-\mathbf{F}$. The moment of the couple is $M = Fa$, so the distance $a = \frac{M}{F}$. Since the $\mathbf{F}$ and $-\mathbf{F}$ applied to the rim of the drum cancel one another, we are left with a resultant force $\mathbf{F}$ at A. $\mathbf{F}$ has magnitude 5 kN and acts vertically downwards. The location of the resultant is the point A which is distance $a = \frac{1.5}{5} = 0.3$ m to the left of the right-hand most point of the rim. 0.3 m is the length of the radius of the drum, so A must be the centre of the drum.

7. Since the sphere is smooth, the forces of contact with the wall and plank must be perpendicular to the surface of contact in each case and must therefore act through the centre of the sphere. Consequently, there are three forces acting on the sphere: $\mathbf{R}_1$ from the wall, $\mathbf{R}_2$ from the plank and $\mathbf{W}$ the weight, i.e. the gravitational force on the sphere. To be in equilibrium they must be concurrent, which they are because they all act through the centre of the sphere, and they must obey the triangle of forces as shown in Figure 2.27. The magnitudes of $\mathbf{R}_1$ and $\mathbf{R}_2$ can be found using elementary trigonometry as

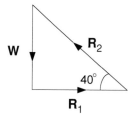

Figure 2.27. Triangle of forces acting on the sphere shown in Figure 2.15.

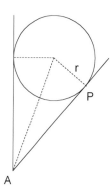

Figure 2.28. Geometry to determine distance AP.

follows:

$$\frac{W}{R_1} = \tan 40°, \quad R_1 = W \cot 40° = 596\,\text{N}, \ \text{since}\ W = 500\,\text{N}.$$

$$\frac{W}{R_2} = \sin 40°, \quad R_2 = W \csc 40° = 778\,\text{N}.$$

Before considering the equilibrium of forces acting on the plank, we need to find where the contact force $-\mathbf{R}_2$ from the sphere acts on the plank. Referring to Figure 2.28:

$$\frac{r}{AP} = \tan 20°, \quad AP = r \cot 20° = 0.549\,\text{m}.$$

Referring now to Figure 2.29, we see that the forces acting on the plank are: its weight $\mathbf{W}_g$ acting like a single force at the centre point G, the tension $\mathbf{T}$ in the cable, the contact force $-\mathbf{R}_2$ from the sphere acting at P and a reactive force $\mathbf{R}$ at the hinge A. The hinge is smooth so there is no reactive couple. Let $\mathbf{R}$ act at an angle α to the upward vertical as shown in the diagram.

There are three equations of equilibrium and they are all needed to calculate the three unknowns T, R and α. By taking moments about A, $\mathbf{R}$ is not involved and we can find the tension $\mathbf{T}$ as follows.

$$T \cdot AB \cos 40° - R_2 \cdot AP - W_g \cdot AG \sin 40° = 0$$

$$T = \frac{R_2 \cdot AP + W_g \cdot AG \sin 40°}{AB \cos 40°} = \frac{778 \times 0.549 + 200 \times 0.65 \sin 40°}{1.3 \cos 40°} = 513\,\text{N}.$$

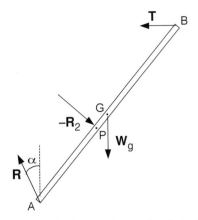

Figure 2.29. Forces acting on the plank shown in Figure 2.15.

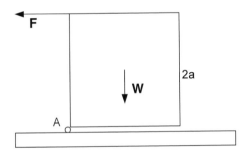

Figure 2.30. Calculation of force required to topple a block.

The other two equations, necessary to find $\mathbf{R}$ and α, are given by equating to zero firstly the sum of the horizontal components of the forces and secondly the sum of the vertical components.

$$-T + R_2 \cos 40° - R \sin \alpha = 0$$
$$-W_g - R_2 \sin 40° + R \cos \alpha = 0.$$

Hence $R \sin \alpha = -T + R_2 \cos 40° = 83$
and $R \cos \alpha = W_g + R_2 \sin 40° = 700.$

Square the two equations and then add to give:

$$R^2 = (83)^2 + (700)^2, \quad R = 705 \, \text{N}.$$

Next, divide the first equation by the second to give:

$$\tan \alpha = \frac{83}{700}, \quad \alpha = 6.76°.$$

Notice that learning the skills for solving problems in mechanics is a cumulative process. For instance, in the problem just solved, we have used the triangle of forces from Chapter 1 and three equations for equilibrium from Chapter 2.

8.   Referring to Figure 2.30, if the force $\mathbf{F}$ increased gradually from zero, the cube would eventually topple when the moment of $\mathbf{F}$ about the hinge A exceeded the moment of the weight $\mathbf{W}$ in the

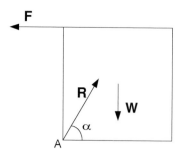

Figure 2.31. Reactive force **R** when the block is on the verge of toppling.

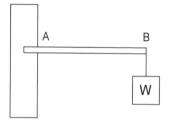

Figure 2.32. A cantilever beam projecting from a wall.

opposite direction. At that stage the only reaction from the supporting surface would be a force acting through A. There would be no reactive couple since the hinge is smooth.

Let the side of the cube be $2a$ so that by taking moments about A, the cube will topple when:

$$F \cdot 2a > W \cdot a, \text{ i.e. } F > \frac{W}{2}.$$

When the cube is just about to topple, it will be in equilibrium with the magnitude of **F** given by $F = W/2$. Denote the reaction at A by the force **R** acting at an angle α to the horizontal as shown in Figure 2.31. The sum of components of force in the horizontal direction must be zero and so must the sum of components in the vertical direction. Hence:

$R \cos \alpha - F = 0, \text{ i.e. } R \cos \alpha = W/2,$

$R \sin \alpha - W = 0, \text{ i.e. } R \sin \alpha = W.$

Square and add: $R^2 = \frac{5}{4} W^2$, $R = \frac{\sqrt{5}}{2} W$.
Divide: $\tan \alpha = 2$, $\alpha = 63.4°$.

9.   Figure 2.32 is a repeat of Figure 2.17 for ease of reference. Let $AB = a$. Then the force **W** acting on the beam at B is equivalent to a force **W** acting at A together with a couple of moment $-Wa$. (The minus sign indicates that the moment is clockwise.) Hence, the reaction at A must be a force $-\mathbf{W}$ (vertically upwards) together with a couple af moment Wa (anti-clockwise), i.e. a vertically upward force of 50 N and couple of moment 100 N m.

10.   Figure 2.33 corresponds to Figure 2.18 with the supports at A and C and the suspended weight replaced by the corresponding forces acting on the beam. Let $AC = a = 0.2$ m and $CB = b = 1.8$ m. Take

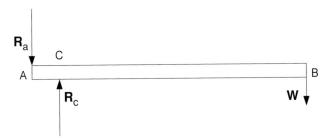

Figure 2.33. Forces acting on a cantilever beam.

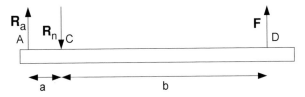

Figure 2.34. Forces on one handle of a nutcracker.

moments firstly about C and secondly about A as follows:

$$R_a \cdot a - W \cdot b = 0, \quad R_a = \frac{b}{a} W = \frac{1.8}{0.2} \times 50 = 450 \, \text{N}$$

$$R_c \cdot a - W(a + b) = 0, \quad R_c = \frac{a + b}{a} W = \frac{2}{0.2} \times 50 = 500 \, \text{N}.$$

11.   Because of the symmetry, we only need consider the equilibrium of one handle of the nutcracker under the action of three parallel forces as shown in Figure 2.34. These are the applied force **F** of magnitude 80 N, the reaction $\mathbf{R}_n$ from the nut (assuming that the nut has not broken at this stage) and the tension $\mathbf{R}_a$ in the linkage AB. Label the application points of $\mathbf{R}_n$ and **F** as C and D, respectively, with $AC = a$ and $CD = b$. Now, take moments about A and C as follows:

$$(a + b)F - aR_n = 0, \quad R_n = \frac{a + b}{a} F = \frac{14}{2} \times 80 = 560 \, \text{N}$$

$$bF - aR_a = 0, \quad R_a = \frac{b}{a} F = \frac{12}{2} \times 80 = 480 \, \text{N}.$$

Hence, with the application of a squeezing force of 80 N, the force transmitted to the nut is 560 N and the tension in the linkage is 480 N.

12.   Referring to the diagram of the boom hoist in Figure 2.20, we note firstly that the hinge A is smooth and so is the pulley at B. Hence, no couple is applied to the beam. Also, by neglecting the weight of the boom, the latter is acted on by only two forces: the hinge reaction at A and a force resulting from the cable tensions at B. For the boom to be in equilibrium, the two forces acting at A and B must be equal, opposite and collinear. They must therefore act along the beam and form the compressive force in the beam.

Figure 2.35 shows the forces acting at B which must be in equilibrium. The two **W** forces correspond to the tensions in the hoisting cable on either side of the pulley. **T** is the tension in the supporting cable CB and $\mathbf{R}_b$ corresponds to the reaction force from the boom itself (which is equal and opposite to the compressive force). Resolving, i.e. taking components of forces, perpendicular to the direction

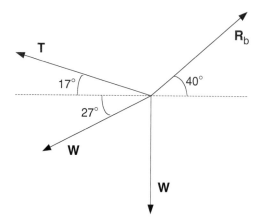

Figure 2.35. Forces acting on the end B of the boom shown in Figure 2.20.

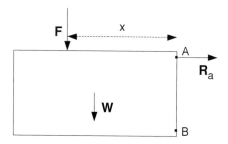

Figure 2.36. Forces on a gate which have moment about hinge B.

AB gives:

$$T \cos 33° + W \cos 77° - W \cos 40° = 0$$

$$T = \frac{W(\cos 40° - \cos 77°)}{\cos 33°} = 4 \times 0.6452 = 2.58 \, \text{kN}.$$

Resolving in the direction *AB* gives:

$$R_b - T \sin 33° - W \sin 77° - W \sin 40° = 0$$

$$R_b = T \sin 33° + W(\sin 77° + \sin 40°) = 7.87 \, \text{kN}.$$

13.   In this problem, we simply take moments about B so that the reaction at hinge B is not involved. Hence, referring to Figure 2.36, with **F** representing the weight of the person,

$$F \cdot x + W \cdot 1 - R_a \cdot 1 = 0 \quad \text{or} \quad xF = R_a - W$$

for equilibrium. Substituting values and noting that $R_a \leq 1550 \, \text{N}$ gives:

$$700x \leq 1550 - 500 = 1050, \quad x \leq \frac{1050}{700} = 1.5 \, \text{m}.$$

The maximum safe distance from A is thus 1.5 m.

3 Centre of gravity

3.1 Coplanar parallel forces

The *centre of gravity* of a body is the point in the body through which the resultant gravitational force acts regardless of the orientation of the body. To show that such a unique point exists and then to identify the position of the point, it is helpful to start by considering the resultant of a few parallel forces acting at given points.

Figure 3.1 indicates two parallel forces $\mathbf{F}_1$ and $\mathbf{F}_2$ acting at points A_1 and A_2, respectively. We know that the resultant $\mathbf{R}_2$ will be such that $R_2 = F_1 + F_2$ and

$$\frac{A_1 C_2'}{C_2' A_2'} = \frac{F_2}{F_1} = \frac{A_1 C_2'}{C_2' A_2''} = \frac{A_1 C_2 \cos\alpha}{C_2 A_2 \cos\alpha} = \frac{A_1 C_2}{C_2 A_2}.$$

Thus we may regard the resultant $\mathbf{R}_2$ as being a force acting at the point C_2 on the line $A_1 A_2$ and such that $\frac{A_1 C_2}{C_2 A_2} = \frac{F_2}{F_1}$. Furthermore, this is true regardless of the angle α, i.e. for any direction of the parallel forces in this plane of action.

If now we have a third parallel force $\mathbf{F}_3$ acting at another point A_3, as shown in Figure 3.2, the three forces $\mathbf{F}_1$, $\mathbf{F}_2$ and $\mathbf{F}_3$ will have a resultant $\mathbf{R}_3$ with $R_3 = R_2 + F_3 = F_1 + F_2 + F_3$ and acting through the point C_3 on the line $C_2 A_3$ such that $\frac{C_2 C_3}{C_3 A_3} = \frac{F_3}{R_2}$. Again, this resultant $\mathbf{R}_3$ will act through the point C_3 regardless of the direction of the parallel forces.

Obviously, this procedure may be continued for any number n of parallel forces which then have a resultant $\mathbf{R}_n$ such that $R_n = \sum_{i=1}^{n} F_i$ and which acts through a unique point C_n regardless of the direction of the parallel forces. C_n may be called the *centre of action* of the parallel forces $\mathbf{F}_1, \mathbf{F}_2, \ldots, \mathbf{F}_n$.

Although the above procedure shows that a unique centre of action C_n does exist, to locate it requires a more direct method. An easy way of doing this uses Cartesian coordinates of the points of action together with the concept of moments discussed in Chapter 2. Figure 3.3 shows just two of the n force vectors, firstly pointing in the positive y-direction and secondly pointing in the negative x-direction. For n such forces, there will be a resultant of magnitude R_n acting in the same direction and a unique centre of

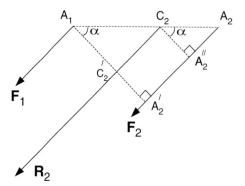

Figure 3.1. Resultant of two parallel forces.

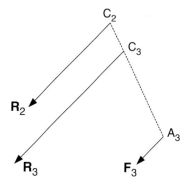

Figure 3.2. Resultant of three parallel forces.

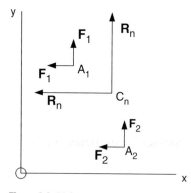

Figure 3.3. Using moments to find the centre of action of parallel forces.

action C_n. Let the coordinates of the point of action of F_i, $i = 1, 2, \ldots, n$, be denoted by $A_i(x_i, y_i)$ and for the centre of action by $C_n(x_c, y_c)$.

Now, the turning effect of the resultant about any point must be the same as the sum of the turning effects of the separate forces. Hence, taking moments about O for the forces acting in the positive y-direction gives:

$$R_n x_c = \sum_{i=1}^{n} F_i x_i$$

$$R_n = \sum_{i=1}^{n} F_i, \text{ so } x_c = \frac{\sum_{i=1}^{n} F_i x_i}{\sum_{i=1}^{n} F_i}.$$

Next, take moments about O of the forces acting in the negative x-direction to obtain:

$$R_n y_c = \sum_{i=1}^{n} F_i y_i, \text{ i.e. } y_c = \frac{\sum_{i=1}^{n} F_i y_i}{\sum_{i=1}^{n} F_i}.$$

EXERCISE 1

Assuming consistent units of distance and force, find the centre of action of four parallel forces of magnitudes $F_1 = 1$, $F_2 = 2$, $F_3 = 3$ and $F_4 = 4$ acting at points with Cartesian coordinates $A_1(1, 2)$, $A_2(-2, 1)$, $A_3(-1, 2)$ and $A_4(1, -2)$, respectively.

Problems 21 and 22.

3.2 Non-coplanar parallel forces

Consider n parallel forces $\mathbf{F}_1, \mathbf{F}_2, \ldots, \mathbf{F}_n$ which act at non-coplanar points A_1, $A_2, \ldots, A_n$, respectively. By the same reasoning as before, the resultant $\mathbf{R}_2$ of $\mathbf{F}_1$ and $\mathbf{F}_2$ has magnitude $R_2 = F_1 + F_2$ and acts through a point C_2 on the line $A_1 A_2$ for any direction of the two parallel forces.

Although not necessarily lying in the plane of $\mathbf{F}_1$ and $\mathbf{F}_2$, the two parallel forces $\mathbf{R}_2$ and $\mathbf{F}_3$ must be coplanar. Therefore, $\mathbf{R}_2$ and $\mathbf{F}_3$ have a resultant $\mathbf{R}_3$ of magnitude $R_3 = R_2 + F_3 = F_1 + F_2 + F_3$ and acting through a point C_3 on the line $C_2 A_3$ for any direction of the three parallel forces.

This procedure may be continued, bringing in one extra force at a time, until we arrive at the resultant $\mathbf{R}_n$ of magnitude $R_n = F_1 + F_2 + \cdots + F_n$ and passing through a unique point C_n for any direction of the n parallel forces. Hence, C_n is the *centre of action* of the set of n non-coplanar forces.

The location of C_n is found using three-dimensional Cartesian coordinates. Each point of action A_i has its position specified by its x-, y- and z-coordinates, i.e. $A_i(x_i, y_i, z_i)$. Similarly, we denote the position of C_n as $C_n(x_c, y_c, z_c)$. Also, since we now have a problem in three-dimensional space, moments of forces must be taken about an axis rather than a point.

Figure 3.4 shows $\mathbf{F}_i$ as a representative of n parallel forces acting at a point $A_i(x_i, y_i, z_i)$ and $\mathbf{R}_n$ as the resultant force acting at $C_n(x_c, y_c, z_c)$. To aid the following analysis, the forces are shown in turn parallel to the x-, y- and z-axes.

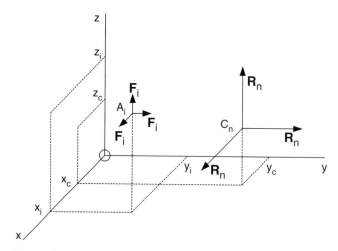

Figure 3.4. Parallel forces in three-dimensional space.

We start by considering the parallel forces acting in the positive y-direction and take moments about the z-axis. The moment of the resultant must equal the sum of the moments of the separate forces, i.e. $R_n x_c = \sum_{i=1}^{n} F_i x_i$. Hence, it follows that:

$$x_c = \frac{\sum_{i=1}^{n} F_i x_i}{R_n} = \frac{\sum_{i=1}^{n} F_i x_i}{\sum_{i=1}^{n} F_i}.$$

Next, let the forces act in the positive z-direction and take moments about the x-axis. This gives:

$$R_n y_c = \sum_{i=1}^{n} F_i y_i \quad \text{and} \quad y_c = \frac{\sum_{i=1}^{n} F_i y_i}{\sum_{i=1}^{n} F_i}.$$

Finally, let the forces act in the positive x-direction and take moments about the y-axis to give:

$$R_n z_c = \sum_{i=1}^{n} F_i z_i \quad \text{and} \quad z_c = \frac{\sum_{i=1}^{n} F_i z_i}{\sum_{i=1}^{n} F_i}.$$

EXERCISE 2

Assuming consistent units of distance and force, find the centre of action of four parallel forces of magnitudes $F_1 = 2$, $F_2 = 3$, $F_3 = 1$ and $F_4 = 4$ acting at points with Cartesian coordinates $A_1(1, 2, -2)$, $A_2(2, -1, -2)$, $A_3(-1, 1, 2)$ and $A_4(0, 2, 1)$, respectively.

Problem 23.

3.3   Finding c.g. positions of uniform plane laminas without using calculus

By 'uniform', we mean that the plane lamina has constant density, i.e. constant weight per unit area. Sometimes the position of the centre of gravity (c.g.) is obvious. For instance, if the lamina is a circular disc, the c.g. is at its centre. If it is in the shape of a parallelogram, as in Figure 3.5, the c.g. must be at G, the intersection of the lines joining the mid-points of opposite sides.

Also, if the lamina can be divided into sections with known c.g. positions, the c.g. for the whole may be found using the method developed in Section 3.1. Consider the lamina in Figure 3.6. It may be divided into equal square sections which have centres of gravity at $A_1(1, 3)$, $A_2(1, 1)$ and $A_3(3, 1)$. Finding the c.g. for the whole lamina is the equivalent to finding the centre of action of three equal parallel forces acting at A_1, A_2 and A_3. The c.g. position is $G(x_g, y_g) = C_3(x_c, y_c)$. If we assume that the weight of each section is W, then $\sum_{i=1}^{3} F_i = 3W$. It follows that:

$$x_g = x_c = \frac{\sum_{i=1}^{3} F_i x_i}{\sum_{i=1}^{3} F_i} = \frac{W(1 + 1 + 3)}{3W} = \frac{5}{3}$$

$$\text{and } y_g = y_c = \frac{\sum_{i=1}^{3} F_i y_i}{\sum_{i=1}^{3} F_i} = \frac{W(3 + 1 + 1)}{3W} = \frac{5}{3}.$$

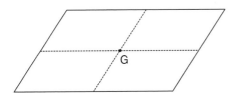

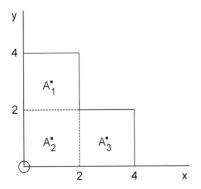

Figure 3.5. Centre of gravity position of uniform parallelogram lamina.

Figure 3.6. Finding the c.g. position of an L-shaped lamina.

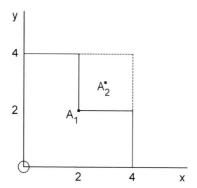

Figure 3.7. Finding the c.g. position using negative weight.

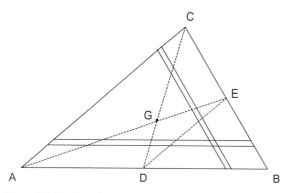

Figure 3.8. Finding the c.g. of a uniform triangular lamina.

Another trick which may be used is to imagine that a hole in a lamina contributes a negative force. With the same example but now referring to Figure 3.7, we can imagine a positive force $F_1 = 4W$ at A_1 and a negative force $F_2 = -W$ at A_2. Then for the L-shaped lamina, $\sum_{i=1}^{2} F_i = 4W - W = 3W$ and apply the formulae as before.

Hence, $x_g = x_c = \dfrac{\sum_{i=1}^{2} F_i x_i}{\sum_{i=1}^{2} F_i} = \dfrac{4W \cdot 2 - W \cdot 3}{3W} = \dfrac{5}{3}$

and $y_g = y_c = \dfrac{\sum_{i=1}^{2} F_i y_i}{\sum_{i=1}^{2} F_i} = \dfrac{4W \cdot 2 - W \cdot 3}{3W} = \dfrac{5}{3}$.

Another method, which will be particularly useful when using calculus, is to divide the lamina into parallel narrow strips. Consider a uniform triangular lamina ABC as shown in Figure 3.8. If it is divided into narrow strips parallel to side AB, the c.g. of each strip will be at its mid-point. Therefore, the c.g. of the whole lamina must lie on the straight line from D, the mid-point of AB, to C since CD passes through the mid-point of all the strips.

Similarly, if the lamina is divided into narrow strips parallel to side BC, the c.g. must lie on the line AE, where E is the mid-point of CB. Since, the c.g. is also on CD, it must

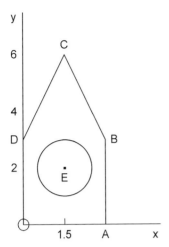

Figure 3.9. Finding the c.g. of a lamina with a circular hole.

be at G, the point of intersection of CD and AE. The location of G will now be found by the application of some simple Euclidean geometry.

The $\triangle$DBE is similar to the $\triangle$ABC since $DB = AB/2$, $EB = CB/2$ and the angle at B is common to both triangles. Therefore $DE \parallel AC$. It follows that $\triangle$GDE is similar to $\triangle$GCA since $\angle$ACD $=$ $\angle$CDE, $\angle$CAE $=$ $\angle$AED and $\angle$AGC $=$ $\angle$EGD. On comparing corresponding sides: $\frac{DG}{GC} = \frac{DE}{AC} = \frac{1}{2}$. Therefore, G lies one third of the way up the *median* from the base to the top corner.

EXERCISE 3

Find the position of the centre of gravity of the uniform lamina shown in Figure 3.9 as OABCD with a circular hole centred at $E(1.5, 2)$ and of radius 1. OABD is a square of side 3 and DBC is a triangle with apex at $C(1.5, 6)$.

Problems 24 and 25.

3.4 Using calculus to find c.g. positions of uniform plane laminas

Let the weight per unit area of the lamina be w. Then if the area is A, the total weight is $W = Aw$. Referring to Figure 3.10, let the bottom and top edges of the lamina be curves defined by the equations $y = y_1(x)$ and $y = y_2(x)$, where $y_1(x)$ and $y_2(x)$ are given functions of x. The area of a narrow strip, parallel to the y-axis and of width dx, will be $dA = [y_2(x) - y_1(x)]\,dx$. Adding up for all such strips from the left at $x = x_1$ to the right at $x = x_2$ gives the total area which is expressed as the integral:

$$A = \int_{x_1}^{x_2} [y_2(x) - y_1(x)]\,dx.$$

If we imagine gravity acting down on the lamina, perpendicular to the lamina and to the x, y plane, the moment about the y-axis of the total weight acting at G must be the

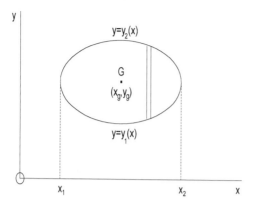

Figure 3.10. Finding the c.g. of a uniform lamina by calculus.

same as the sum of the moments about the y-axis of the weights of the strips. The strip is distance x from the y-axis, so the moment of its weight is $xw[y_2(x) - y_1(x)]\,dx$. w is constant, so summing over all the strips gives:

$$Awx_g = w\int_{x_1}^{x_2} x[y_2(x) - y_1(x)]\,dx$$

and dividing by Aw gives x_g.

To find the y-coordinate of G, we take moments about the x-axis. The centre of gravity of the strip is at its mid-point with y-coordinate $[y_1(x) + y_2(x)]/2$. Hence, the moment about the x-axis of the weight of the strip is:

$$\frac{1}{2}[y_1(x) + y_2(x)]w[y_2(x) - y_1(x)]\,dx = \frac{w}{2}\left[y_2^2(x) - y_1^2(x)\right]dx.$$

Summing over all the strips gives:

$$Awy_g = \frac{w}{2}\int_{x_1}^{x_2}\left[y_2^2(x) - y_1^2(x)\right]dx$$

and dividing by Aw gives y_g.

Let us illustrate this technique with a uniform triangular lamina bounded by the x- and y-axes and the line $y = (6 - x)/2$, as shown in Figure 3.11. Thus, $y_1(x) = 0$, $y_2(x) = (6 - x)/2$, $x_1 = 0$ and $x_2 = 6$. The area:

$$A = \int_0^6 \frac{6 - x}{2}\,dx = \frac{1}{2}\left[6x - \frac{x^2}{2}\right]_0^6 = (36 - 18)/2 = 9,$$

which agrees with the formula $\frac{1}{2}$ base $\times$ height.

$$9wx_g = w\int_0^6 x\frac{6 - x}{2}\,dx = \frac{w}{2}\left[3x^2 - \frac{x^3}{3}\right]_0^6 = \frac{w}{2}(108 - 72) = 18w, \ x_g = 2.$$

$$9wy_g = \frac{w}{2}\int_0^6 \frac{(6 - x)^2}{4}\,dx = \frac{w}{8}\left[-\frac{(6 - x)^3}{3}\right]_0^6 = 9w, \ y_g = 1.$$

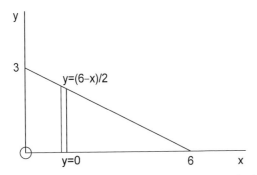

Figure 3.11. Finding the c.g. of a uniform triangular lamina.

Hence, the coordinates of the c.g. are $G(2,1)$, which corresponds to the position of the c.g. being one third the way up the median from the base.

EXERCISE 4

Find the position of the centre of gravity of a uniform semi-circular plane lamina of radius a.

EXERCISE 5

Find the position of the centre of gravity of a uniform plane lamina bounded by the x- and y-axes and the curve: $y^2 = 4 - x$ with $y > 0$.

Problems 26 and 27.

3.5   Centre of gravity positions of uniform solid bodies of revolution

The curved surface of a body of revolution corresponds to a curve in the x, y plane, $y = f(x)$, which is then rotated about the x-axis. Since the solid body has uniform density, its centre of gravity must lie on its axis, which corresponds to the x-axis. Hence, all we need to find is the x-coordinate of the c.g.

Take a thin slice through the body as indicated in Figure 3.12. This slice will be a circular lamina of area πy^2, where $y = f(x)$. If w is the weight per unit volume of the body, the weight of the slice is $w\pi y^2 \, dx$, where dx is the thickness of the slice. The total weight W of the body is given by adding up the weight of all such slices from one end $x = x_1$ to the other $x = x_2$, i.e.

$$W = w\pi \int_{x_1}^{x_2} y^2 \, dx,$$

with $y = f(x)$.

To find the c.g. position, imagine gravity acting perpendicular to the x, y plane and take moments about the y-axis. The moment of the total weight, i.e. $W x_g$, will be equal

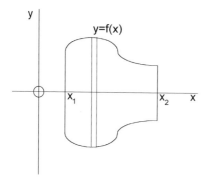

Figure 3.12. Slice through a body of revolution.

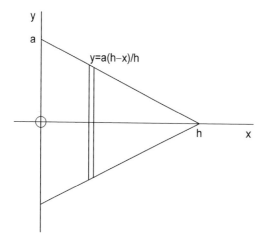

Figure 3.13. Finding the c.g. of a uniform solid cone.

to the sum of the moments of the slices. Hence,

$$W x_g = w\pi \int_{x_1}^{x_2} x y^2 \, dx.$$

Let us illustrate the method by finding the c.g. position of a uniform solid cone of height h and circular base of radius a. The curved surface is generated by rotating the line $y = \frac{a}{h}(h - x)$ about the x-axis with $x_1 = 0$ and $x_2 = h$, as indicated in Figure 3.13.

$$W = \frac{w\pi a^2}{h^2} \int_0^h (h - x)^2 \, dx = \frac{w\pi a^2}{h^2} \left[-\frac{1}{3}(h - x)^3 \right]_0^h = w\pi a^2 h/3$$

$$W x_g = \frac{w\pi a^2}{h^2} \int_0^h x(h - x)^2 \, dx = \frac{w\pi a^2}{h^2} \int_0^h (h^2 x - 2hx^2 + x^3) \, dx$$

$$= \frac{w\pi a^2}{h^2} \left[\frac{h^2 x^2}{2} - \frac{2hx^3}{3} + \frac{x^4}{4} \right]_0^h = w\pi a^2 h^2 \left(\frac{3}{4} - \frac{2}{3} \right) = \frac{1}{12} w\pi a^2 h^2.$$

$$x_g = \frac{1}{12} w \pi a^2 h^2 \times \frac{3}{w \pi a^2 h} = h/4.$$

Hence, the c.g. position of a uniform solid circular cone is one quarter the way up the axis from its base.

EXERCISE 6

Find the c.g. position of a uniform solid hemisphere of radius a.

EXERCISE 7

Find the c.g. position for a uniform solid body of revolution when its curved surface is generated by rotating about the x-axis the curve $y^2 = 4 - x$, with $0 \le x \le 4$, in appropriate units.

Problems 28 and 29.

3.6 Answers to exercises

1.   If the centre of action of the parallel forces has coordinates $C_4(x_c, y_c)$, then

$$x_c = \frac{\sum_{i=1}^{4} F_i x_i}{\sum_{i=1}^{4} F_i} = \frac{1 - 4 - 3 + 4}{10} = \frac{-2}{10} = -0.2$$

and $y_c = \dfrac{\sum_{i=1}^{4} F_i y_i}{\sum_{i=1}^{4} F_i} = \dfrac{2 + 2 + 6 - 8}{10} = \dfrac{2}{10} = 0.2.$

2.   The Cartesian coordinates of the centre of action $C_4(x_c, y_c, z_c)$ are found by applying the formulae derived in Section 3.2. Firstly, $\sum_{i=1}^{4} F_i = 2 + 3 + 1 + 4 = 10$. Then

$$x_c = \frac{1}{10} \sum_{i=1}^{4} F_i x_i = (2 + 6 - 1 + 0)/10 = 0.7,$$

$$y_c = \frac{1}{10} \sum_{i=1}^{4} F_i y_i = (4 - 3 + 1 + 8)/10 = 1.0,$$

$$z_c = \frac{1}{10} \sum_{i=1}^{4} F_i z_i = (-4 - 6 + 2 + 4)/10 = -0.4.$$

3.   The lamina (see Figure 3.14) is symmetrical about the line $x = 1.5$. Therefore, the x-coordinate of the centre of gravity must be $x_g = 1.5$.

Let the weight of the square part without a hole be $9W$. Then the weight of the triangular part will be $4.5W$ and the weight of the piece which is removed, to make a circular hole of radius 1, is $\pi r^2 W = \pi W$. The centre of gravity of the lamina is now equivalent to the centre of action of the parallel forces $F_1 = 9W$ at $A_1(1.5, 1.5)$, $F_2 = 4.5W$ at $A_2(1.5, 4)$ and $F_3 = -\pi W$ at $A_3(1.5, 2)$. Thus:

$$y_g = y_c = \frac{\sum_{i=1}^{3} F_i y_i}{\sum_{i=1}^{3} F_i} = \frac{(9 \times 1.5 + 4.5 \times 4 - \pi \times 2)W}{(9 + 4.5 - \pi)W} \approx 2.43.$$

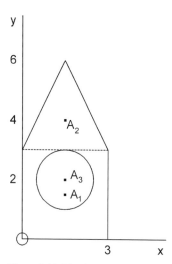

Figure 3.14. Finding the c.g. of a lamina with a circular hole.

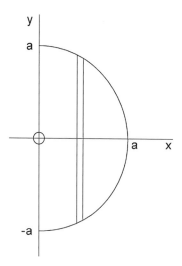

Figure 3.15. Finding the c.g. of a plane semi-circular lamina.

4. Position the semi-circle with base along the y-axis and centre at O, as shown in Figure 3.15. By symmetry, the c.g. must lie on the x-axis. For radius a, the area of the semi-circle is $A = \frac{1}{2}\pi a^2$.

Since the equation for the semi-circle is $x^2 + y^2 = a^2$, $x \geq 0$, it follows that:

$$y_1(x) = -(a^2 - x^2)^{1/2}, \quad y_2(x) = (a^2 - x^2)^{1/2}, \quad x_1 = 0 \quad \text{and} \quad x_2 = a.$$

Thus, $Awx_g = w \displaystyle\int_0^a x[(a^2 - x^2)^{1/2} + (a^2 - x^2)^{1/2}]\,dx$

$$= 2w \int_0^a x(a^2 - x^2)^{1/2}\,dx = 2w \left[-\frac{1}{3}(a^2 - x^2)^{3/2}\right]_0^a = \frac{2}{3}wa^3.$$

Therefore, $x_g = \dfrac{2}{3}wa^3 \times \dfrac{2}{\pi wa^2} = \dfrac{4a}{3\pi}$.

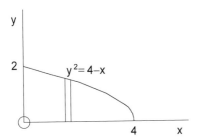

Figure 3.16. Lamina for finding the c.g. in Exercise 5.

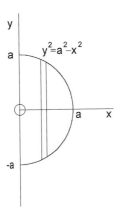

Figure 3.17. Finding the c.g. of a uniform solid hemisphere.

5. The lamina, as shown in Figure 3.16, is such that $y_1 = 0$, $y_2 = (4 - x)^{1/2}$, $x_1 = 0$ and $x_2 = 4$. The area of the lamina is:

$$A = \int_0^4 (4 - x)^{1/2}\, dx = \left[-\frac{2}{3}(4 - x)^{3/2} \right]_0^4 = \frac{16}{3}.$$

Then, $Awx_g = w \int_0^4 x(4 - x)^{1/2}\, dx$ and, integrating by parts,

$$= w \left[-\frac{2}{3}x(4 - x)^{3/2} \right]_0^4 + w \int_0^4 \frac{2}{3}(4 - x)^{3/2}\, dx$$

$$= w \left[-\frac{4}{15}(4 - x)^{5/2} \right]_0^4 = \frac{128}{15}w.$$

Therefore, $x_g = \frac{128w}{15} \times \frac{3}{16w} = \frac{8}{5}$.

$$Awy_g = \frac{w}{2}\int_0^4 (4 - x)\, dx = \frac{w}{2}\left[4x - \frac{x^2}{2} \right]_0^4 = \frac{w}{2}(16 - 8) = 4w \quad \text{and} \quad y_g = 4w \times \frac{3}{16w} = \frac{3}{4}.$$

6. The curved surface of the hemisphere is generated by rotating about the x-axis the curve $y = f(x) = (a^2 - x^2)^{1/2}$, with $0 \le x \le a$, as shown in profile in Figure 3.17. The weight of the hemisphere is:

$$W = w\pi \int_0^a y^2\, dx = w\pi \int_0^a (a^2 - x^2)\, dx = w\pi \left[a^2 x - \frac{x^3}{3} \right]_0^a = \frac{2}{3}w\pi a^3.$$

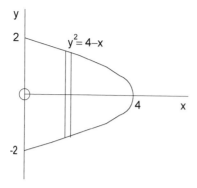

Figure 3.18. Body of revolution in Exercise 7.

Then, $Wx_g = w\pi \int_0^a xy^2\,dx = w\pi \int_0^a x(a^2 - x^2)\,dx = w\pi \left[\dfrac{a^2 x^2}{2} - \dfrac{x^4}{4}\right]_0^a = \dfrac{w\pi a^4}{4}.$

Therefore, $x_g = \dfrac{w\pi a^4}{4} \times \dfrac{3}{2w\pi a^3} = \dfrac{3}{8}a.$

7.   Referring to Figure 3.18,

$$W = w\pi \int_0^4 (4 - x)\,dx = w\pi \left[4x - \dfrac{x^2}{2}\right]_0^4 = 8w\pi.$$

$$Wx_g = w\pi \int_0^4 x(4 - x)\,dx = w\pi \left[2x^2 - \dfrac{x^3}{3}\right]_0^4 = w\pi \left(32 - \dfrac{64}{3}\right) = \dfrac{32w\pi}{3}.$$

$$x_g = \dfrac{32w\pi}{3} \times \dfrac{1}{8w\pi} = \dfrac{4}{3}.$$

4 Distributed forces

4.1 Distributed loads

The gravitational force acting on a body is distributed over the whole volume of the body. However, if the body is rigid, we can replace this system of distributed gravitational forces by a single resultant gravitational force (weight) acting through the centre of gravity of the body.

It is helpful sometimes to represent a system of distributed parallel forces by a so-called *load diagram*. Consider a horizontal beam AB with sand piled on it to a constant depth d, as illustrated in Figure 4.1. We can then represent the load due to the sand and the weight of the beam by a load diagram with constant intensity q, as shown in Figure 4.2. If the total weight of the sand over the beam is S, the weight of the beam is W and the length of the beam is a, then $qa = S + W$. Measuring distance in metres and force in newtons, the intensity q has units N/m.

If the sand is heaped with a varying depth d, as shown in Figure 4.3a, then the corresponding load diagram, shown in Figure 4.3b, reduces at the ends A and B to an intensity equal to the weight per unit length of the beam. As in the case of uniform loading, the magnitude of the resultant or total load $Q = S + W$ must equal the area of the load diagram ABDC. If q is written as a function $q(x)$ of x, the distance along the beam from A, then the resultant load:

$$Q = \int_0^a q(x)\,dx.$$

If $q(x)$ is symmetrical about the mid-point of AB, the resultant load Q will act through that point. If it is not symmetrical, we must take moments about A of slices of the load diagram of width dx and add from A to B, i.e. integrate from 0 to a. Then, if the resultant load Q acts through a point in AB at distance x_c from A, we have:

$$Q x_c = \int_0^a x q(x)\,dx.$$

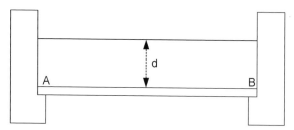

Figure 4.1. A beam supporting a depth d of sand.

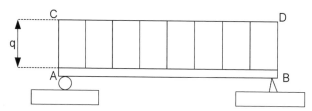

Figure 4.2. Load diagram with constant intensity q.

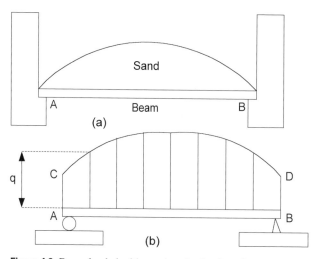

Figure 4.3. Beam loaded with varying depth of sand.

EXERCISE 1

A light cantilever beam AC has a length $AB = a$ set into a supporting wall. A weight W is suspended from the other end C of the beam and the length $BC = b$. Assuming linear load diagrams (as shown in Figure 4.4) for the reactions from the wall onto the top and bottom of the beam between A and B, find the maximum intensity of these reactions: q_a at A and q_b at B.

Problems 30 and 31.

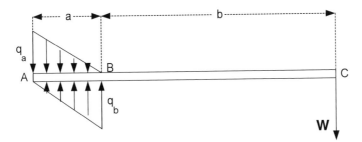

Figure 4.4. Load diagram where a wall supports a cantilever beam.

4.2 Hydrostatics

Hydrostatics is concerned with water or other incompressible fluids in statical equilibrium. The force between such a fluid and a surface with which it is in contact is always perpendicular to the surface. The force is distributed over the surface and is measured in terms of pressure, i.e. force per unit area.

Suppose we isolate a horizontal cylindrical volume in the fluid as indicated in Figure 4.5. Since the volume of fluid is in equilibrium, the force at one end must be equal and opposite to the force at the other end. Therefore, the pressure is the same at either end. Consequently, the pressure must be the same over any horizontal area in the fluid.

Next, we consider a vertical cylindrical volume in the fluid. For equilibrium, the force P_b on the bottom must balance the force P_t on the top together with the weight of the fluid, as indicated in Figure 4.6. Let the cross-sectional area of the fluid cylinder be A and its length be a. If the weight per unit volume is w, then:

$$P_b = p_b A = P_t + W = p_t A + aAw.$$

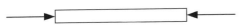

Figure 4.5. Horizontal cylindrical volume of fluid with forces acting at either end.

Figure 4.6. Vertical forces acting on a vertical cylinder of fluid.

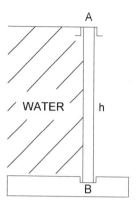

Figure 4.7. Sluice gate of height h.

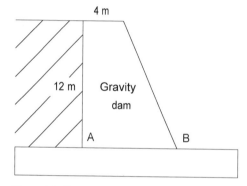

Figure 4.8. Cross-sectional view of a gravity dam.

Hence, the pressures p_t and p_b at the top and bottom, respectively, are related by the equation:

$$p_b = p_t + aw.$$

Thus, we see that the pressure increases with depth at a rate equal to w, the weight per unit volume of the fluid.

EXERCISE 2

If w is the weight per unit volume of water, which is held back by a sluice gate of height h, as shown in Figure 4.7, find the reactive forces per unit length of the supports A at the top and B at the bottom. Hint: start by drawing the load diagram corresponding to the water pressure on a vertical section of the sluice gate.

EXERCISE 3

A gravity dam has trapezoidal cross-section, as shown in Figure 4.8, and holds back water against a vertical face. If the masonry of the dam has weight per unit volume of $2.5w$, where w is the weight per unit volume of the water, find the minimum width AB of the base of the dam to have a safety factor of 2 against overturning about B.

Problems 32 and 33.

4.3 Buoyancy

When a body is totally or partially immersed in a liquid, buoyancy is the upthrust exerted on the body by the liquid. All we need to know about buoyancy is contained in the famous principle of Archimedes. This states that the upthrust on a body is equal and opposite to the weight of the liquid displaced by that body and its resultant acts upwards through the centre of gravity of the displaced liquid.

Archimedes' principle may be reasoned as follows. Consider a body totally immersed in a liquid, as indicated in Figure 4.9. The downward thrust on the top surface S_1 of the body is equal to the weight of the liquid vertically above S_1. If the bottom surface is S_2 and the body were not there, the downward thrust on an imaginary surface S_2 would be the weight of liquid vertically above S_2. This would be balanced by an equal and opposite upthrust on the imaginary S_2. The latter upthrust must be the actual upthrust on the real S_2, i.e. the bottom surface of the body.

Combining the upthrust on S_2 with the downthrust on S_1, we arrive at a resultant upthrust on the body which is equal and opposite to the weight of liquid displaced. Furthermore, the resultant upthrust must act through the centre of gravity of the liquid displaced.

The same argument may be used for a body which is only partially immersed. In the case shown in Figure 4.10a, there is still some downthrust on S_1, which is the part of the upper surface immersed, and that downthrust is equal to the weight of water vertically above S_1. The total upthrust is the upthrust on S_2 minus the downthrust on S_1, which together equal the total weight of liquid displaced by S_1 and S_2.

In the case of Figure 4.10b, only the part S_2 of the bottom surface is immersed and the upthust on it is the weight of liquid displaced.

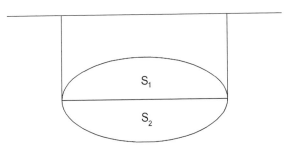

Figure 4.9. A body totally immersed in a liquid.

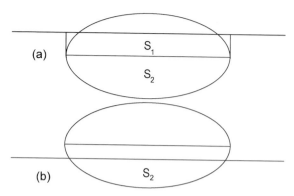

Figure 4.10. A body partially immersed in a liquid.

EXERCISE 4

Find the limiting ratio of thickness to radius in order that a hollow spherical shell of metal may float in water, given that the density of the metal is four times that of water.

EXERCISE 5

A 1 m diameter sphere floats half submerged in water when free. Given that the weight per unit volume of water is 9.81 kN/m^3, find what force is needed to keep the sphere completely submerged.

Problems 34 and 35.

4.4 Centre of pressure on a plane surface

The centre of pressure of a liquid on a plane surface is the point where the resultant pressure force acts. As with finding the centre of gravity, the centre of pressure is found by taking moments.

Consider the example of the vertical face of a dam holding back water in a V-shaped valley, as shown in Figure 4.11. The pressure increases with depth x and is constant across any horizontal strip of the surface. The centre of pressure on a horizontal strip is at its mid-point and therefore, the centre of pressure for the whole triangular surface must lie on the line joining the mid-point of the top water line to the point at the bottom of the dam.

The force of water pressure on the horizontal strip at depth x is:

$$xw\frac{h-x}{h}l\,dx = \frac{wl}{h}(h-x)x\,dx,$$

where l is the length of the top water line on the dam, w is the weight per unit volume of water and dx is the width of the strip. Then, the total pressure force on the dam is:

$$P = \int_0^h \frac{wl}{h}(h-x)x\,dx = \frac{wl}{h}\left[\frac{hx^2}{2} - \frac{x^3}{3}\right]_0^h = \frac{wlh^2}{6}.$$

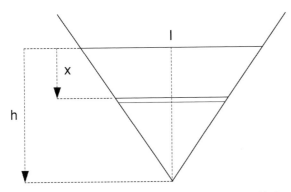

Figure 4.11. Finding the centre of water pressure on a V-shaped dam.

Let x_c be the depth of the centre of pressure and take moments about the top water line:

$$P x_c = \int_0^h x \frac{wl}{h}(h - x)x\, dx = \frac{wl}{h}\int_0^h (h - x)x^2\, dx = \frac{wl}{h}\left[\frac{hx^3}{3} - \frac{x^4}{4}\right]_0^h = \frac{wlh^3}{12}.$$

Therefore, $x_c = \dfrac{wlh^3}{12} \times \dfrac{6}{wlh^2} = \dfrac{h}{2}.$

EXERCISE 6

A right circular cylindrical tank, with its axis horizontal, is completely filled with a liquid. Find the position of the centre of pressure of the liquid on an end face of the tank.

Problems 36 and 37.

4.5 Answers to exercises

1.   By analogy with the problem of finding the position of the centre of gravity of a uniform triangular lamina, we can see that the resultant reaction for the top load diagram acts at a distance $a/3$ from A (see Figure 4.12). Similarly, the resultant reaction for the bottom load diagram acts at a distance $a/3$ from B. Their magnitudes are equal to the areas of the load diagrams, i.e. $Q_a = aq_a/2$ and $Q_b = aq_b/2$, where q_a and q_b are the maximum intensities of the load diagrams.

Now, consider the equilibrium of the beam under the action of Q_a, Q_b and W. If Q_a acts through A' and Q_b through B', then $AA' = A'B' = B'B = a/3$. Take moments about B':

$$Q_a \frac{a}{3} - W\left(\frac{a}{3} + b\right) = 0, \quad Q_a = \left(1 + \frac{3b}{a}\right)W.$$

Take moments about A':

$$Q_b \frac{a}{3} - W\left(\frac{2a}{3} + b\right) = 0, \quad Q_b = \left(2 + \frac{3b}{a}\right)W.$$

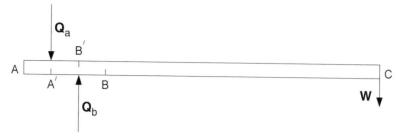

Figure 4.12. Resultant reactions from wall for cantilever beam.

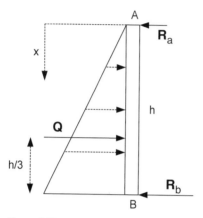

Figure 4.13. Horizontal forces acting on a sluice gate.

Referring back to the equations involving the maximum intensities of load q_a and q_b, we have:

$$q_a = \frac{2}{a}Q_a = \frac{2}{a}\left(1 + \frac{3b}{a}\right)W \quad \text{and} \quad q_b = \frac{2}{a}Q_b = \frac{2}{a}\left(2 + \frac{3b}{a}\right)W.$$

2. Figure 4.13 shows the horizontal forces acting on the sluice gate, including the load diagram for the water pressure. The water pressure (above atmospheric) at depth x is wx. Therefore, for a unit width vertical strip of the sluice gate, the intensity of the load diagram increases from zero at A to wh at B. The total force on the strip is the area of the load diagram, i.e. $Q = wh^2/2$, and this resultant acts at a height $h/3$ above B. Let the reactions per unit length of the support be R_a at A and R_b at B.

Taking moments about B:

$$R_a h - Qh/3 = 0, \quad R_a = Q/3 = wh^2/6.$$

Taking moments about A:

$$-R_b h + 2Qh/3 = 0, \quad R_b = 2Q/3 = wh^2/3.$$

3. Since the width of the dam would be the same factor in both the total water pressure on the face of the dam and in the total weight of the dam, we only need consider a unit width vertical strip through the dam. The resultant water pressure force will act at one third the height of the dam (as in the sluice gate problem) and to allow a safety factor of 2, we make $Q = 2 \times wh^2/2$, where $h = 12$ m. Referring to

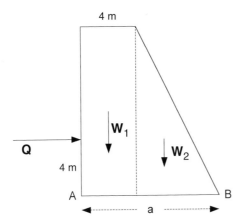

4 m

W_1

Q

W_2

4 m

A

B

a

Figure 4.14. Active forces on a gravity dam.

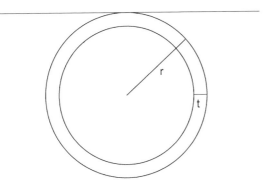

r

t

Figure 4.15. Spherical shell floating completely submerged.

Figure 4.14, the moment of Q about B will then balance the opposite moment about B of the weight for minimum a, i.e.

$$W_1(a-2) + W_2 \tfrac{2}{3}(a-4) = Q \cdot 4.$$

Now, $W_1 = 2.5w \times 4 \times 12 = 120w$, $W_2 = 2.5w \times (a-4)/2 \times 12 = 15w(a-4)$.

Therefore, $120w(a-2) + 10w(a-4)^2 = 576w$,

i.e. $10a^2 + 40a - 656 = 0$ or $a^2 + 4a - 65.6 = 0$.

Thus, $a = \dfrac{-4 \pm \sqrt{16 + 262.4}}{2} = 6.34$ m (a must be positive).

4.   The limiting case is where the sphere floats completely submerged, i.e. when the buoyancy equals the weight. Referring to Figure 4.15, let r = outer radius, t = thickness and w = weight per unit volume of water.

Weight of shell = $\tfrac{4}{3}\pi[r^3 - (r-t)^3]4w$.

Buoyancy = $\tfrac{4}{3}\pi r^3 w$.

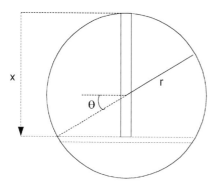

Figure 4.16. Finding the centre of pressure on the end of a tank filled with liquid.

These are equal if: $4[r^3 - (r - t)^3] = r^3$.

Divide by r^3: $4\left[1 - \left(1 - \dfrac{t}{r}\right)^3\right] = 1$.

$\left(1 - \dfrac{t}{r}\right)^3 = \dfrac{3}{4}$, $\dfrac{t}{r} = 1 - \left(\dfrac{3}{4}\right)^{1/3} = 0.09144$.

5.   If the sphere floats half submerged, the buoyancy is equal to the weight of water displaced, i.e. to:

$$\frac{2}{3}\pi r^3 \times w = \frac{2}{3}\pi \left(\frac{1}{2}\right)^3 \times 9.81 \times 10^3 = 2568\,\text{N}.$$

If the sphere were completely submerged, the buoyancy would be doubled and an extra downward force of 2568 N would be required to achieve total submersion.

6.   The rules for the pressure of a liquid in a tank are basically the same as before, provided no extra pressurization is applied. In this example, we can assume that the pressure at the top of the tank is atmospheric. Since we are concerned with pressure in excess of atmospheric (since that is applied to the outside of the tank), we can let the pressure in the liquid be zero at the top.

     For equilibrium, the pressure in the liquid over any horizontal plane is constant. Considering a column of liquid down from the highest point, as shown in Figure 4.16, the same reasoning as given in Section 4.2 shows that the pressure at depth x is xw, where w is the weight per unit volume of liquid.

     Measuring the angle θ as shown in Figure 4.16, we can write $x = r(1 + \sin\theta)$ and $dx = r\cos\theta\,d\theta$. Then the pressure force on the strip of width dx is:

$$dP = wx \cdot 2r\cos\theta \cdot dx = 2wr(1 + \sin\theta) \cdot r\cos\theta \cdot r\cos\theta\,d\theta$$
$$= 2wr^3(1 + \sin\theta)\cos^2\theta\,d\theta = wr^3(1 + \cos 2\theta + 2\cos^2\theta\,\sin\theta)d\theta.$$

Integrating from $\theta = -\pi/2$ to $\theta = \pi/2$ gives the total pressure force over the circular end of the tank:

$$P = \int_{-\pi/2}^{\pi/2} wr^3(1 + \cos 2\theta + 2\cos^2\theta\,\sin\theta)\,d\theta$$
$$= wr^3\left[\theta + \frac{1}{2}\sin 2\theta - \frac{2}{3}\cos^3\theta\right]_{-\pi/2}^{\pi/2} = w\pi r^3.$$

As might be expected, this is the average pressure wr times the area πr^2.

The centre of pressure will lie on the vertical centre line at a depth $x = x_c$. To find x_c, we take moments about the top of the tank. The moment of the pressure on the horizontal strip at depth x is:

$$dM = x\,dP = wx^2 \cdot 2r\cos\theta \cdot dx = wr^2(1 + \sin\theta)^2 \cdot 2r\cos\theta \cdot r\cos\theta\,d\theta$$

$$= 2wr^4(1 + 2\sin\theta + \sin^2\theta)\cos^2\theta\,d\theta$$

$$= wr^4(1 + \cos 2\theta + 4\cos^2\theta\sin\theta + \tfrac{1}{4} - \tfrac{1}{4}\cos 4\theta)d\theta.$$

Integrating from $\theta = -\pi/2$ to $\theta = \pi/2$ gives the total moment:

$$M = Px_c = \int_{-\pi/2}^{\pi/2} wr^4 \left(\frac{5}{4} + \cos 2\theta + 4\cos^2\theta\sin\theta - \frac{1}{4}\cos 4\theta \right) d\theta$$

$$= wr^4 \left[\frac{5}{4}\theta + \frac{1}{2}\sin 2\theta - \frac{4}{3}\cos^3\theta - \frac{1}{16}\sin 4\theta \right]_{-\pi/2}^{\pi/2} = \frac{5}{4}wr^4\pi.$$

Then, $x_c = M/P = \frac{5}{4}wr^4\pi/w\pi r^3 = \frac{5}{4}r.$

5 Trusses

5.1 Method of sections

A truss is a vertical framework of struts connected together at their ends so as to form a rigid structure, even when the connections are smooth hinge points. We shall assume that the structure is light compared with any supported loads. Thus, in the truss shown in Figure 5.1, we will neglect the weight of the struts which is assumed to be small compared with the load L. Furthermore, all the joints will be regarded as hinge points, i.e. any moments exerted at the joints are small enough to be neglected. This means that a strut will exert a force at a joint in the direction of the length of the strut. This force will be a push if the strut is in compression or a pull if it is in tension.

Finally, we shall only consider trusses with no redundant struts. In other words, the truss would collapse under the action of the load if any one strut were removed. Such a structure may be built up as a series of triangles. The struts need not have the same length, so the triangles need not be equilateral, as in Figure 5.1. However, there is a relation between the number s of struts and the number j of joints. For one triangle, $s = j = 3$. Then for each triangle added after that, there are two extra struts and one extra joint. It follows that $s = 2j - 3$.

The *method of sections*, referred to in the title to this section of the book, may be used to find the tension or compression forces in the struts. It is particularly useful when we want to examine only one particular strut.

Firstly, referring to Figure 5.2, we must find the reactions R_a and R_d at the supports A and D. These are obviously forces acting upwards to counter balance the load L. Taking moments about D gives: $-3aR_a + La = 0$, where a is the length of a strut. Therefore, $R_a = L/3$. Taking moments about A gives: $3aR_d - 2aL = 0$ and $R_d = 2L/3$.

Next, we make an imaginary cut through the structure and passing through the strut in which we are particularly interested. Let this strut be FG but notice that the cut must go through two other struts as well, i.e. through BF and BC in the way that the cut has been made in Figure 5.2. Now, replace these struts by their tension forces T_1, T_2 and T_3, acting at joints G and B to the left of the cut and at F and C to the right of the cut,

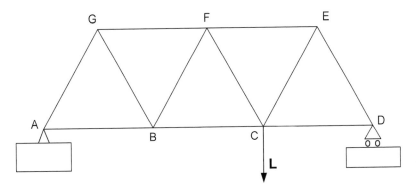

Figure 5.1. A light truss supporting a load L.

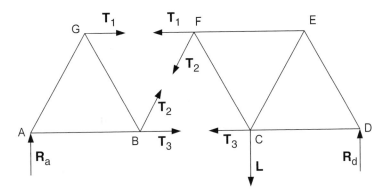

Figure 5.2. Imaginary cut through the structure.

as shown in Figure 5.2. Of course, some of the struts may be in compression but this will eventually be indicated by the corresponding tensions turning out to be negative.

Consider the equilibrium of the part of the truss ABG to the left of the cut. Taking moments about B will only involve T_1, since T_2 and T_3 both act through B.

Thus: $T_1 \dfrac{\sqrt{3}}{2}a + R_a a = 0$ and $T_1 = -\dfrac{2}{\sqrt{3}}R_a = -\dfrac{2}{3\sqrt{3}}L$.

Since T_1 is negative, it means that the strut FG is in compression. It exerts a push to the left at G and a push to the right at F, each push being of magnitude $\frac{2}{3\sqrt{3}}L$.

If instead of T_1, we had wanted to find T_2, we could have done this without involving T_1 or T_3 by resolving in the vertical direction the forces acting on the part of the truss ABG. Thus, $R_a + T_2 \cos 30° = 0$ and

$$T_2 = -R_a \sec 30° = -\frac{L}{3}\frac{2}{\sqrt{3}} = -\frac{2}{3\sqrt{3}}L.$$

Again, the negative sign implies compression so the strut BF exerts a push at B and F, each push having magnitude $\frac{2}{3\sqrt{3}}L$.

Alternatively, if we had wanted T_3 rather than T_1 or T_2, we could have taken moments of forces on ABG about F, the point through which both T_1 and T_2 act. Thus:

$$-R_a\frac{3}{2}a + T_3\frac{\sqrt{3}}{2}a = 0 \quad \text{and} \quad T_3 = \frac{2}{\sqrt{3}}\frac{3}{2}R_a = \sqrt{3}\frac{L}{3} = L/\sqrt{3}.$$

The positive sign of T_3 implies that it is a tension force. Hence, the strut BC exerts a pull at B and C, each pull being of magnitude $L/\sqrt{3}$.

EXERCISE 1

Use the method of sections to find T_1, T_2 and T_3 again in the above example but this time by considering equilibrium of the part of the truss CDEF to the right of the cut.

Problems 38 and 39.

5.2 Method of joints

The *method of joints* considers the equilibrium of the forces acting at each joint in turn. Starting from a joint where there is a known supporting force and only two struts, the triangle of forces may be constructed to determine the forces in the struts.

Considering next an adjacent joint where three struts meet, a triangle of forces can be constructed on the strut force already determined. This will give the forces in the other two struts. By constructing a polygon of forces at each successive adjacent joint, it is possible to find all the strut forces.

Consider again the example of Section 5.1, in which a truss made up of 11 equal light struts carries a load L at C, as shown again in Figure 5.3, but this time with the supporting forces inserted. If we had not already done so, we would have to start by finding the supporting forces R_a and R_d.

Starting with joint A, we have the force $R_a = L/3$ acting vertically upwards. The other two forces are in the direction of the struts, so we can construct the $30°$ right-angle

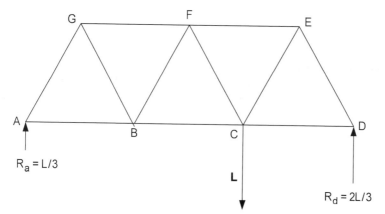

Figure 5.3. A light truss carrying a load L with supporting forces included.

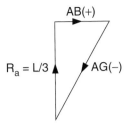

Figure 5.4. Triangle of forces at joint A.

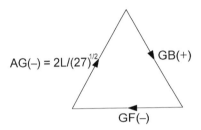

Figure 5.5. Triangle of forces at joint G.

triangle of forces, as shown in Figure 5.4. Since the AB strut force must act to the right, it must be a tension and has been indicated as such with a plus sign in the diagram. Also, the AG strut force must be downwards to complete the triangle of forces. The latter force must therefore be a compression and this is indicated by a minus sign in the diagram. The ratios of the magnitudes of **AB**$(+)$ to **AG**$(-)$ to $R_a = L/3$ are $1 : 2 : \sqrt{3}$, from Pythagoras' theorem applied to the $30°$ right angle triangle.

Therefore, $AB(+) = \dfrac{L}{3\sqrt{3}}$ and $AG(-) = \dfrac{2L}{3\sqrt{3}}$.

In moving now to an adjacent joint, we have the choice of B or G. The former, i.e. B, is of no use at this stage because it is a joint of four struts and we only know the strut force in one of them. We must therefore consider the equilibrium of the three strut forces acting at G. One of these is known; that is the compressive force $AG(-) = \frac{2L}{3\sqrt{3}}$ from the strut AG. Starting with this, we can then complete the triangle of forces, as shown in Figure 5.5. The directions must be all the same way round the triangle. Thus the **GB** strut force is a pull down from G, i.e. tension, and the **GF** strut force is a push from the right, i.e. compression. The triangle is equilateral, so the magnitudes of the forces are equal, i.e.

$$GB(+) = \frac{2L}{3\sqrt{3}} \quad \text{and} \quad GF(-) = \frac{2L}{3\sqrt{3}}.$$

Now that we know the strut forces **AB**$(+)$ and **GB**$(+)$, we can consider the four strut forces acting at B. We start to draw the polygon of forces in Figure 5.6 with two sides, corresponding to the strut forces (both tensions) from the struts AB and GB of lengths

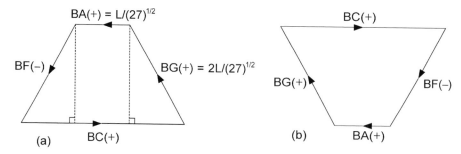

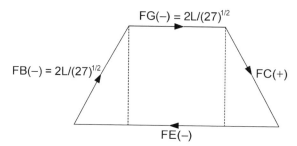

Figure 5.6. Polygons of forces at joint B.

Figure 5.7. Polygon of forces acting at joint F.

proportional to $BA(+) = \frac{L}{3\sqrt{3}}$ and $BG(+) = \frac{2L}{3\sqrt{3}}$. The polygon is then completed with two sides having the directions of the struts BC and BF. Either of the two polygons, Figures 5.6a or 5.6b, will serve our purpose. The first one started with **BG**(+) and went on to **BA**(+), whereas the second started with **BA**(+) followed by **BG**(+).

It follows from the directions of the arrows round the polygon that the force from strut CB pulls away from B, whereas that from strut FB pushes down on B. Thus CB is in tension and FB is in compression. From the shape of the polygon, it is obvious that the magnitude:

$$BF(-) = BG(+) = \frac{2L}{3\sqrt{3}}.$$

The two sides of the polygon are each at angle of 30° to the vertical. Drawing in vertical dotted lines, through the top corners of the first polygon and perpendicular to the base, forms two 30° right-angled triangles. In each of the latter, the length of the base is half that of the hypotenuse. It follows that:

$$BC(+) = \frac{L}{3\sqrt{3}} + \frac{L}{3\sqrt{3}} + \frac{L}{3\sqrt{3}} = L/\sqrt{3}.$$

Proceeding now to joint F, we start the polygon of forces in Figure 5.7 with sides corresponding to the compression forces $FB(-) = \frac{2L}{3\sqrt{3}}$ and $FG(-) = \frac{2L}{3\sqrt{3}}$. Then, complete the polygon with sides having directions of the struts CF and EF as shown. From the

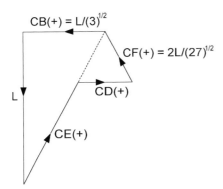

Figure 5.8. Polygon of forces acting at joint C.

directions of the arrows, we see that strut CF is in tension and strut EF is in compression. Hence, we have:

$$FC(+) = \frac{2L}{3\sqrt{3}} \quad \text{and} \quad FE(-) = \frac{L}{3\sqrt{3}} + \frac{2L}{3\sqrt{3}} + \frac{L}{3\sqrt{3}} = \frac{4L}{3\sqrt{3}}.$$

Moving on to joint C, we now have not only four strut forces but also the downward load **L**. Hence, the polygon of forces for equilibrium of the forces acting at C is a pentagon. The latter is started by sides corresponding to $CF(+) = \frac{2L}{3\sqrt{3}}$, $CB(+) = \frac{L}{\sqrt{3}}$ and the downward load **L**. We then complete the pentagon, as in Figure 5.8, with sides in the directions of struts EC and DC. The pentagon may take different shapes depending on the order in which we draw sides corresponding to the five forces. For instance, the pentagon would have been completely convex had we drawn **CD**(+) onto the end of **L** and finished with **CE**(+). The direction of the arrows indicates a pull from C towards D and a pull from C towards E. This means that struts DC and EC are in tension as already indicated by the plus signs.

Referring to Figure 5.8, the $30°$ right-angle triangle, with **L** on the left-hand side, has hypotenuse of magnitude $2L/\sqrt{3}$. The magnitude of **CE**(+) is two thirds of this, i.e.

$$CE(+) = \frac{4L}{3\sqrt{3}} \quad \text{and finally,} \quad CD(+) = CF(+) = \frac{2L}{3\sqrt{3}}.$$

We now have the choice of joint D or joint E, in order to find the forces exerted by strut DE. In fact, we shall do both since this will act as a check on previous working.

Consider joint E. We start by drawing in sides corresponding to $EF(-) = \frac{4L}{3\sqrt{3}}$ and $EC(+) = \frac{4L}{3\sqrt{3}}$, as shown in Figure 5.9. Then complete the triangle of forces with a side corresponding to an upward force, i.e. a compression: $ED(-) = EC(+) = \frac{4L}{3\sqrt{3}}$.

Finally, as a check, we consider joint D. Start the triangle of forces in Figure 5.10 with sides corresponding to $DC(+) = \frac{2L}{3\sqrt{3}}$ horizontally to the left followed by the supporting force $2L/3$ vertically upwards. Then complete the triangle with the hypotenuse corresponding to $DE(-) = 2DC(+) = \frac{4L}{3\sqrt{3}}$, which agrees with our previous calculation for $ED(-)$.

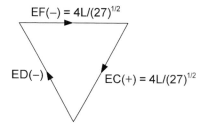

Figure 5.9. Triangle of forces acting at joint E.

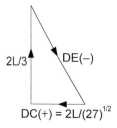

Figure 5.10. Triangle of forces acting at joint D.

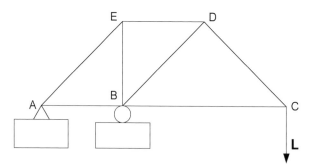

Figure 5.11. A truss, supported at A and B, carrying a load L at C.

EXERCISE 2

The light truss, shown in Figure 5.11, is supported at joints A and B, and it carries a load L at joint C. The lengths of the struts are such that $AB = BE = ED$ and $BC = 2AB$. Use the method of joints to find the tension or compression forces in all of the struts.

Problems 40 and 41.

5.3 Bow's notation

Bow's notation is a way of putting all the information contained in the method of joints into one diagram. Instead of labelling the joints with capital letters, we use the capital letters to label the spaces in between the struts and also the spaces between external loads and support forces.

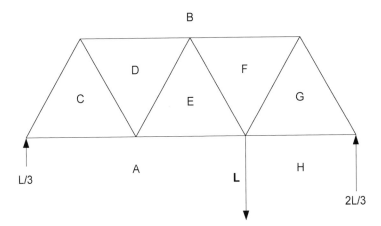

Figure 5.12. Labelling of truss for Bow's notation.

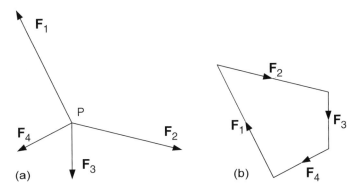

Figure 5.13. Forces acting at point P taken clockwise in order to form the polygon of forces.

Consider the truss used as an example in Sections 5.1 and 5.2. This is now labelled as shown in Figure 5.12. A, B and H are the spaces outside the truss between load L and support $L/3$, from $L/3$ round the top of the truss to support $2L/3$ and between $2L/3$ and L, respectively. The other letters C, D, E, F and G label the spaces between the struts.

We proceed as with the method of joints, drawing the polygon of forces for each joint but doing it in such a way that the straight line corresponding to a strut serves for the force at either end of the strut. To do this, it is necessary to take each force successively as one goes round each joint in the same direction. For example, the forces shown in Figure 5.13a, acting at point P, would be represented by the polygon in Figure 5.13b by taking the forces in order clockwise about the point P.

Finally, the corners of each polygon are labelled with small letters corresponding to the capital letters used for the spaces in the truss diagram. Thus, going clockwise around the first joint gives the triangle of forces abc in Figure 5.14, where ab corresponds to the supporting force $L/3$, bc to the compression in the strut between B and C, and ca to the

Figure 5.14. Triangle of forces at first joint on left of truss.

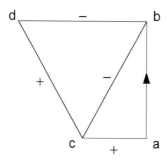

Figure 5.15. Triangle of forces added for the second joint.

tension in the strut between C and A. Arrows are not entered on the sides corresponding to strut forces because the same sides will be re-used for the polygons of forces of the adjacent joints when the forces are in the opposite directions.

Now go to the next joint circled in the clockwise direction by the spaces CBD. We already have the strut force cb. We then complete the triangle of forces with two sides having the directions of the other two struts, as shown in Figure 5.15. bd is to the left corresponding to compression and dc is downwards corresponding to tension. In fact, it is useful to indicate tension and compression with + and −, respectively, on the diagram as shown in Figure 5.15.

We now proceed to successive joints as in the method of joints. Each time, we start with the sides of the polygon already determined and remember to order the sides according to a clockwise rotation around the joint. In our example, we go next to the joint circled by ACDE. We start with the sides ac and cd already in the diagram and then complete the polygon with sides de and ea having the directions of the corresponding struts, as shown in Figure 5.16. The directions of the forces are also that way round, so de is compression (−) and ea is tension (+).

The next joint is that encircled by EDBF. The polygon is started with the sides ed and db already there and completed with sides bf and fe drawn in the directions of the corresponding struts, as shown in Figure 5.17. The forces go from e to d to b and hence

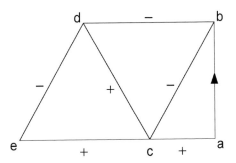

Figure 5.16. Polygon of forces added for the third joint.

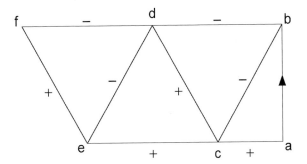

Figure 5.17. Polygon of forces added for the fourth joint.

continue from b to f to e. Referring to the original truss, we see that bf is compression (−) and fe is tension (+).

The next joint is encicled clockwise by HAEFG. Since the first side of the polygon is ha, which corresponds to the downward force L, ha will be three times the length of ab, so we need plenty of room to expand the diagram. Having drawn ha, we continue round the polygon, as shown in Figure 5.18, with the sides ae and ef already drawn. We complete the polygon with the sides fg and gh in the directions of the corresponding struts. The new strut forces are from f to g and from g to h, i.e. both away from the joint and hence they are both tensions (+).

The only strut force which remains to be dealt with is that for the strut between B and G. Considering the joint encircled by GFB, we already have two sides gf and fb for the triangle of forces, so this is completed by bg, as shown in Figure 5.19. The force is in the direction from b to g at this joint and is therefore a compression (−).

The only force which has not been included is the supporting force $2L/3$. This lies between the spaces B and H, and therefore corresponds to the side bh in Figure 5.19. Since it is vertically upwards, a corresponding arrow may be inserted in the diagram on the line bh.

The process using Bow's notation may seem as long-winded as the original method of joints. However, part of the reason for this is that separate diagrams have been drawn

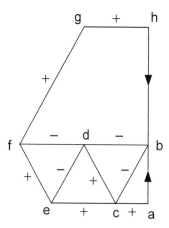

Figure 5.18. Polygon of forces added for the fifth joint.

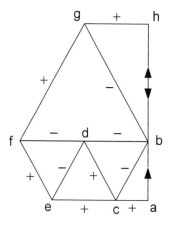

Figure 5.19. Triangle of forces added for the sixth joint.

to illustrate each stage in the construction. In practice, we would only draw one diagram, i.e. the final one.

It is easy to deduce the magnitudes of the strut forces from the final diagram. In our example, it is particularly simple, starting with $ha = L$ and $ab = L/3$, and noticing that all of the pentagon sides are either horizontal, vertical or at $30°$ to the vertical. It will be a useful exercise to deduce from the diagram the magnitudes of all the strut forces and check that they agree with what we obtained using the original method of joints.

Remember: when considering each joint, the forces must take the order which follows by going clockwise around the joint, starting with the forces already known, i.e. with the sides of the polygon already drawn. (You can go anti-clockwise for each joint if you like but you must be consistent.)

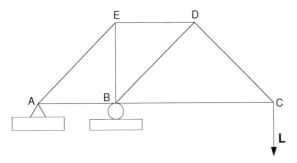

Figure 5.20. A light truss carrying a load L.

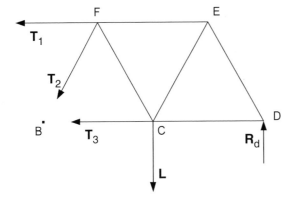

Figure 5.21. Finding T_1, T_2 and T_3 by the method of sections.

EXERCISE 3

Use Bow's notation to find the strut forces in the light truss shown in Figure 5.20, which is the same as that in Exercise 2 of Section 5.2.

Problems 42 and 43.

5.4 Answers to exercises

1.   Referring to Figure 5.21, take moments about B so as not to involve T_2 or T_3:

$$T_1 \frac{\sqrt{3}}{2} a - La + R_d 2a = 0.$$

Thus: $T_1 = \dfrac{2}{\sqrt{3}}(L - 2R_d) = \dfrac{2}{\sqrt{3}}\left(1 - \dfrac{4}{3}\right)L = -\dfrac{2}{3\sqrt{3}}L.$

Resolve vertically so as not to involve T_1 or T_3:

$$-T_2 \cos 30° - L + R_d = 0.$$

Thus: $T_2 = \dfrac{2}{\sqrt{3}}(-L + R_d) = \dfrac{2}{\sqrt{3}}\left(-1 + \dfrac{2}{3}\right)L = -\dfrac{2}{3\sqrt{3}}L.$

Take moments about F so as not to involve T_1 or T_2:

$$-T_3 \frac{\sqrt{3}}{2}a - L\frac{1}{2}a + R_d \frac{3}{2}a = 2.$$

Thus: $T_3 = \frac{2}{\sqrt{3}}\left(-\frac{1}{2}L + \frac{3}{2}R_d\right) = \frac{2}{\sqrt{3}}\left(-\frac{1}{2}+1\right)L = \frac{1}{\sqrt{3}}L.$

2.   Referring to Figure 5.22, by taking moments about B, we see that there is a downward supporting force of $2L$ at A. It follows that there is an upward balancing supporting force of $3L$ at B.

Consider the equilibrium of forces at the joints, starting from A, using mainly a triangle but also a polygon of forces, as shown successively in Figures 5.23a, 5.23b, 5.24a and 5.24b. As a final check, we consider joint C forces in Figure 5.25.

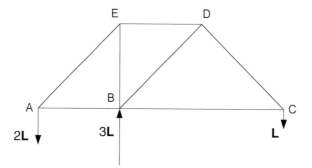

Figure 5.22. External forces acting on a light truss.

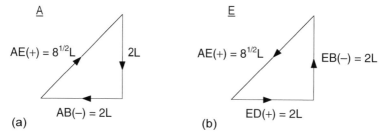

Figure 5.23. Triangles of forces at joints (a) A and (b) E.

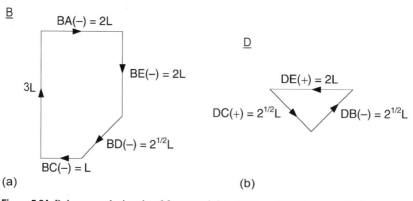

Figure 5.24. Polygon and triangle of forces at joints (a) B and (b) D, respectively.

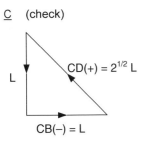

$\underline{C}$ (check)

L

CD(+) = $2^{1/2}$ L

CB(−) = L

Figure 5.25. Triangle of forces at joint C.

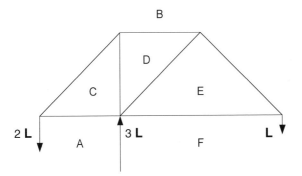

B

D

C

E

2 **L**

A

3 **L**

F

L

Figure 5.26. A light truss with external forces and with spaces labelled A . . . F.

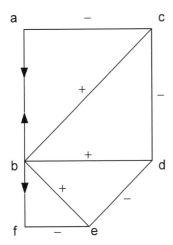

a

−

c

+

−

b

+

d

+

−

f

−

e

Figure 5.27. Using Bow's notation to find the strut forces in the truss shown in Figure 5.26.

3.   By taking moments about the support points, we find as before that the support forces are $2L$ down and $3L$ up, as shown in Figure 5.26. Label the spaces as in Figure 5.26 and start with the joint to the left of the diagram.

Referring to Figure 5.27, we draw ab vertically downward and corresponding to a force $2L$. Complete the triangle of forces with sides bc and ca. bc corresponds to a force away from the joint, i.e. tension, and ca corresponds to compression. Hence, they are labelled + and −, respectively.

Moving to joint CBD, start with side cb and complete the triangle with sides bd($+$) and dc($-$), as shown. As indicated, the latter correspond to tension and compression, respectively. Next, to joint DBE, start with side db and complete triangle with sides be($+$) and ed($-$). Finally, for joint EBF, start with side eb and complete triangle with sides bf (for load L) and fe($-$).

The magnitudes of the strut forces are deduced readily from Figure 5.27, noting that the magnitudes of ab and bf are $2L$ and L, respectively. We find that we have tensions: $bc = 2\sqrt{2}L$, $bd = 2L$ and $be = \sqrt{2}L$. The compressions are: $ac = 2L$, $cd = 2L$, $de = \sqrt{2}L$ and $ef = L$. Finally, it should be noted that fa in the diagram corresponds to the upward supporting force $3L$ and that the force polygon for the joint ACDEF is acdef in Figure 5.27.

6 Beams

6.1 Shearing force and bending moment

In Chapter 5, the struts in trusses were assumed to be subject to only axial loads of either tension or compression. In contrast, beams are subject mainly to lateral loads. These loads cause what are called *shearing forces* and *bending moments* in a beam; we shall investigate them in this chapter.

Consider a horizontal beam AB supported at either end A and B, as shown in Figure 6.1. Neglect the weight of the beam but let it be subject to a downward load L at distance a from A and b from B. To support the beam, there will be upward forces $R_a = \frac{b}{a+b}L$ at A and $R_b = \frac{a}{a+b}L$ at B. The magnitudes R_a and R_b are found by taking moments about B and A, respectively.

Let us now examine the equilibrium conditions for a section of the beam from A to a point at distance $x < a$ from A. The action of the beam to the right of this section must be such as to keep this section in equilibrium when it is subject to the supporting force R_a at A. Now, force R_a at A has the same effect as R_a at the other end of the section together with a couple of moment $-R_a x$. This must be cancelled by the action of the beam to the right of this section. This action must therefore be a downward force $F = R_a$ together with a couple of moment $M = R_a x$, as shown in Figure 6.2. The force F is called the shearing force and the couple of moment M is called the bending moment.

Next, let us consider the equilibrium of a section of the beam extending from A to a point beyond where the load acts. Thus, the length of the section is now x such that $a < x < a + b$. Referring to Figure 6.3, the shearing force is F upwards with $F = L - R_a = R_b$. The bending moment is $M = R_a x - L(x - a) = La - R_b x$. Notice that when $x = a$, $M = (L - R_b)a = R_a a$, as would be given by the shorter section formula $M = R_a x$. Also, when $x = a + b$, since $R_b = \frac{a}{a+b}L$, $M = La - R_b(a + b) = 0$.

We can now draw graphs called the shearing force and bending moment diagrams as shown in Figure 6.4.

Suppose the beam were an I beam as illustrated in cross-section in Figure 6.5. Then, the shearing force would be mainly taken by the vertical web of the I section. On the

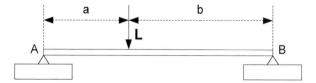

Figure 6.1. A horizontal beam supported at both ends.

Figure 6.2. A section of the beam kept in equilibrium by the shearing force and bending moment.

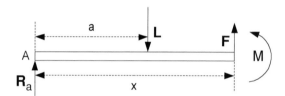

Figure 6.3. Section of beam extending beyond the load point.

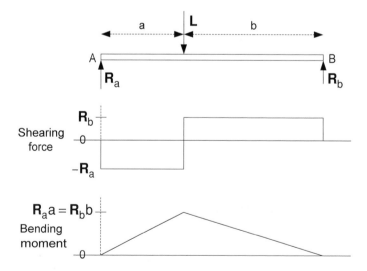

Figure 6.4. Graphs of shearing force and bending moment.

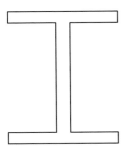

Figure 6.5. Cross-section of an I beam.

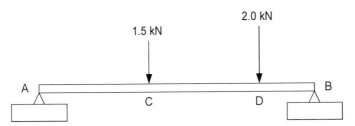

Figure 6.6. A light beam supporting two loads.

other hand, the bending moment would be mainly taken by a compression in the top flange of the I section and a tension in the bottom flange of the I section.

EXERCISE 1

A light beam AB of length 5 m rests horizontally on supports at A and B, as shown in Figure 6.6. The beam is subject to loads of 1.5 kN at C and 2 kN at D, where $AC = 2$ m and $CD = 2$ m. Draw the shearing force and bending moment diagrams for the beam, i.e. plot the graphs of shearing force and bending moment against distance along the beam.

EXERCISE 2

Referring to Figure 6.7, a light cantilever beam, i.e. one which is supported at one end only, carries a load L at its free end. Draw the shearing force and bending moment diagrams for the beam.

Problems 44 and 45.

6.2   Uniformly distributed beam loading

Having considered the shearing force and bending moment in a beam when subjected to point loads, let us now proceed to examine the effect of a load which is continuously distributed along the beam. Start with the simplest situation where a horizontal beam AB is supported at A and B, and has a uniform load along its length. If w is the load

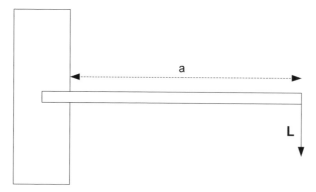

Figure 6.7. A cantilever beam supporting a load L.

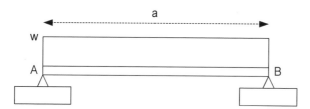

Figure 6.8. A beam subjected to a uniform load.

intensity in downward force per unit length of the beam, then w is constant, as indicated in Figure 6.8. The total load is wa, where a is the length of the beam.

The supporting forces R_a and R_b at A and B, respectively, obviously share the load equally so that $R_a = R_b = wa/2$. Consider the equilibrium of a section of the beam from A of length $x < a$. In this case, the resultant load on this section is a force wx acting at a distance $x/2$ from A, as indicated in Figure 6.9. Then, the shearing force at the right-hand end of the section is $F = R_a - wx = \frac{wa}{2} - wx$, downwards. Thus, when $x = 0$, i.e. at A, $F = wa/2$ downwards. F varies linearly with x until $x = a$, i.e. at B, where $F = -wa/2$ downwards or $wa/2$ upwards.

For zero moment about the right-hand end of the section of beam, we must have a bending moment:

$$M = R_a x - wx \cdot \frac{x}{2} = \frac{w}{2}(ax - x^2) = -\frac{w}{2}\left(x - \frac{a}{2}\right)^2 + \frac{wa^2}{8}$$

or $M - \frac{wa^2}{8} = -\frac{w}{2}\left(x - \frac{a}{2}\right)^2$.

The equation has been expressed in this form to help in drawing the bending moment diagram. If M is the y-coordinate, we see that the equation in coordinate geometry defines a parabola, which is concave downwards with axis parallel to the y-axis and apex at the point $(\frac{a}{2}, \frac{wa^2}{8})$. Also, M is zero at $x = 0$ and $x = a$, i.e. at the ends A and B of the beam.

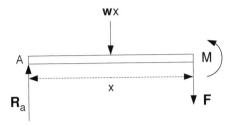

Figure 6.9. Section of beam of length x from A.

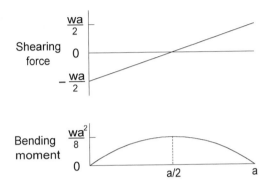

Figure 6.10. Shearing force and bending moment for a beam with uniform load.

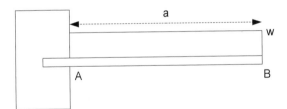

Figure 6.11. A cantilever beam with uniformly distributed load.

The shearing force and bending moment diagrams can now be drawn as shown in Figure 6.10

Notice that, in previous examples with point loads, we assumed that the weight of the beam was small enough to be neglected. In this last example, the continuously distributed load could be a load on top of a light beam or just the weight of the beam itself or both.

EXERCISE 3

A cantilever beam projects a distance a from its support, as shown in Figure 6.11. Formulate and draw the shearing force and bending moment diagrams for the beam when the load is uniformly distributed along its length with intensity w.

Problem 46.

6.3   Using calculus

In order to use calculus in the shearing force and bending moment analysis, we start by examining the equilibrium conditions for a small segment of the beam, illustrated in Figure 6.12. Let the segment be distance x from the left-hand end A and its thickness be δx. For a load intensity w, the load on the segment is of magnitude $w\delta x$. Let the shearing force and bending moment have magnitudes F and M on the left of the segment. Note that these are both drawn on the diagram with negative directions since they must oppose the corresponding force and moment acting in the beam to the right of the segment. On the right of the segment, the shearing force and bending moment become $F + \delta F$ and $M + \delta M$, respectively, drawn with positive directions.

Resolving the forces on the segment in the upward vertical direction, for equilibrium, we have:

$$F + \delta F - F - w\delta x = 0, \text{ i.e. } \frac{\delta F}{\delta x} = w.$$

Taking moments about the centre point O of the segment, for equilibrium, we have:

$$(F + \delta F)\frac{\delta x}{2} + F\frac{\delta x}{2} + M + \delta M - M = 0, \text{ i.e. } \frac{\delta M}{\delta x} = -F - \frac{\delta F}{2}.$$

Finally, taking the limit as $\delta x \to 0$, the shearing force and bending moment equations become:

$$\frac{dF}{dx} = w \quad \text{and} \quad \frac{dM}{dx} = -F.$$

These are simple differential equations which can be used to derive shearing force and bending moment diagrams if the load intensity w is given as a function of x.

Let AB be a light horizontal beam of length a supported at the ends A and B, as shown in Figure 6.13. Suppose the load L is distributed with intensity w, which varies

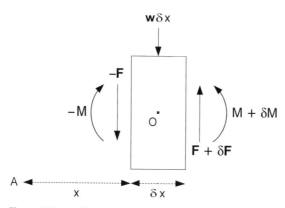

Figure 6.12. Small segment of a beam.

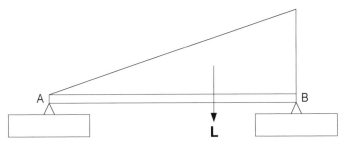

Figure 6.13. Light horizontal beam subjected to a distributed load L.

linearly from zero at A to its maximum value at B. The centre of gravity of the load is above a point which is 2/3 of the way along the beam from A. It follows that the supporting forces at A and B must be $L/3$ and $2L/3$, respectively.

If x is the distance along the beam from A, then $w = cx$, where c is a constant and the total load

$$L = \int_0^a w\,dx = \int_0^a cx\,dx = c\left[\frac{x^2}{2}\right]_0^a = ca^2/2 \quad \text{and} \quad c = 2L/a^2.$$

Then, since

$$\frac{dF}{dx} = w, \quad F = \int w\,dx = \int cx\,dx = cx^2/2 + C,$$

where C is an integrating constant. From previous work, we know that the shearing force F is the negative of the supporting force at A, i.e. when $x = 0$. Thus,

$$-\frac{1}{3}L = 0 + C \quad \text{and} \quad F = \frac{L}{a^2}x^2 - \frac{L}{3} = \left(\frac{x^2}{a^2} - \frac{1}{3}\right)L.$$

It follows that $F = 0$ when $x = a/\sqrt{3}$ and at B where $x = a$, $F = 2L/3$, which is the supporting force at B.

Next, we use the equation $\frac{dM}{dx} = -F$ to find M as a function of x:

$$M = \int -F\,dx = -L\int \left(\frac{x^2}{a^2} - \frac{1}{3}\right) dx = -L\left(\frac{x^3}{3a^2} - \frac{x}{3}\right) + K,$$

where K is an integrating constant. However, $M = 0$ at A, i.e. where $x = 0$ and therefore $K = 0$.

Hence, $M = \dfrac{L}{3}\left(1 - \dfrac{x^2}{a^2}\right)x$,

which, as expected, is also zero at $x = a$, i.e. at end B. M takes its maximum value where $\frac{dM}{dx} = -F = 0$, i.e. at $x = a/\sqrt{3}$ and the maximum value is $2aL/9\sqrt{3}$.

The shearing force and bending moment diagrams can now be drawn as shown in Figure 6.14.

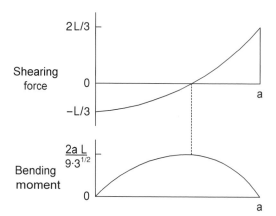

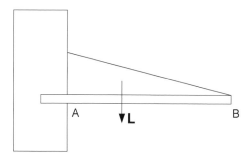

Figure 6.14. Shearing force and bending moment diagrams for beam in Figure 6.13.

Figure 6.15. A cantilever beam bearing a load L varying linearly in intensity.

EXERCISE 4

A light cantilever beam, of length a from the support at A to the free end at B, bears a load L varying linearly in intensity w from its maximum value at A to zero at B (see Figure 6.15). Use calculus to formulate and draw the shearing force and bending moment diagrams for the projecting part of the beam from A to B.

Problems 47 and 48.

6.4 Answers to exercises

1.   Referring to Figure 6.16, take moments about B:

$$1.5 \times 3 + 2 \times 1 - R_a \times 5 = 0, \ 5R_a = 6.5, \ R_a = 1.3 \, \text{kN}.$$

Take moments about A:

$$R_b \times 5 - 1.5 \times 2 - 2 \times 4 = 0, \ 5R_b = 11, \ R_b = 2.2 \, \text{kN}.$$

Consider a section of the beam of length from A of $x < 2\,\text{m}$ (see Figure 6.17). For equilibrium, the shearing force has magnitude $F = R_a = 1.3\,\text{kN}$ downwards. The bending moment $M = R_a x =$

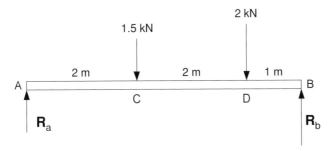

Figure 6.16. Forces acting on a light horizontal beam.

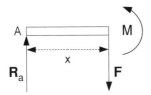

Figure 6.17. Section of beam with $x < 2$ m.

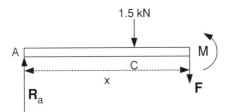

Figure 6.18. Section of beam with 2 m $< x < 4$ m.

$1.3x$ kN m, assuming that x is measured in metres. For this section of the beam, M varies linearly between 0 at A and 2.6 kN m at C.

Next, consider a section of the beam from A of length x, where 2 m $< x < 4$ m (see Figure 6.18). Now, the shearing force $F = R_a - 1.5 = 1.3 - 1.5 = -0.2$ kN downwards or 0.2 kN upwards. The bending moment $M = R_a x - 1.5(x - 2) = 3 - 0.2x$. Thus, for this section of the beam, M varies linearly between 2.6 kN m at C and 2.2 kN m at D.

Finally, consider a section of beam from A of length x, where 4 m $< x < 5$ m (see Figure 6.19). With F now shown upwards in direction in the diagram, the shearing force $F = 3.5 - R_a = 2.2$ kN. The bending moment

$$M = R_a x - 1.5(x - 2) - 2(x - 4) = 11 - 2.2x,$$

which varies linearly between 2.2 kN m at D and 0 at B.

Now, the shearing force and bending moment diagrams can be drawn, as shown in Figure 6.20.

Note: (a) The shearing force is constant between the ends and successive load points. It starts with a negative value equal in magnitude to the supporting force at the left end. It then changes positively at each load point by an amount equal in magnitude to the load.

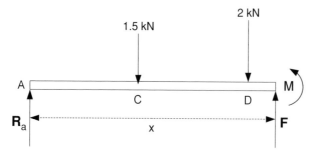

Figure 6.19. Section of beam with $4 \text{ m} < x < 5 \text{ m}$.

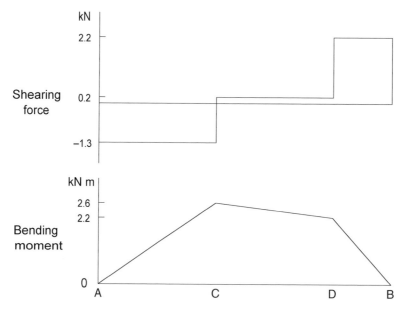

Figure 6.20. Shearing force and bending moment diagrams for the beam shown in Figure 6.16.

(b) The bending moment is continuous but changes slope at each load point. The slope of the bending moment diagram is the negative of shearing force. (The bending moment is often shown to be negative and in that case the bending moment slope equals the shearing force.)

2.   In this case, the support provides at A an upward force $R_a = L$ together with a couple M_a (see Figure 6.21). The latter must balance out the couple formed by the supporting force R_a at A and the load L at B. Hence, $M_a = La$.

Now consider a section of beam from A of length $x < a$ (see Figure 6.22). The shearing force F and bending moment M must balance out the supporting force R_a and moment M_a. Therefore, $F = R_a = L$ downwards and

$$M = R_a x - M_a = Lx - La = L(x - a) = -L(a - x).$$

Thus the shearing force is a constant L downwards throughout the projecting length of the beam and the bending moment varies linearly between $-La$ at A (the support) and 0 at B (the free end where the load is attached). Hence, the shearing force and bending moment diagrams are as shown

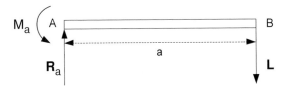

Figure 6.21. Supporting force R_a and couple M_a for a cantilever beam.

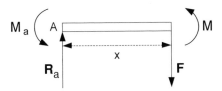

Figure 6.22. Section of beam of length $x < a$.

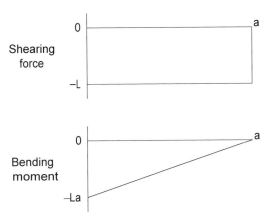

Figure 6.23. Shearing force and bending moment diagrams for the cantilever beam of Figure 6.21.

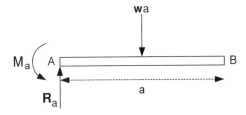

Figure 6.24. External forces and couple acting on a cantilever beam.

in Figure 6.23. Notice again that the slope of the bending moment is the negative of the shearing force.

3.   Consider the projecting part of the beam AB (see Figure 6.24). The resultant load is wa acting at the mid-point of the beam, i.e. at distance $a/2$ from A. The reactions at the support will be a vertical force $R_a = wa$ together with a couple of moment $M_a = wa \cdot \frac{a}{2} = wa^2/2$ to keep the beam in equilibrium.

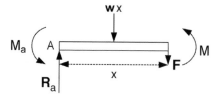

Figure 6.25. Section of cantilever beam of length $x < a$.

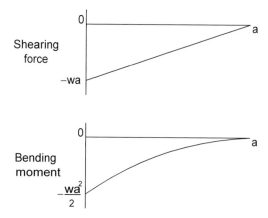

Figure 6.26. Shearing force and bending moment diagrams for the cantilever beam of Figure 6.24.

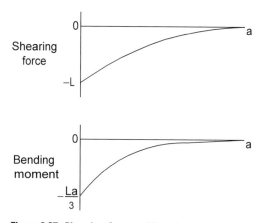

Figure 6.27. Shearing force and bending moment diagrams for the cantilever beam of Figure 6.15.

Next, consider the equilibrium of the section of beam from A of length $x < a$ (see Figure 6.25). The downward shearing force at the right-hand end of this section is:

$$F = R_a - wx = wa - wx = w(a - x).$$

The corresponding bending moment is:

$$M = R_a x - wx \cdot \frac{x}{2} - M_a = wax - \frac{wx^2}{2} - \frac{wa^2}{2} = -w(x - a)^2/2.$$

If M is the y-coordinate, this equation is that of a parabola which is concave downwards with axis parallel to the y-axis. The apex is at $(a, 0)$ and $y = -wa^2/2$ at $x = 0$.

The shearing force and bending moment diagrams can now be drawn as shown in Figure 6.26.

4.   Referring to Figure 6.15, the load intensity $w = c(a - x)$, where c is a constant. The total load is:

$$L = \int_0^a w \, dx = c \int_0^a (a - x) \, dx = c \left[ax - \frac{x^2}{2} \right]_0^a = ca^2/2.$$

Hence, $c = 2L/a^2$.

Now, $\dfrac{dF}{dx} = w$, so $F = \displaystyle\int w \, dx = c(a - x/2)x + K$,

where K is an integrating constant. The shearing force is zero at a free end, i.e. at $x = a$ in this case.

Thus, $\dfrac{ca^2}{2} + K = 0$ and $F = -\dfrac{c}{2}(x^2 - 2ax + a^2) = -\dfrac{c}{2}(x - a)^2$.

Therefore, $F = -\dfrac{L}{a^2}(x - a)^2$.

If F is the y-coordinate, this is the equation of a parabola which is concave downwards, has its axis parallel to the y-axis and apex at $(a, 0)$. $F = -L$ at $x = 0$ and $F = 0$ at $x = a$.

Now, $\dfrac{dM}{dx} = -F = \dfrac{c}{2}(x - a)^2$, so $M = -\displaystyle\int F \, dx = \dfrac{c}{6}(x - a)^3 + J$,

where J is an integrating constant. $M = 0$ at a free end, i.e. at $x = a$ and therefore $J = 0$.

Hence, $M = \dfrac{c}{6}(x - a)^3 = \dfrac{L}{3a^2}(x - a)^3$.

This is zero at $x = a$ and $-La/3$ at $x = 0$.

The shearing force and bending moment diagrams can now be drawn, as shown in Figure 6.27.

7 Friction

7.1 Force of friction

Friction between two solid surfaces is that which tries to prevent one surface from sliding over the other. It is caused by a roughness in the surfaces so that protruding particles which form part of one surface interlock with particles which protrude from the other surface. Such surfaces would quickly wear down if forced to slide over one another. To prevent this, a lubricant oil is used to separate bearing surfaces. However, friction is often an important factor in statics for maintaining equilibrium.

Friction between two dry surfaces is sometimes called *Coulomb friction* since Coulomb performed many experiments to establish some empirical laws (C.A. Coulomb, *Theorie des machines simples*, Paris, 1821). However, much was known about friction before then, as may be seen from the information on friction in a physics textbook published in 1740 (Pieter Van Musschenbroek, *A treatise on natural philosophy for the use of students in the university*, translated into English by John Colson, Lucasian Professor of Mathematics in the University of Cambridge, 2nd edn, London, 1740).

Let us regard friction as being the tangential component of the reaction force, between two surfaces in contact, which tries to prevent sliding. If the force which is trying to produce sliding is gradually increased, friction will increase to maintain equilibrium until a maximum value is reached, after which sliding commences. We shall refer to this maximum value as *limiting friction*.

The laws of friction for dry surfaces in contact are as follows.

1. Limiting friction is proportional to the normal component (perpendicular to the surface) of reaction between the surfaces in contact.
2. Limiting friction is independent of the area in contact.
3. The first two laws still apply for friction when there is sliding. Referring to friction when there is sliding as *kinetic friction* and to friction when there is no sliding as *static friction*, for any two dry surfaces in contact, the magnitude of kinetic friction is less than that of limiting static friction. For a low velocity of slide, the kinetic friction is independent of velocity.

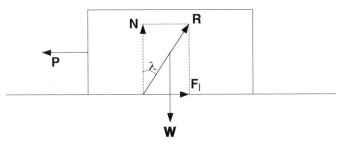

Figure 7.1. A block on the point of slipping.

The first law can be expressed simply by the equation $F_1 = \mu N$, where F_1 is limiting friction, N is the normal component of reaction and μ is the proportionality constant called the *coefficient of friction*. The value of μ depends on the nature and materials of the surfaces of contact. The value of μ is often about 0.5, as in the case of leather on wood, metal on wood and masonry on dry clay. It is less for metal on metal, more like 0.2, and with some materials it can be much smaller, e.g. for steel on ice it is about 0.03. The value of μ for a motor car's tyres on dry concrete is about 1.0, so the maximum braking force available is likely to be about the same as the weight of the vehicle. Since kinetic friction is less than limiting static friction, a driver will not bring a vehicle to a halt in the shortest possible distance if the brakes are applied so hard that the tyres skid. It is for this reason that many cars are equipped with ABS (antilock braking system).

Let a block of weight W rest on a horizontal surface, as shown in Figure 7.1. Apply a pull P which is just sufficient for the block to be on the point of slipping. The total force of reaction from the surface is represented by R, which has a normal component N and a limiting static frictional component F_1.

If μ is the coefficient of friction, then $F_1 = \mu N$. Moreover, if λ is the angle between the total surface reaction R and the normal component N, then $\mu = F_1/N = \tan \lambda$, and λ is called the *angle of friction*.

EXERCISE 1

In the example just considered (see Figure 7.1), let the pull P be removed but let the surface on which the block rests be tilted until the block is on the verge of slipping. Find the relation between the angle of tilt of the surface and the angle of friction.

Problems 49 and 50.

7.2 Sliding or toppling?

In Figure 7.1 applied to Exercise 1, the base of the block was deliberately taken to be long compared to its height. However if the opposite were the case, as shown in

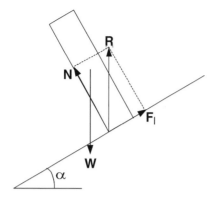

Figure 7.2. A block about to topple.

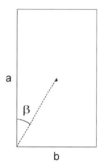

Figure 7.3. The angle β which determines whether a tilted block will topple or slide.

Figure 7.2, the block would topple over rather than slide down the incline. If F_1 is the limiting static friction, then α is the angle of friction and the block is on the verge of sliding. However, the reaction R must act through the base of the block. If the resultant weight force W acts outside the base as shown, then R and W form a couple which topples the block.

To decide whether a block will eventually topple or slide when the surface on which it rests is tilted, we need to know a and b, the height and the base length, respectively, and also μ, the coefficient of static friction. In the absence of sliding, the block will topple when the vertical line through the centre of gravity no longer cuts the base of the block, i.e. when $\alpha > \beta$, where β is the angle indicated in Figure 7.3. In the absence of toppling, the block will slide when $\alpha > \lambda$. It follows that the block will slide rather than topple if $\beta > \lambda$, i.e. if

$$\mu = \tan \lambda < \tan \beta = b/a.$$

EXERCISE 2

Suppose a rectangular packing case, with height a and base length b, has its centre of gravity at the centre point C of the case. If the case rests on a horizontal surface and is subjected to a horizontal pull P at the top, as shown in Figure 7.4, what is the condition on the coefficient of static friction μ between the case and the surface for the case to slide rather than topple as P is increased?

Problems 51 and 52.

7.3   Direction of minimum pull

In Exercise 2, the pull P required to slide the case could have been reduced by raising the direction of pull above the horizontal, as shown in Figure 7.5. If the coefficient of static friction between the case and the horizontal surface is $\mu = \tan \lambda$, let us find θ, the angle of pull P to the horizontal, to minimize the pull required to slide the case and also find the magnitude of the minimum P.

If the case is on the verge of sliding,

$$F_1 = \mu N = P \cos \theta \quad \text{and} \quad N + P \sin \theta = W.$$

Thus, $P \cos \theta = \mu W - \mu P \sin \theta$ and $P = \dfrac{\mu W}{\cos \theta + \mu \sin \theta}.$

We can now simplify the problem by writing $\mu = \tan \lambda$, so that:

$$P = \frac{W \tan \lambda}{\cos \theta + \sin \theta \tan \lambda} = \frac{W \sin \lambda}{\cos \theta \cos \lambda + \sin \theta \sin \lambda} = \frac{W \sin \lambda}{\cos(\theta - \lambda)}.$$

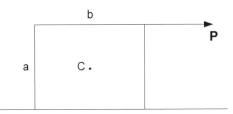

Figure 7.4. A packing case subjected to a horizontal pull P.

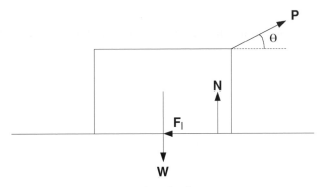

Figure 7.5. Raising the direction of pull.

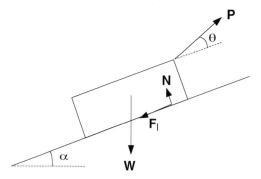

Figure 7.6. Finding the minimum force P to pull a case up an inclined plane.

P is minimum when $\cos(\theta - \lambda)$ is maximum, i.e. when $\theta = \lambda$.

Then, $P_{min} = W \sin \lambda$.

EXERCISE 3

Now, let the packing case be placed on an inclined plane, as shown in Figure 7.6. Find the direction and magnitude for the minimum pull P to slide the case up the plane.

Problem 53.

7.4   Ladder leaning against a wall

If a ladder AB of length $2a$ is leant against a wall, what is the minimum angle θ to the horizontal for the ladder to remain in equilibrium? Let the frictional and normal components of reaction from wall and ground be F_1, N_1 at A and F_2, N_2 at B, as shown in Figure 7.7. Since friction tries to prevent sliding, F_1 is upward at A and F_2 is to the left at B.

If the coefficients of friction between the ladder and the wall and ground are μ_1 and μ_2, respectively, then:

$$F_1 < \mu_1 N_1 \quad \text{and} \quad F_2 < \mu_2 N_2.$$

Resolving horizontally: $N_1 = F_2$ and vertically: $F_1 + N_2 = W$, where W is the weight of the ladder.

Taking moments about the mid-point C:

$$N_2 a \cos \theta - F_2 a \sin \theta - F_1 a \cos \theta - N_1 a \sin \theta = 0.$$

Hence, it follows that:

$$\tan \theta = \frac{N_2 - F_1}{F_2 + N_1} \geq \frac{N_2 - \mu_1 N_1}{2N_1} \geq \frac{F_2 - \mu_1 \mu_2 N_1}{2\mu_2 N_1} = \frac{N_1 - \mu_1 \mu_2 N_1}{2\mu_2 N_1} = \frac{1 - \mu_1 \mu_2}{2\mu_2}.$$

Therefore, $\theta \geq \tan^{-1} \left(\dfrac{1 - \mu_1 \mu_2}{2\mu_2} \right)$.

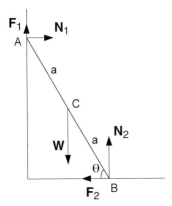

Figure 7.7. A ladder leaning against a wall.

EXERCISE 4

If a person climbs a ladder that is leaning against a wall, show that it is less likely to slip when the person is on the lower half but more likely to slip when on the upper half.

Problems 54 and 55.

7.5   Motor vehicle clutch

The purpose of a motor vehicle clutch is to connect or disconnect the engine from the drive wheels in a smooth manner. Torque is transmitted by friction between one or more pairs of co-axial annular faces. Friction, and hence maximum transmitted torque, is proportional to the axial thrust provided by springs which push the faces together. Disconnection is achieved by pulling the faces apart. Let us examine the relationship between the maximum torque T and the axial thrust P for just two faces in contact.

Figure 7.8 represents two aspects of the annular friction plates. One shows the axial thrust P. The other shows the inner and outer circular edges to the faces of radii r_1 and r_2, respectively, and an intermediate co-axial circular strip of radius r and width dr.

Let p be the pressure intensity (force per unit area) between the two friction surfaces. Then, the normal force on the narrow strip is $p \times 2\pi r dr$. It follows that:

$$P = 2\pi \int_{r_1}^{r_2} pr\, dr.$$

If μ is the coefficient of limiting static friction, the friction force around the strip is $\mu p \times 2\pi r dr$ and its moment about the axis is $r \times \mu p \times 2\pi r dr$. Hence, the maximum transmitted torque is:

$$T = 2\pi \mu \int_{r_1}^{r_2} pr^2\, dr.$$

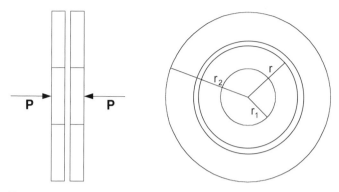

Figure 7.8. Diagrammatic representation of two clutch plates.

If the pressure intensity does not vary with r, then p is constant and:

$$P = 2\pi p \int_{r_1}^{r_2} r \, dr = \pi p \left(r_2^2 - r_1^2 \right),$$

$$T = 2\pi \mu p \int_{r_1}^{r_2} r^2 \, dr = 2\pi \mu p \left(r_2^3 - r_1^3 \right) \big/ 3 = \frac{2\mu P \left(r_2^3 - r_1^3 \right)}{3 \left(r_2^2 - r_1^2 \right)}.$$

EXERCISE 5

If it is assumed that wear occurs at a constant rate over the area of the surfaces in contact, since wear speed is proportional to pressure × velocity, which is proportional to pressure × radius, it follows that: $pr = $ constant $(= c$, say$)$. Find the relationship between the maximum transmitted torque T and the axial thrust P in this case.

Problem 56.

7.6 Capstan

The principle of a capstan is to have a drum rotated at constant speed by powerful machinery. A rope is wrapped around the drum so that a relatively small force applied to one end of the rope induces a relatively large force at the other end. In this way, heavy loads may be shifted by pulling the loose end of the capstan rope.

Consider a small element of the rope in contact with the drum and subtending an angle $d\theta$ (radians) at the centre of the drum, as indicated in Figure 7.9. Let R be the component of force acting on the element of rope from the drum and normal to its surface. Assume that the drum is slipping against the rope with coefficient of kinetic friction μ so that the frictional force on the element of the rope is μR. Let the tension on one side of the element be T and on the other side $T + dT$. Finally, let the tension in the rope in contact with the drum change from T_0 to T_1 and the angle subtended at the centre by that section of the rope be θ_1 radians.

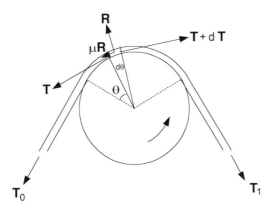

Figure 7.9. Analysing the action of a capstan.

Resolving forces on the small element of rope and using the fact that, since $d\theta$ is small, we may take $\sin d\theta = d\theta$ and $\cos d\theta = 1$:

normally: $R = T d\theta/2 + (T + dT)d\theta/2 = T d\theta$ (neglecting $dT d\theta$),

tangentially: $T + dT = T + \mu R$, hence, $dT = \mu T d\theta$.

Now integrate over the length of rope in contact with the drum:

$$\int_{T_0}^{T_1} \frac{dT}{T} = \int_0^{\theta_1} \mu \, d\theta,$$

$$\ln T_1 - \ln T_0 = \ln \frac{T_1}{T_0} = \mu \theta_1$$

and the tension amplification factor is:

$$\frac{T_1}{T_0} = \exp \mu \theta_1,$$

where the angle θ_1 is measured in radians.

EXERCISE 6

Evaluate the force amplification factor T_1/T_0 when the rope has a coefficient of kinetic friction with the drum of 0.3 and the rope is wrapped round the drum either once, twice, three times or four times.

Problem 57.

7.7 Answers to exercises

1.   The answer to this question is provided immediately by examining Figure 7.10. Since the block is on the verge of slipping, it is in equilibrium with limiting static friction being applied. This is indicated by the arrow F_1 in the diagram, while N is the normal component of reaction from the surface. The

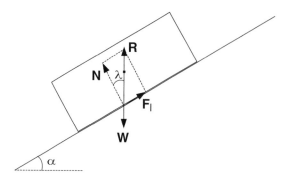

Figure 7.10. A block on the verge of slipping.

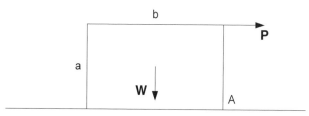

Figure 7.11. Toppling the case about edge A.

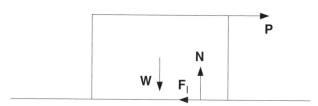

Figure 7.12. Sliding the case.

resultant R of F_1 and N must be equal, opposite and collinear to the weight of the block, indicated by the arrow W.

As already defined, the angle of friction is the angle λ between the directions of R and N. Thus, λ is the angle between the vertical and the normal to the plane. Since N is perpendicular to the line of tilt of the plane and R is vertical, it follows that $\lambda = \alpha$, the angle of tilt of the plane to the horizontal.

2. If the case were to topple, it would do so about the bottom edge A (see Figure 7.11), with the moment of P given by Pa being greater than the opposing moment of the weight $Wb/2$, i.e. $P > \frac{Wb}{2a}$.

On the other hand, if the case were to slide, we must have $P > F_1 = \mu N = \mu W$ (see Figure 7.12).

The case will slide rather than topple if the force P necessary for sliding is less than that for toppling. Thus, the case will slide rather than topple if:

$$\frac{Wb}{2a} > \mu W, \text{ i.e. } \mu < \frac{b}{2a}.$$

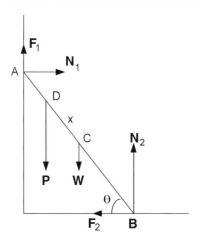

Figure 7.13. A ladder with person's weight P at D.

3. Referring to Figure 7.6 with the case on the verge of sliding, i.e. with $F_1 = \mu N$, resolve forces parallel and perpendicular to the plane:

$$P \cos \theta = W \sin \alpha + F_1 = W \sin \alpha + \mu N,$$

$$P \sin \theta + N = W \cos \alpha.$$

Therefore, $P \cos \theta = W \sin \alpha + \mu W \cos \alpha - \mu P \sin \theta$ or $P = \dfrac{W(\sin \alpha + \mu \cos \alpha)}{\cos \theta + \mu \sin \theta}$.

Write $\mu = \tan \lambda = \sin \lambda / \cos \lambda$, so that:

$$P = \frac{W(\sin \alpha \cos \lambda + \cos \alpha \sin \lambda)}{\cos \theta \cos \lambda + \sin \theta \sin \lambda} = \frac{W \sin(\alpha + \lambda)}{\cos(\theta - \lambda)}.$$

P is minimum when $\theta = \lambda$ and

$$P_{\min} = W \sin(\alpha + \lambda).$$

4. The only difference from the case of the ladder discussed in Section 7.4 is the person's weight P at D, as indicated in Figure 7.13. The position D is distance x up the ladder from its centre C. If x were negative, D would be down the ladder from C.

    Again, $F_1 \le \mu_1 N_1$ and $F_2 \le \mu_2 N_2$.
    Resolving horizontally and vertically:

$$N_1 = F_2 \quad \text{and} \quad F_1 + N_2 = P + W.$$

    Taking moments about C:

$$N_2 a \cos \theta + P x \cos \theta = F_2 a \sin \theta + F_1 a \cos \theta + N_1 a \sin \theta.$$

Hence, $\tan \theta = \dfrac{(N_2 - F_1)a + Px}{(F_2 + N_1)a} = \dfrac{N_2 - F_1 + P\alpha}{2N_1}, \quad \alpha = \dfrac{x}{a},$

and therefore $\tan \theta \ge \dfrac{1 - \mu_1 \mu_2}{2\mu_2} + \dfrac{P\alpha}{2N_1}.$

Thus, if x is positive, α is positive and a larger θ is required to ensure equilibrium. On the other hand, if x is negative, α is negative and a smaller θ will ensure equilibrium.

5. If $pr = c$,

$$P = 2\pi \int_{r_1}^{r_2} pr \, dr = 2\pi c(r_2 - r_1).$$

Then, $T = 2\pi\mu \int_{r_1}^{r_2} pr^2 \, dr = 2\pi\mu c \int_{r_1}^{r_2} r \, dr$

$$= \pi\mu c\left(r_2^2 - r_1^2\right) = \mu P(r_2 + r_1)/2 = \mu P R,$$

where R is the mean radius of the friction surface.

6. Evaluating $T_1/T_0 = \exp\mu\theta_1$, we obtain the following result:

θ_1	2π	4π	6π	8π
T_1/T_0	6.6	43	286	1881

8 Non-coplanar forces and couples

8.1 Coplanar force and couple

We have already seen in Section 2.4 that a force acting at one point of a rigid body is equivalent to the same force with a different (but parallel) line of action together with a couple. However, we shall study this in greater detail before considering several non-coplanar forces and couples.

Firstly, let us see with the aid of diagrams how a force and a coplanar couple may be replaced by a single force. Figure 8.1a shows a force and a couple both acting in the x, y plane. The force is $\mathbf{F}$ acting at P and the couple is made up from the two forces $\mathbf{G}$ and $-\mathbf{G}$ with their lines of action a distance d_1 apart. A couple has the same effect wherever it acts in the plane. Hence, we can replace it by a couple at P, indicated in Figure 8.1b by a double-line vector, using the right-hand thread rule. The magnitude of the couple vector is $|\mathbf{G}|d_1$. Since it acts in the positive z-direction, the vector moment of the couple is $|\mathbf{G}|d_1\mathbf{k}$, where $\mathbf{k}$ is the unit vector in the z-direction.

Finally, the force and couple at P is equivalent to just the force $\mathbf{F}$ at the point Q, still in the x, y plane, as shown in Figure 8.2. Q is positioned so that the moment of $\mathbf{F}$ about P is the same as the moment of the original couple. The shift in the line of action of $\mathbf{F}$ is d_2 such that:

$$|\mathbf{F}|d_2 = |\mathbf{G}|d_1 \text{ or } d_2 = |\mathbf{G}|d_1/|\mathbf{F}|.$$

Next, suppose that we have a force $\mathbf{F} = F_x\mathbf{i} + F_y\mathbf{j} + F_z\mathbf{k}$ acting at a point $P(x_p, y_p, z_p)$, see Figure 8.3, and we wish to replace it by the equivalent force and couple at the origin $O(0, 0, 0)$. Let us use a three-lined equals sign $\equiv$ to indicate equivalence. Then $F_x\mathbf{i}$ at $P \equiv F_x\mathbf{i}$ at B with couple of moment $F_x z_p\mathbf{j} \equiv F_x\mathbf{i}$ at A with couples $F_x z_p\mathbf{j}$ and $-F_x y_p\mathbf{k} \equiv F_x\mathbf{i}$ at O with couples $F_x z_p\mathbf{j}$ and $-F_x y_p\mathbf{k}$.

$F_y\mathbf{j}$ at $P \equiv F_y\mathbf{j}$ at $D \equiv F_y\mathbf{j}$ at Q with couple $F_y x_p\mathbf{k} \equiv F_y\mathbf{j}$ at O with couples $F_y x_p\mathbf{k}$ and $-F_y z_p\mathbf{i}$.

$F_z\mathbf{k}$ at $P \equiv F_z\mathbf{k}$ at $B \equiv F_z\mathbf{k}$ at A with couple $F_z y_p\mathbf{i} \equiv F_z\mathbf{k}$ at O with couples $F_z y_p\mathbf{i}$ and $-F_z x_p\mathbf{j}$.

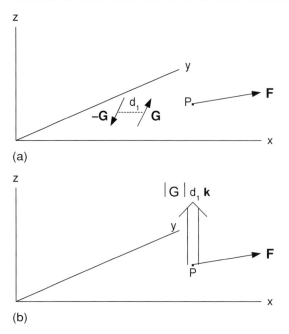

(a)

(b)

Figure 8.1. Shifting a couple to the point where a force acts.

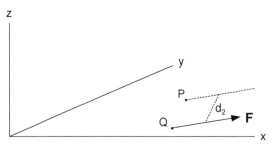

Figure 8.2. Replacing force and couple at P by a force at Q.

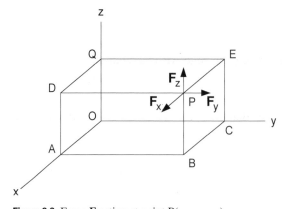

Figure 8.3. Force **F** acting at point $P(x_p, y_p, z_p)$.

We shall illustrate in Section 8.2 that two couples are equivalent to one couple with a moment obtained by vectorial addition of the vector moments of the original couples. Thus the six moments at the origin obtained by shifting the components of $\mathbf{F}$ at P to the origin are equivalent to one couple with moment:

$$\mathbf{C} = (F_z y_\mathrm{p} - F_y z_\mathrm{p})\mathbf{i} + (F_x z_\mathrm{p} - F_z x_\mathrm{p})\mathbf{j} + (F_y x_\mathrm{p} - F_x y_\mathrm{p})\mathbf{k}.$$

This may be written more briefly as the *determinant* of a 3×3 matrix as follows:

$$\begin{vmatrix} \mathbf{i} & \mathbf{j} & \mathbf{k} \\ x_\mathrm{p} & y_\mathrm{p} & z_\mathrm{p} \\ F_x & F_y & F_z \end{vmatrix}.$$

Since $\mathbf{F} = F_x\mathbf{i} + F_y\mathbf{j} + F_z\mathbf{k}$ and if we denote the position vector of the point P from the origin O as:

$$\mathbf{p} = x_\mathrm{p}\mathbf{i} + y_\mathrm{p}\mathbf{j} + z_\mathrm{p}\mathbf{k},$$

then $\mathbf{C}$ becomes the vector product of $\mathbf{p}$ with $\mathbf{F}$ in that order, i.e.

$$\mathbf{C} = \mathbf{p} \times \mathbf{F}.$$

EXERCISE 1

Let P_1 and P_2 be two points in a rigid body with position vectors $\mathbf{r}_1 = (-2\mathbf{i} + \mathbf{j} + \mathbf{k})\,\mathrm{cm}$ and $\mathbf{r}_2 = (2\mathbf{i} - 3\mathbf{j} + 3\mathbf{k})\,\mathrm{cm}$, respectively. Find the force and couple at P_2 equivalent to the force $\mathbf{F} = (10\mathbf{i} + 5\mathbf{j} + 3\mathbf{k})\,\mathrm{N}$ at P_1.

Problem 58.

8.2   Effect of two non-coplanar couples

Let two couples be applied to adjacent sides of a cube as shown in Figure 8.4. We can move couples around in their planes, so shift the top one so that one of its forces lies along AB. Then do the same with the side couple. This leaves us with the situation shown in Figure 8.5a, in which only the edge AB is drawn.

Next, we replace the forces 6 N placed 1.5 cm apart by forces 10 N placed 0.9 cm apart, as shown in Figure 8.5b. This is still a couple of moment 9 N cm. Now, when we examine the diagram, we see that the two forces on the edge AB cancel each other, leaving one force 10 N on top of the cube and in the opposite direction to the 10 N on the side of the cube. They are parallel to each other and form a new couple.

The plane of the new couple passes through two lines parallel to the edge AB of the cube. One line is in the top of the cube, 2 cm away from the edge AB, and the other line is in the side of the cube, 0.9 cm away from the edge AB. The two forces, each of 10 N,

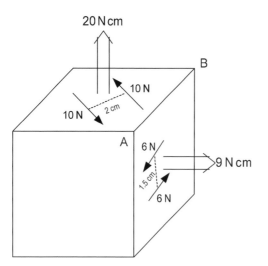

Figure 8.4. Couples applied to adjacent sides of a cube.

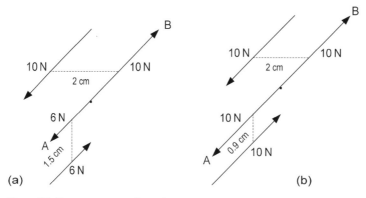

Figure 8.5. Rearrangement of couples.

which form the new couple lie along these lines, are distance $\sqrt{2^2 + 0.9^2} = 2.19$ cm apart (see Figure 8.6). Hence, if we represent the new resultant couple by the moment vector $\mathbf{C}$, it has magnitude:

$$|\mathbf{C}| = 10 \times 2.19 = 21.9\,\text{N cm} = 0.219\,\text{N m}.$$

Also, we see from Figure 8.6 that $\mathbf{C}$ makes an angle:

$$\alpha = \tan^{-1}(0.9/2) = 24.2°$$

with the vertical.

Using Cartesian coordinates, if the original couple moment vectors were $9\,\text{N cm}$ parallel to the x-axis and $20\,\text{N cm}$ parallel to the z-axis, and the edge of the cube AB were parallel to the y-axis, then the two couple moment vectors would be $\mathbf{C}_1 = 9\mathbf{i}\,\text{N cm}$ and $\mathbf{C}_2 = 20\mathbf{k}\,\text{N cm}$. Now, the resultant couple moment vector as already calculated would be:

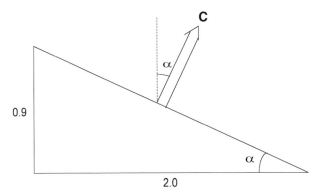

Figure 8.6. Couple **C** formed by forces in the top and side of a cube.

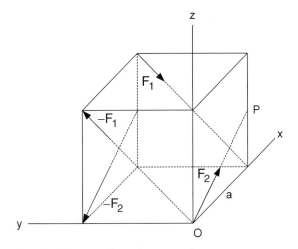

Figure 8.7. Two couples acting on a cube.

$$\mathbf{C} = (21.9\sin 24.2°)\mathbf{i} + (21.9\cos 24.2°)\mathbf{k} = (9\mathbf{i} + 20\mathbf{k})\,\mathrm{N\,cm} = \mathbf{C}_1 + \mathbf{C}_2.$$

In other words, the two couples with moment vectors $\mathbf{C}_1$ and $\mathbf{C}_2$ have the same effect on the rigid body as a single resultant couple with moment vector $\mathbf{C} = \mathbf{C}_1 + \mathbf{C}_2$.

In fact, this is a general rule for finding the resultant of two couples. It has only been proved here for a particular example. (See page 105 of *Mechanics*, by J. P. Den Hartog, published by Dover, New York, 1961 for a general proof.)

EXERCISE 2

Suppose two couples act on a cube as shown in Figure 8.7. Let $|\mathbf{F}_1| = 200\,\mathrm{N}$, $|\mathbf{F}_2| = 100\,\mathrm{N}$, the length of the edge of the cube be $a = 0.5\,\mathrm{m}$ and P be the mid-point of a vertical edge. With coordinate axes as shown, find the resultant couple acting on the cube expressed as a moment vector **C**.

Problems 59 and 60.

8.3   The wrench

In Section 8.1, we found that a force and a coplanar couple acting on a rigid body was equivalent to the force alone with a different but appropriately chosen line of action. Let us now examine the situation when the force and couple are not coplanar.

Let a force $\mathbf{F}$ act at a point P which has position vector $\mathbf{r}_p$ (see Figure 8.8). This is combined with a couple with moment vector $\mathbf{C}$, which is not perpendicular to $\mathbf{F}$, i.e. the couple and the force $\mathbf{F}$ are not coplanar.

Firstly, we replace the couple by two couples, one with moment vector $\mathbf{C}_f$ along the line of $\mathbf{F}$ and the other with moment vector $\mathbf{C}_p$ perpendicular to $\mathbf{F}$ (see Figure 8.9). Thus, $\mathbf{C}_p \perp \mathbf{C}_f$ and $\mathbf{C}_p + \mathbf{C}_f = \mathbf{C}$.

Secondly, we move the point of application of the force $\mathbf{F}$ to another point Q in order to eliminate the couple with moment vector $\mathbf{C}_p$ (see Figure 8.10). The position vector

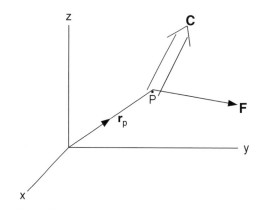

Figure 8.8. Force $\mathbf{F}$ and couple $\mathbf{C}$ combination.

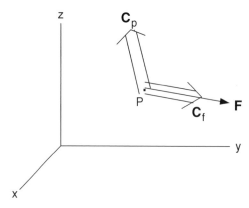

Figure 8.9. Couple $\mathbf{C}$ replaced by two couples $\mathbf{C}_f$ and $\mathbf{C}_p$ at right angles.

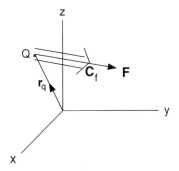

Figure 8.10. Point of application of $\mathbf{F}$ moved to eliminate couple $\mathbf{C}_p$.

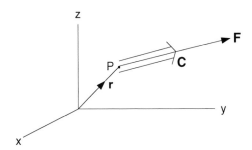

Figure 8.11. Positive wrench acting at P.

$\mathbf{r}_q$ of Q must be such that $\mathbf{F}$ acting at Q has moment $\mathbf{C}_p$ about P, i.e.

$$\mathbf{C}_p = (\mathbf{r}_q - \mathbf{r}_p) \times \mathbf{F}.$$

What we are left with is the combination of a force $\mathbf{F}$ together with a couple in a plane perpendicular to $\mathbf{F}$. This combination is called a *wrench*. If $\mathbf{C}_f$ is in the same direction as $\mathbf{F}$, as shown in Figure 8.10, then we have a *positive wrench*. If $\mathbf{C}_f$ had been in the opposite direction, we would have had a *negative wrench*.

EXERCISE 3

Referring to Figure 8.11, a positive wrench has a couple of moment $100\,\mathrm{N\,m}$ and force component $\mathbf{F} = (-50\mathbf{i} + 40\mathbf{j} + 30\mathbf{k})\,\mathrm{N}$. If its point of application P has position vector $\mathbf{r} = (\mathbf{i} + \mathbf{j} + 2\mathbf{k})\,\mathrm{m}$, find the moment of the wrench about the origin and its moment about the y-axis.

EXERCISE 4

Suppose that a force $\mathbf{F}$ and a couple with moment vector $\mathbf{C}$ act on a rigid body. Coordinate axes are drawn so that $\mathbf{F}$ acts at the origin. If $\mathbf{F} = (30\mathbf{i} + 50\mathbf{j} + 40\mathbf{k})\,\mathrm{N}$ and $\mathbf{C} = (10\mathbf{i} + 5\mathbf{j} - 10\mathbf{k})\,\mathrm{N\,m}$, find the equivalent wrench acting on the body.

Problems 61 and 62.

8.4　Resultant of a system of forces and couples

Let us recap on a result obtained in Section 8.1. Suppose we have a Cartesian coordinate system with its origin O at a point in a rigid body. Then a force $\mathbf{F}$ acting at another point P of the body will have a moment $\mathbf{p} \times \mathbf{F}$ about O, where $\mathbf{p}$ is the position vector of P (see Figure 8.12). Calling this moment $\mathbf{C}$, i.e. $\mathbf{C} = \mathbf{p} \times \mathbf{F}$, it follows that a force $\mathbf{F}$ at P is equivalent to a force $\mathbf{F}$ at O together with a couple with moment vector $\mathbf{C}$.

Now, suppose that we have several forces $\mathbf{F}_i$ acting at points P_i with position vectors $\mathbf{p}_i$, $i = 1, 2, \ldots, n$. Each force $\mathbf{F}_i$ is equivalent to the same force acting at O together with a couple with moment vector $\mathbf{C}_i = \mathbf{p}_i \times \mathbf{F}_i$. We can then find the resultant force $\mathbf{F}_R$ acting at O from the vector sum:

$$\mathbf{F}_R = \sum_{i=1}^{n} \mathbf{F}_i$$

together with a resultant couple with moment vector given by the vector sum:

$$\mathbf{C}_R = \sum_{i=1}^{n} \mathbf{C}_i = \sum_{i=1}^{n} \mathbf{p}_i \times \mathbf{F}_i.$$

Also, there may be couples with moment vectors $\mathbf{C}_j$, $j = 1, 2, \ldots, m$, acting on the body independently of the forces $\mathbf{F}_i$. In this case, there will be a resultant couple with moment vector:

$$\mathbf{C}_R = \sum_{i=1}^{n} \mathbf{p}_i \times \mathbf{F}_i + \sum_{j=1}^{m} \mathbf{C}_j.$$

EXERCISE 5

If a rigid body is acted on by two forces $\mathbf{F}_1$ and $\mathbf{F}_2$ at points P_1 and P_2, respectively, and also by a couple with moment vector $\mathbf{C}$, find the resultant force and couple at the origin of coordinates

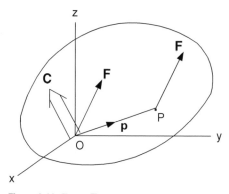

Figure 8.12. Force $\mathbf{F}$ at P replaced by $\mathbf{F}$ at O together with a couple of moment $\mathbf{C}$.

given that:

$$F_1 = (80\mathbf{i} - 40\mathbf{j} + 50\mathbf{k})\,\text{N} \quad \text{and} \quad F_2 = (-30\mathbf{i} + 20\mathbf{j} - 40\mathbf{k})\,\text{N},$$

position vectors of P_1 and P_2 are:

$$\mathbf{p}_1 = (\mathbf{i} - \mathbf{j} + 2\mathbf{k})\,\text{m} \quad \text{and} \quad \mathbf{p}_2 = (-\mathbf{i} + 2\mathbf{j} - \mathbf{k})\,\text{m},$$

respectively, and $\mathbf{C} = (40\mathbf{i} + 30\mathbf{j} - 50\mathbf{k})\,\text{N m}$.

Problems 63 and 64.

8.5 Equations of equilibrium

We showed in Section 8.4 that a system of forces and couples acting on a rigid body can be reduced to a single resultant force acting at the origin together with a single couple. Equilibrium will exist if the resultant force and couple are both zero, i.e.:

$$\sum_{i=1}^{n} \mathbf{F}_i = 0 \quad \text{and} \quad \sum_{i=1}^{n} \mathbf{p}_i \times \mathbf{F}_i + \sum_{j=1}^{m} \mathbf{C}_j = 0.$$

For the resultant force $\mathbf{F}_R$ to be zero, all three of its Cartesian components must be zero. Hence, the vector equation $\sum_{i=1}^{n} \mathbf{F}_i = 0$ splits into three scalar equations:

$$\sum_{i=1}^{n} (F_x)_i = 0, \; \sum_{i=1}^{n} (F_y)_i = 0 \quad \text{and} \quad \sum_{i=1}^{n} (F_z)_i = 0,$$

where $(F_x)_i$, $(F_y)_i$ and $(F_z)_i$ are the x-, y- and z-components of $\mathbf{F}_i$.

Similarly the couple moment equation:

$$\mathbf{C}_R = \sum_{i=1}^{n} \mathbf{p}_i \times \mathbf{F}_i + \sum_{j=1}^{m} \mathbf{C}_j = 0$$

can be split into three scalar equations by taking the x-, y- and z-components:

$$\sum_{i=1}^{n} (\mathbf{p}_i \times \mathbf{F}_i) \cdot \mathbf{i} + \sum_{j=1}^{m} \mathbf{C}_j \cdot \mathbf{i} = 0$$

$$\sum_{i=1}^{n} (\mathbf{p}_i \times \mathbf{F}_i) \cdot \mathbf{j} + \sum_{j=1}^{m} \mathbf{C}_j \cdot \mathbf{j} = 0$$

$$\sum_{i=1}^{n} (\mathbf{p}_i \times \mathbf{F}_i) \cdot \mathbf{k} + \sum_{j=1}^{m} \mathbf{C}_j \cdot \mathbf{k} = 0.$$

Since the x-, y- and z-components of a couple moment vector correspond to the moments about the x-, y- and z-axes, respectively, the three scalar moment equations may be interpreted as equating to zero the sum of the moments about the x-, y- and z-axes, in turn.

Altogether, we have six independent scalar equations which must be satisfied to ensure equilibrium. In the coplanar case (Section 2.5), we were able to replace two force and one moment equations by three moment equations about non-collinear points. In the non-coplanar case, we can replace the three force and three moment equations by six moment equations provided there is no straight line intersecting all six moment axes. Otherwise, a non-zero resultant force along that straight line would not affect any of the moments.

EXERCISE 6

Three pieces of string are attached to points A, B and C in a horizontal plane. The other ends of the strings are joined together at a point O from which a weight W is suspended. Find the tensions in the pieces of string given that OA, OB and OC are mutually perpendicular and have lengths in the ratio $3 : 4 : 5$, respectively. (Hint: let OA, OB and OC lie along Cartesian axes, find the equation of the plane through A, B and C, and hence, the direction of the perpendicular to the plane.)

EXERCISE 7

A rectangular table top $2\,\text{m} \times 1\,\text{m}$ is supported horizontally $1\,\text{m}$ from the ground by six light struts, as shown in Figure 8.13. The struts are smoothly jointed to the corners of the table top and to the ground. The table top weighs $200\,\text{N}$ and is also subjected to two horizontal forces, each of $200\,\text{N}$, as shown. Calculate the compression or tension in each of the struts. By taking them all as compression, as indicated in the diagram, any negative result will correspond to tension.

Problems 65 and 66.

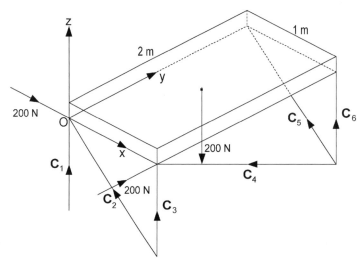

Figure 8.13. A horizontal table supported on six light struts.

8.6 Answers to exercises

1. Referring to Figure 8.14, force $\mathbf{F}$ at P_1 is equivalent to force $\mathbf{F}$ at P_2 together with a couple of moment:

$$\mathbf{C} = \mathbf{r} \times \mathbf{F}.$$

$$\mathbf{r} = \mathbf{r}_1 - \mathbf{r}_2 = (-2\mathbf{i} + \mathbf{j} + \mathbf{k}) - (2\mathbf{i} - 3\mathbf{j} + 3\mathbf{k}) = -4\mathbf{i} + 4\mathbf{j} - 2\mathbf{k}.$$

$$\mathbf{C} = \begin{vmatrix} \mathbf{i} & \mathbf{j} & \mathbf{k} \\ -4 & 4 & -2 \\ 10 & 5 & 3 \end{vmatrix}$$

$$= (22\mathbf{i} - 8\mathbf{j} - 60\mathbf{k})\,\mathrm{N\,cm} = (0.22\mathbf{i} - 0.08\mathbf{j} - 0.6\mathbf{k})\,\mathrm{N\,m}.$$

2. Let the couple moment vectors corresponding to the forces $\pm\mathbf{F}_1$ and $\pm\mathbf{F}_2$ be $\mathbf{C}_1$ and $\mathbf{C}_2$, respectively. Using the right-hand thread rule, we see that $\mathbf{C}_1$ is parallel to the y, z plane with equal components in the positive y and negative z directions. $|\mathbf{C}_1| = |\mathbf{F}_1|a = 100\,\mathrm{N\,m}$. It follows that:

$$\mathbf{C}_1 = 100\left(\frac{1}{\sqrt{2}}\mathbf{j} - \frac{1}{\sqrt{2}}\mathbf{k}\right)\,\mathrm{N\,m}.$$

$\mathbf{C}_2$ is parallel to the x, z plane and has components in the negative x and positive z directions proportional to 1 and 2, respectively. $|\mathbf{C}_2| = |\mathbf{F}_2|a = 50\,\mathrm{N\,m}$. It follows that:

$$\mathbf{C}_2 = 50\left(-\frac{1}{\sqrt{5}}\mathbf{i} + \frac{2}{\sqrt{5}}\mathbf{k}\right)\,\mathrm{N\,m}.$$

The resultant couple has moment vector:

$$\mathbf{C} = \mathbf{C}_1 + \mathbf{C}_2 = -10\sqrt{5}\mathbf{i} + 50\sqrt{2}\mathbf{j} + (20\sqrt{5} - 50\sqrt{2})\mathbf{k}$$
$$= (-22.4\mathbf{i} + 70.7\mathbf{j} - 26.0\mathbf{k})\,\mathrm{N\,m}.$$

3. Referring to Figure 8.11, the direction of $\mathbf{C}$ is that of $\mathbf{F}$ or that of $-5\mathbf{i} + 4\mathbf{j} + 3\mathbf{k}$. Now, $|\mathbf{C}| = 100\,\mathrm{N\,m}$, so:

$$\mathbf{C} = \frac{100(-5\mathbf{i} + 4\mathbf{j} + 3\mathbf{k})}{\sqrt{25 + 16 + 9}} = 10\sqrt{2}(-5\mathbf{i} + 4\mathbf{j} + 3\mathbf{k})\,\mathrm{N\,m}.$$

The moment of $\mathbf{F}$ about the origin is:

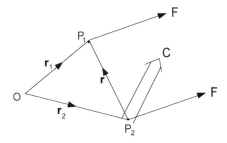

Figure 8.14. Force at P_1 replaced by a force and couple at P_2.

$$C_f = r \times F = \begin{vmatrix} i & j & k \\ 1 & 1 & 2 \\ -50 & 40 & 30 \end{vmatrix} = (-50i - 130j + 90k)\,N\,m.$$

The moment of the wrench about the origin is:

$$C_w = C + C_f = -50(1 + \sqrt{2})i - 10(13 - 4\sqrt{2})j + 30(3 + \sqrt{2})k$$
$$= (-120.7i - 73.4j + 132.4k)\,N\,m.$$

The moment of the wrench about the y-axis is simply the y-component of its moment about the origin, i.e. $-73.4\,N\,m$.

4.   Comparing our general theory with this particular example, we see that P is now the origin and therefore $r_p = 0$. $C = (10i + 5j - 10k)\,N\,m$, so the component of C in the direction of $F = 30i + 50j + 40k$ or of $3i + 5j + 4k$ is:

$$|C_f| = \frac{(10i + 5j - 10k) \cdot (3i + 5j + 4k)}{\sqrt{9 + 25 + 16}} = \frac{30 + 25 - 40}{5\sqrt{2}} = \frac{3}{\sqrt{2}}\,N\,m.$$

It follows that:

$$C_f = \frac{3}{\sqrt{2}} \cdot \frac{3i + 5j + 4k}{5\sqrt{2}} = (0.9i + 1.5j + 1.2k)\,N\,m.$$

Hence, the wrench is the force $F = (30i + 50j + 40k)\,N$ combined with the couple $(0.9i + 1.5j + 1.2k)\,N\,m$.

However, we still have to find the position vector r_q of the point of application Q of the wrench. The component of C perpendicular to F is:

$$C_p = C - C_f = 9.1i + 3.5j - 11.2k.$$

Since, $r_p = 0$, we have the vector equation:

$$C_p = r_q \times F.$$

Thus, if we write $r_q = r_x i + r_y j + r_z k$, the vector equation becomes:

$$9.1i + 3.5j - 11.2k = \begin{vmatrix} i & j & k \\ r_x & r_y & r_z \\ 30 & 50 & 40 \end{vmatrix}.$$

Comparing coefficients of i, j and k in turn gives us the three equations:

$$40r_y - 50r_z = 9.1$$
$$-40r_x + 30r_z = 3.5$$
$$50r_x - 30r_y = -11.2.$$

These have no unique solution but if we put $r_y = 0$, then $r_x = -0.224$ and $r_z = -0.182$.
Hence, one possible point of application Q of the wrench has position vector r_q given by:

$$r_q = (-0.224i - 0.182k)\,m.$$

5.   The resultant force is:

$$F_R = \sum_{i=1}^{2} F_i = (80i - 40j + 50k) + (-30i + 20j - 40k) = (50i - 20j + 10k)\,N.$$

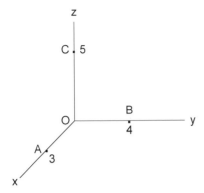

Figure 8.15. Points A, B and C as suggested in the 'Hint' in Exercise 6.

The couples due to the forces are:

$$C_1 = p_1 \times F_1 = \begin{vmatrix} i & j & k \\ 1 & -1 & 2 \\ 80 & -40 & 50 \end{vmatrix} = 30i + 110j + 40k,$$

$$C_2 = p_2 \times F_2 = \begin{vmatrix} i & j & k \\ -1 & 2 & -1 \\ -30 & 20 & -40 \end{vmatrix} = -60i - 10j + 40k.$$

Also, $C = 40i + 30j - 50k$. Therefore, the resultant couple is:

$$C_R = C_1 + C_2 + C = (10i + 130j + 30k)\,N\,m.$$

6. The general equation for a plane is: $lx + my + nz = p$. Thus, if points A, B and C, as shown in Figure 8.15, lie on the plane, then: for A(3, 0, 0), $3l = p$; for B(0, 4, 0), $4m = p$; and for C(0, 0, 5), $5n = p$. Therefore, the plane through A, B and C has the equation:

$$\frac{x}{3} + \frac{y}{4} + \frac{z}{5} = 1 \text{ or } 20x + 15y + 12z = 60.$$

Now, the coefficients of x, y and z are direction ratios of a line perpendicular to the plane. Since $\sqrt{20^2 + 15^2 + 12^2} = 27.73$, the corresponding direction cosines are:

$$\cos\alpha = \frac{20}{27.73}, \quad \cos\beta = \frac{15}{27.73} \quad \text{and} \quad \cos\gamma = \frac{12}{27.73}.$$

Returning to the physical problem, as illustrated in Figure 8.16, since the strings are mutually perpendicular, resolving along:

OA gives $T_1 = W\cos\alpha = 0.721W$,
OB gives $T_2 = W\cos\beta = 0.541W$ and
OC gives $T_3 = W\cos\gamma = 0.433W$.

7. Referring to Figure 8.13, we see that C_2 and C_5 act at $\alpha = 45°$ to the vertical; $\cos\alpha = \sin\alpha = 0.707$. C_4 acts at angle β to the vertical with $\cos\beta = 1/\sqrt{5} = 0.447$ and $\sin\beta = 2/\sqrt{5} = 0.894$. To evaluate C_i, $i = 1, \ldots, 6$, we write down six equilibrium equations obtained by resolving the forces in the x-, y- and z-directions and by taking moments about the x-, y- and z-axes. These equations are numbered (1)–(6) respectively, as follows:

(1) $0.707C_2 + 0.707C_5 = 200$
(2) $0.894C_4 = 200$

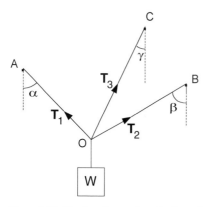

Figure 8.16. Finding the tensions in the strings in Exercise 6.

(3) $C_1 + 0.707C_2 + C_3 + 0.447C_4 + 0.707C_5 + C_6 = 200$
(4) $1.414C_5 + 2C_6 = 200$
(5) $C_3 + 0.447C_4 + C_6 = 100$
(6) $0.894C_4 - 1.414C_5 = 200$

As we shall see, these equations are particularly easy to solve. In a more difficult case, we may use a computer program such as that provided by MATLAB. Here, referring to the equations by numbers in brackets, we have:

(2) $C_4 = 223.7$, (6) $C_5 = 0$, (1) $C_2 = 282.8$,
(4) $C_6 = 100$, (5) $C_3 = -100$, (3) $C_1 = -100$.

Referring to the struts by the subscripts shown in Figure 8.13, we have the following results:
1, tension 100 N; 2, compression 283 N; 3, tension 100 N; 4, compression 224 N; 5, zero tension/compression; 6, compression 100 N.

9 Virtual work

9.1 Work done by a force

The definition of work done by a force is the scalar quantity given by the force multiplied by the distance moved by its point of application in the direction of the force. Thus, if you carry the shopping home along a horizontal path, the force from your hand which holds up the basket does no work since there is no movement in the vertical direction. The unit of work is the joule, which is the same as a newton metre.

If the force is $\mathbf{F}$ and the distance moved by its point of application is $\mathbf{a}$ at angle α to $\mathbf{F}$, as shown in Figure 9.1, then the work done by $\mathbf{F}$ is $W = Fa\cos\alpha$. In fact, this is the scalar product of the two vectors $\mathbf{F}$ and $\mathbf{a}$, i.e. $W = \mathbf{F}.\mathbf{a}$.

EXERCISE 1

Find the work done by a force $\mathbf{F} = (20\mathbf{i} + 30\mathbf{j} + 40\mathbf{k})\,\mathrm{N}$ when its point of application moves from A(4, 3, 2) to B(1, 2, 3) where the Cartesian coordinate distances are given in metres.

Problems 67 and 68.

9.2 Work done by a couple

Firstly, consider movement of a body in the plane of a couple acting on it which involves a translation $\mathbf{a}$ and a small rotation through angle $\delta\phi$ radians. Referring to Figure 9.2, let the couple consist of two forces $\mathbf{F}$ and $-\mathbf{F}$, a perpendicular distance b apart. In the translation, the work done by the couple is $W = \mathbf{F}.\mathbf{a} - \mathbf{F}.\mathbf{a} = 0$. However, the work done in the rotation $\delta\phi$ is:

$$W = F\frac{b}{2}\delta\phi + F\frac{b}{2}\delta\phi = Fb\delta\phi = M\delta\phi,$$

where M is the moment of the couple.

Moving into three dimensions, still no work is done by the couple in a translation since movement perpendicular to the plane of the couple is perpendicular to $\mathbf{F}$ and $-\mathbf{F}$. Also, no work is done by the couple in rotations of the body about axes which lie in the

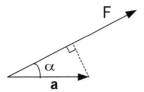

Figure 9.1. Force **F** and movement **a** of its point of application.

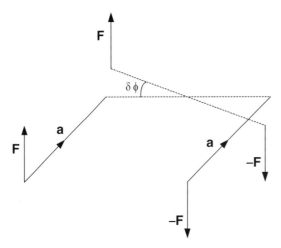

Figure 9.2. Work done by a couple in translation and rotation.

plane of the couple. Using the right-hand thread rule, we can specify a small rotation of a body by the vector $\delta\phi$. Then the work done by a couple with moment vector **C** in this rotation is $W = \mathbf{C}.\delta\phi$. The components of **C** will be measured in newton metres and the components of $\delta\phi$ in radians, in order to give W in joules.

EXERCISE 2

Find the work done by a couple of moment $200(\mathbf{i} + \mathbf{j} + \mathbf{k})\,\mathrm{N\,m}$ when the body on which it acts turns through $+1°$ about the x-axis.

Problems 69 and 70.

9.3 Virtual work for a single body

Consider a rigid body in equilibrium when acted on by various forces and couples. Equilibrium implies that if we find the resultant force at a given point together with the resultant couple, then both will be zero. Consequently, if we imagine a small movement of the body, the total work done by the forces and couples equals the total work done by the resultant force and resultant couple. Then, since the resultant force and resultant couple are both zero, the total work done must be zero. Since the movement is imagined, it is called a *virtual displacement* and the corresponding work done is called *virtual work*.

The converse also applies. If the total virtual work done in any small virtual displacement of the rigid body is zero, then the body must be in equilibrium. We must include the word 'any' because otherwise we could imagine a small displacement perpendicular to the direction of a non-zero resultant and not detect the latter because no work would be done.

EXERCISE 3

A uniform straight rod of weight W and length $2a$ is freely hinged at one end so that it can turn in a vertical plane. A *spiral spring* is attached to the hinge mounting and to the rod at that end, so that the spring is in its natural state when the rod is horizontal. Its weight will cause the rod to droop by an angle θ. Use virtual work to find the equation which determines the value of θ for equilibrium, given a *spring constant* of k N m per radian. Check that the same equation is obtained by taking moments. (Hint: find the virtual work done by the spring and gravity in a small virtual displacement $\delta\theta$ from θ.)

Problems 71 and 72.

9.4 Virtual work for a system of bodies

Exercise 3 leaves us with the impression that no advantage is gained through the use of the concept of virtual work. This is true when considering the equilibrium of a single rigid body. However, if we have a system of rigid bodies connected together, no total work is done by connecting forces in any displacement since action and reaction are always equal and opposite. Thus, the system is in equilibrium if the total virtual work of external forces is zero for any small virtual displacement of the system. Considering the system as a whole in this way gives a simple approach for some problems.

One of the first applications of virtual work was to rope and pulley systems. In the two-pulley system shown in Figure 9.3, let us make a virtual downward displacement δx of the end of the rope where force F is applied. This will shorten the loop of the

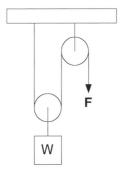

Figure 9.3. A two-pulley system.

rope passing under the pulley which supports the weight W by δx and hence will lift W by $\delta x/2$. Neglecting friction and any weights other than W, the virtual work done by external forces is: $F\delta x - W\delta x/2 = 0$ for equilibrium. The second term is negative since the gravitational force W is downwards while the weight moves upwards by a distance $\delta x/2$.

The equation shows that $F = W/2$ for equilibrium.

Next consider a pin-jointed framework ABCDEF, as shown in Figure 9.4, which is made up of seven vertical or horizontal struts, each of weight w and length a. A weight W is suspended from C, while the framework is attached to fixed mountings at A and B, and collapse is prevented by two light cables AE and EC. The problem is to find the tension T in the cable AE.

In order to solve for T, we must devise a virtual work equation which involves T. To do this, we replace the cable AE by the force T which it applies to the framework at A and E. Then make a small virtual displacement which leaves the lengths of the struts and the other cable EC unaltered.

Make a virtual displacement as shown in Figure 9.5, rocking the sides of ABEF over by the small angle $2\delta\theta$ (radians) and moving E and F over to E' and F', respectively.

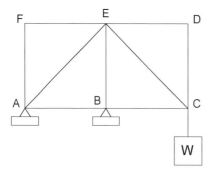

Figure 9.4. Vertical pin-jointed framework prevented from collapse by light cables AE and EC.

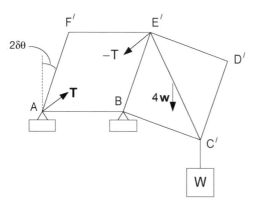

Figure 9.5. Virtual displacement used to calculate T.

The tension T at A does no work, since A does not move, but the T at E does do work: $-T(AE' - AE)$.

Referring to Figure 9.6, the angle $2\theta = \pi/2 + 2\delta\theta$ and therefore, $\theta = \pi/4 + \delta\theta$.

$$AG = a \sin\theta = a\left(\sin\frac{\pi}{4}\cos\delta\theta + \cos\frac{\pi}{4}\sin\delta\theta\right) = a(1 + \delta\theta)/\sqrt{2},$$

since $\delta\theta$ is small. Therefore, $AE' = \sqrt{2}a(1 + \delta\theta)$ and since $AE = \sqrt{2}a$, we have:

$$AE' - AE = \sqrt{2}a\delta\theta.$$

Since $\delta\theta$ is small, the centres of gravity of AF and FE remain at the same height, the centre of the square BCDE drops by $2\delta\theta \cdot a/2 = a\delta\theta$ and W drops by $a \cdot 2\delta\theta$.

Hence, the virtual work done by T and the external forces is:

$$-T\sqrt{2}a\delta\theta + 4wa\delta\theta + W2a\delta\theta = 0$$

for equilibrium.

Therefore, $T = \sqrt{2}(2w + W)$.

Another way of tackling this problem is to make a virtual downward (or upward) displacement of the support at A, as indicated in Figure 9.7. In this case the supporting force R_A does work. Taking moments about B (re. Figure 9.4), we see that $R_A = W$ downwards. In the virtual displacement, nothing moves to the right of BE, so that part may be omitted from the diagram.

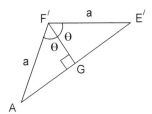

Figure 9.6. Finding the length AE'.

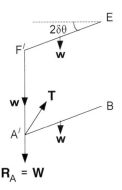

Figure 9.7. Virtual downward displacement of the support at A.

In this case, the virtual work is:

$$-T\sqrt{2}a\delta\theta + R_A a \cdot 2\delta\theta + wa \cdot 2\delta\theta + 2w \cdot \frac{a}{2} \cdot 2\delta\theta$$
$$= -T\sqrt{2}a\delta\theta + W \cdot 2a\delta\theta + 4wa\delta\theta = 0,$$

for equilibrium. This gives $T = \sqrt{2}(2w + W)$ as before.

As a final example, consider a simple truss ABCDE made up of seven pin-jointed struts, each of length a and weight w. The truss has fixed supports at A and C, and a weight W is suspended from B, as shown in Figure 9.8. Suppose we wish to find the tension or compression in strut CD.

Let the strut CD have tension T. To find T by virtual work, we must replace the strut CD by the forces which the strut itself exerts at joints C and D. This will be a pull of T together with half the weight, i.e. $w/2$, acting downwards at each joint. In order to find T by virtual work, we must make a virtual displacement which changes the length of CD so that work is done by T. The most convenient displacement moves C vertically, which means that work will be done by the supporting force R_C. From the symmetry of the diagram, we see immediately that $R_C = (7w + W)/2$, acting upwards.

In our virtual displacement, nothing moves to the left of BD, so we have the particularly simple diagram shown in Figure 9.9. However, the work done by T is

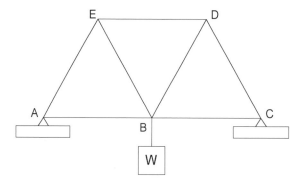

Figure 9.8. A truss made up of seven identical pin-jointed struts.

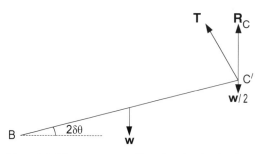

Figure 9.9. Movement caused by virtual reduction in length of strut DC.

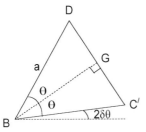

Figure 9.10. Finding the length DC'.

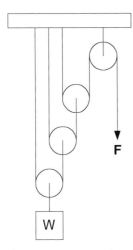

Figure 9.11. A rope and pulley system.

$T(DC - DC')$, so we need to find the length DC'. From Figure 9.10, we see that:

$$2\theta + 2\delta\theta = \frac{\pi}{3}, \text{ i.e. } \theta = \frac{\pi}{6} - \delta\theta.$$

Now, $DG = a\sin\theta = a\left(\frac{1}{2} \cdot 1 - \frac{\sqrt{3}}{2} \cdot \delta\theta\right)$, since $\delta\theta$ is small.

Thus, $DC' = a(1 - \sqrt{3}\delta\theta)$ and $DC - DC' = \sqrt{3}a\delta\theta$.

We can now write the virtual work as:

$$T\sqrt{3}a\delta\theta + R_Ca \cdot 2\delta\theta - w\frac{a}{2} \cdot 2\delta\theta - \frac{w}{2}a \cdot 2\delta\theta = T\sqrt{3}a\delta\theta + (5w + W)a\delta\theta = 0,$$

for equilibrium. Therefore, $T = -(5w + W)/\sqrt{3}$. Since this is negative, the strut CD has a compression of $(5w + W)/\sqrt{3}$

EXERCISE 4

If the rope and pulley system, shown in Figure 9.11, is in equilibrium, use virtual work to find the relationship between the force F and weight W, neglecting any friction and weights other than W.

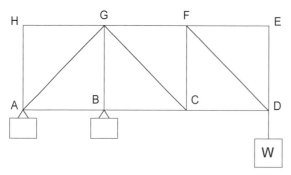

Figure 9.12. A framework of vertical and horizontal struts with light diagonal cables.

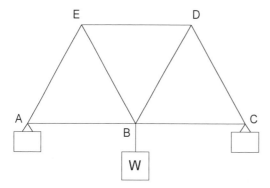

Figure 9.13. A truss with seven identical pin-jointed struts.

EXERCISE 5

A pin-jointed framework ABCDEFGH, as shown in Figure 9.12, consists of ten vertical and horizontal struts, each of length a and weight w. The framework is prevented from collapsing by light cables AG, GC and FD. The framework is attached to fixed supports at A and B, and a weight W is suspended from D. Use virtual work to find the tension in the cable AG.

EXERCISE 6

For the truss discussed in the text with seven struts, each of length a and weight w, and illustrated again in Figure 9.13, use virtual work to find the tension or compression in strut DE.

Problems 73, 74 and 75.

9.5 Stability of equilibrium

In Section 9.3, we found that we could sometimes use virtual work to determine the equilibrium position of a rigid body. However, some equilibrium positions are completely impracticable since they cannot be sustained. An extreme example of this is to

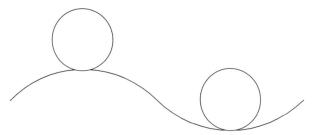

Figure 9.14. Balls on a corrugated roof.

try to balance a pencil on its point. In theory, it is in equilibrium when its centre of gravity is vertically above its point. In practice, it always falls over. Such an equilibrium position is referred to as an unstable position of equilibrium.

Stability of equilibrium is defined as follows. Release a body from a position slightly displaced from its equilibrium position. Then:

1. if it moves back to its equilibrium position, the latter is a *stable* position;
2. if it moves away from its equilibrium position, the latter is *unstable*;
3. if it stays where it is, the equilibrium position is *neutral*.

A simple illustration of this concept is to consider a ball on a horizontal corrugated iron roof, in which case there are two types of equilibrium position, as shown in Figure 9.14. We may be able to balance the ball on top of a ridge but if we move it slightly to one side and release it, it will roll down into the trough. On the other hand, if we release it near to the bottom of the trough, it will roll back to the bottom of the trough. Hence, the top of a ridge is a position of unstable equilibrium and the bottom of a trough is a position of stable equilibrium.

However, if we consider positions along the horizontal bottom of the trough, the ball will be in equilibrium wherever it is placed. Hence, in this direction we have neutral stability.

In order to develop a method for analysing such situations, we introduce the concept of *mechanical energy*. The mechanical energy possessed by a body is its capacity for doing work by virtue of either its position or its velocity. The former is called *potential energy* and the latter *kinetic energy*. For instance, if a clock is driven by a suspended weight, the weight is given potential energy when it is wound up since it then has the capacity to drive the clock for a considerable length of time. On the other hand, if a hammer is used to drive in a nail, the hammer is given kinetic energy, in terms of its velocity prior to impact, which is used to drive in the nail. Provided that no heat is generated by friction, the total mechanical energy of a body, i.e. the sum of its potential energy and its kinetic energy, remains constant. For example, the weight used in a pile-driver has potential energy due to its height above the pile when it is dropped and this is converted into kinetic energy due to its velocity immediately prior to impact with the pile.

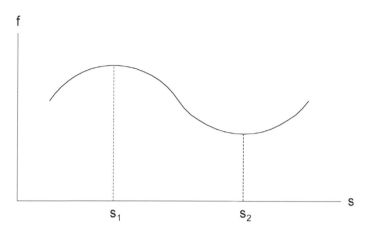

Figure 9.15. Potential energy f plotted against position s.

The potential energy of a body is defined as the work done against the field of force, e.g. gravity, in moving the body from a reference position to its present position. Alternatively, it is the work which would be done by the field if the body were to move back from its present position to the reference position. Given this definition for the potential energy of a body, we can relate virtual work to virtual change in potential energy.

In a small virtual displacement of the body, the virtual work done by the field of force acting on it is the negative of the virtual change in potential energy and this must therefore be zero for equilibrium. Let us now interpret this mathematically by writing the potential energy as the function $f(s)$, where s is the position variable. Then δf, the change in f due to the small virtual change δs in s is:

$$\delta f = \frac{df}{ds}\delta s = 0 \text{ for equilibrium.}$$

Since $\delta s \neq 0$, $\dfrac{df}{ds} = 0$ for equilibrium.

Plotting potential energy f against position s, as in Figure 9.15, we see that $df/ds = 0$ at $s = s_1$, where f is maximum and at $s = s_2$, where f is minimum. If the body is released from a position away from but near to $s = s_1$, $df/ds \neq 0$ so the body will start to move. In doing so, the kinetic energy increases from zero. Consequently, the potential energy f must decrease. It can only do this if the body moves further away from $s = s_1$. From this we conclude that a position of maximum potential energy is an unstable equilibrium position.

Next, suppose that we release the body at a position away from but near to $s = s_2$. Again, $df/ds \neq 0$ and the body will start to move. As before, the kinetic energy increases from zero and the potential energy must decrease. This can only happen if the body moves towards $s = s_2$. Hence, the position of minimum potential energy is a stable equilibrium position.

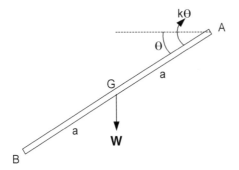

Figure 9.16. Rod AB hinged at A and restrained by a spiral spring.

Usually the function $f(s)$ will have a negative second derivative, i.e. $d^2f/ds^2 < 0$, where f is maximum and a positive second derivative, i.e. $d^2f/ds^2 > 0$, where f is minimum. Hence, our analytical method consists of finding $f(s)$, which is the potential energy as a function of a position variable s, differentiating and putting the derivative equal to zero, i.e. $df/ds = 0$, to find the equilibrium position and then differentiating again to test for stability.

Examine again the example given as Exercise 3 in Section 9.3 and illustrated now in Figure 9.16. The spiral spring is in its natural state when $\theta = 0$. Let the potential energy in the spring be zero in this position. Then, the potential energy due to the spring at angle θ is:

$$\int_0^\theta k\phi \, d\phi = \frac{1}{2}k\theta^2.$$

The potential energy due to gravity is: $-Wa \sin\theta$. Hence, the total potential energy is:

$$f(\theta) = \frac{1}{2}k\theta^2 - Wa \sin\theta.$$

Then, $\dfrac{df}{d\theta} = k\theta - Wa \cos\theta = 0$

for equilibrium and this is the same equation as obtained by virtual work.

Next, $\dfrac{d^2f}{d\theta^2} = k + Wa \sin\theta > 0,$

since the equilibrium θ will be in the range $0 < \theta < \pi/2$. This implies that the equilibrium θ is a position of minimum potential energy and is therefore stable.

EXERCISE 7

A body with a convex lower surface rests in equilibrium on top of a convex supporting surface. At the point of contact the radii of curvature are r_1 and r_2 for the body and supporting surfaces, respectively, and h is the height of the centre of gravity of the body above the point of contact. If the body is rolled to one side so that the arc followed by the point of contact subtends an angle θ at the centre of curvature of the supporting surface, find an expression for the total potential energy of the body as a function of θ. Hence, show that the equilibrium position ($\theta = 0$) is stable if:

$$\frac{1}{h} > \frac{1}{r_1} + \frac{1}{r_2}.$$

Problems 76 and 77.

9.6 Answers to exercises

1. The distance moved by the point of application of the force is:

$$\mathbf{a} = (1-4)\mathbf{i} + (2-3)\mathbf{j} + (3-2)\mathbf{k} = -3\mathbf{i} - \mathbf{j} + \mathbf{k}.$$

Then, the work done by $\mathbf{F}$ is:

$$W = \mathbf{F}.\mathbf{a} = (20\mathbf{i} + 30\mathbf{j} + 40\mathbf{k}).(-3\mathbf{i} - \mathbf{j} + \mathbf{k}) = -60 - 30 + 40 = -50\,\mathrm{J}.$$

Notice that the work done by a force may be negative.

2. $\mathbf{C} = 200(\mathbf{i} + \mathbf{j} + \mathbf{k})\,\mathrm{N\,m}$ and $\boldsymbol{\delta\phi} = (\pi/180)\mathbf{i}$.

Therefore, the work done by the couple is:

$$W = \mathbf{C}.\boldsymbol{\delta\phi} = 200(\mathbf{i} + \mathbf{j} + \mathbf{k}).(\pi/180)\mathbf{i} = 200\pi/180 = 10\pi/9\,\mathrm{J}.$$

3. Referring to Figure 9.17, we see that in a small angular displacement $\delta\theta$, the centre of gravity of the rod moves $a\delta\theta$ perpendicular to the rod. Hence, the vertical component is $a\delta\theta\cos\theta$ and the virtual work done by gravity is $Wa\delta\theta\cos\theta$. The virtual work done by the spring is the moment multiplied by the angular displacement, i.e. $-k\theta\delta\theta$. Hence, the total virtual work is: $Wa\delta\theta\cos\theta - k\theta\delta\theta = 0$ for equilibrium. Dividing through by $\delta\theta$ gives: $Wa\cos\theta - k\theta = 0$.

In fact, this equation would have been given directly by taking moments about A.

4. Referring to Figure 9.18, if the end of the rope where force F is applied is pulled down a distance δx, the pulley wheel second from the right will move up by $\delta x/2$. This in turn will move the third pulley wheel up by $(\delta x/2)/2 = \delta x/4$. The latter will move the fourth pulley wheel up by $(\delta x/4)/2 = \delta x/8$. The work done by the external forces is: $F\delta x - W\delta x/8 = 0$ for equilibrium and $F = W/8$.

5. Referring to Figure 9.19, let R_A be the support force from the mounting at A. Taking moments about B: $R_\mathrm{A}a = 10wa/2 + W2a$ so $R_\mathrm{A} = 5w + 2W$.

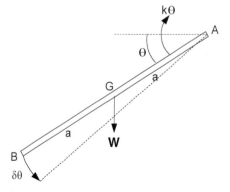

Figure 9.17. Virtual angular displacement $\delta\theta$ of rod about end A.

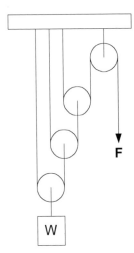

Figure 9.18. A pulley system in which force F holds weight W.

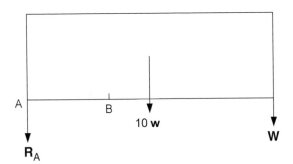

Figure 9.19. Finding the support force R_A by taking moments about B.

Make a virtual displacement to the left of BG, as shown in Figure 9.20. By the same trigonometry as given in the text, we see that: $A'G - AG = \sqrt{2}a\delta\theta$. Then, the virtual work is:

$$-T \cdot \sqrt{2}a\delta\theta + R_A a \cdot 2\delta\theta + wa \cdot 2\delta\theta + 2wa/2 \cdot 2\delta\theta = -T \cdot \sqrt{2}a\delta\theta + (14w + 4W)a\delta\theta = 0$$

for equilibrium. Therefore, $T = \sqrt{2}(7w + 2W)$.

6.   Make a small virtual rotation $2\delta\theta$ about B of the part of the truss BCD, in order to make a virtual change in the length of DE. The only forces which will do work are indicated in Figure 9.21. The strut DE has been replaced by its tension T and half its weight, $w/2$, acting at D'. The supporting force $R_C = (7w + W)/2$, as before.

The length of each strut is a, so we can now write down the virtual work in the small virtual displacement $2\delta\theta$ as:

$$Ta \cdot 2\delta\theta \cos \frac{\pi}{6} + R_C a \cdot 2\delta\theta - w\frac{a}{2} \cdot 2\delta\theta - w\frac{a}{2} \cdot 2\delta\theta \cos \frac{\pi}{3} - w\frac{\sqrt{3}}{2}a \cdot 2\delta\theta \cos \frac{\pi}{6}$$

$$- \frac{w}{2}a \cdot 2\delta\theta \cos \frac{\pi}{3} = T\sqrt{3}a\delta\theta + \left(\frac{7}{2}w + W\right)a\delta\theta = 0$$

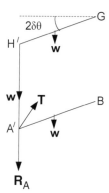

Figure 9.20. Virtual displacement $2\delta\theta$.

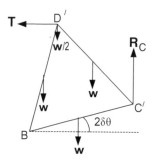

Figure 9.21. Virtual rotation of part of truss BCD.

for equilibrium. Therefore,

$$T = -\left(\frac{7}{2}w + W\right)\Big/\sqrt{3}.$$

This is negative, so strut DE has a *compression* of

$$\left(\frac{7}{2}w + W\right)\Big/\sqrt{3}.$$

7.   **Note:** θ in Figure 9.22 should really be small; it is made large here to facilitate the drawing.

As the body rolls away from the top of the surface, the point of contact moves from P_2 to P, while P_1 is the point of the body originally in contact with P_2. With our exaggerated angles, P_2P and P_1P are taken to be arcs of circles about centres C_2 and C_1, respectively. The lengths of the arcs are the same, so $s = r_1\phi = r_2\theta$ (angles in radians). Let the potential energy of the body be $f(\theta)$ with $f(0) = 0$.

As the body rolls to one side, its centre of gravity G moves upwards relative to C_1 by $(r_1 - h)$ $[1 - \cos(\theta + \phi)]$. In the same movement, C_1 moves downwards by $(r_1 + r_2)(1 - \cos\theta)$. The potential energy in the new position is W times the total upward movement of G. Thus, having substituted $\phi = r_2\theta/r_1$:

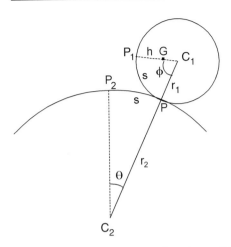

Figure 9.22. A body rolled away from its equilibrium position.

$$f(\theta) = W(r_1 - h)\left[1 - \cos\left(1 + \frac{r_2}{r_1}\right)\theta\right] - W(r_1 + r_2)(1 - \cos\theta).$$

$$\frac{df}{d\theta} = W(r_1 - h)\left(1 + \frac{r_2}{r_1}\right)\sin\left(1 + \frac{r_2}{r_1}\right)\theta - W(r_1 + r_2)\sin\theta = 0$$

when $\theta = 0$, the equilibrium position.

$$\frac{d^2f}{d\theta^2} = W(r_1 - h)\left(1 + \frac{r_2}{r_1}\right)^2\cos\left(1 + \frac{r_2}{r_1}\right)\theta - W(r_1 + r_2)\cos\theta$$

$$= W(r_1 - h)\left(\frac{r_1 + r_2}{r_1}\right)^2 - W(r_1 + r_2) \text{ when } \theta = 0$$

> 0 for stable equilibrium.

Multiply the inequality by the positive quantity: $r_1^2/[W(r_1 + r_2)]$.

Then, $(r_1 - h)(r_1 + r_2) - r_1^2 = r_1 r_2 - h(r_1 + r_2) > 0$.

Moving $h(r_1 + r_2)$ onto the other side of the inequality and dividing by the positive quantity $h r_1 r_2$ gives:

$$\frac{1}{h} > \frac{1}{r_1} + \frac{1}{r_2}$$

for stable equilibrium at $\theta = 0$.

Part II

Dynamics

10 Kinematics of a point

Kinematics of a point

10.1 Rectilinear motion

Kinematics is concerned with motion in an abstract sense without reference to force or mass. We shall start by analysing motion in a straight line, i.e. *rectilinear motion*.

Consider a point P moving along a straight line from O, as indicated in Figure 10.1. If it takes t seconds to travel x metres from O to P, then its average speed or velocity is x/t metres per second or x/t m/s or x/t m s^{-1}. We have to insert the word 'average' because the velocity may not be constant over that period of time.

If the velocity is varying, we need to be able to define it at any instant of time. To do this, we have to use calculus, which involves taking the average velocity over a short interval of time and then taking the limit as this interval tends to zero. Suppose our point P moves a distance δx in time δt. The *velocity* at the start of this interval is then:

$$v = \lim_{\delta t \to 0} \frac{\delta x}{\delta t} = \frac{dx}{dt} = \dot{x}.$$

dx/dt is called the *derivative* of x with respect to t and it is the rate at which x is changing at that instant. The dot notation ($\dot{x}$) was introduced by Newton for the derivative with respect to time and it continues to be used as a shorthand notation.

If the velocity v is varying, then v itself will have a rate of change and this rate is called the *acceleration* of P. Let the velocity v change by δv over a short interval of time δt. Then the acceleration at the start of that interval is:

$$\lim_{\delta t \to 0} \frac{\delta v}{\delta t} = \frac{dv}{dt} = \dot{v} = \frac{d\dot{x}}{dt} = \frac{d^2 x}{dt^2} = \ddot{x}.$$

Hence, we may regard the acceleration as being the *first derivative* with respect to time of the velocity, i.e. $\dot{v}$, or the *second derivative* with respect to time of position, i.e. $\ddot{x}$. Note that two dots are used to denote the second derivative with respect to time.

Although the derivative of velocity with respect to time is the most obvious way to define acceleration, there is another way that is sometimes useful. This involves velocity and distance as follows. In a small interval of time δt, let the velocity change by δv and its position by δx. Then the acceleration is:

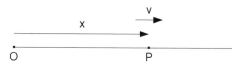

Figure 10.1. Rectilinear motion of point P.

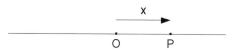

Figure 10.2. P oscillating from side to side of O.

$$\dot{v} = \frac{dv}{dt} = \lim_{\delta t \to 0} \frac{\delta v}{\delta t} = \lim_{\delta t \to 0} \frac{\delta v}{\delta x} \cdot \frac{\delta x}{\delta t} = \lim_{\delta t \to 0} \frac{\delta x}{\delta t} \cdot \lim_{\delta x \to 0} \frac{\delta v}{\delta x} = \frac{dx}{dt} \cdot \frac{dv}{dx} = v\frac{dv}{dx}.$$

Going from position to velocity ($x \to v$) and from velocity to acceleration ($v \to \dot{v}$) involves *differential calculus*. Going the opposite way, i.e. $\dot{v} \to v \to x$, involves *integral calculus*.

$$v = \frac{dx}{dt} \quad \text{so } x = \int v \, dt \quad \text{and} \quad \dot{v} = \frac{dv}{dt} \quad \text{so } v = \int \dot{v} \, dt.$$

EXERCISE 1

Let a stone be dropped from a height above the ground where $x = 0$, positive x being measured vertically downwards from that point. Assume that the stone travels with a constant acceleration $\ddot{x} = g$. Sketch graphs to show how $\ddot{x}, \dot{x}$ and x vary with time t and also how the velocity v varies with x.

Problems 78 and 79.

10.2 Simple harmonic motion

Again, consider P to be moving along a straight line through O, as indicated in Figure 10.2, in such a way that it oscillates from side to side of O. x is the distance of P from O, being positive to the right of O and negative to the left.

This motion is called *simple harmonic motion* when it is described by the equation: $x = a \sin \omega t$. Here, t is time as usual and a and ω are constants. In this case, P oscillates between $x = a$ and $x = -a$, and a is called the *amplitude* of the oscillation.

Plotting x against t gives a sine wave of *period* $2\pi/\omega$ and *frequency* $\omega/2\pi$, as shown in Figure 10.3.

Let us now examine the velocity and acceleration of P. The velocity is:

$$\dot{x} = a\omega \cos \omega t = a\omega \sin\left(\omega t + \frac{\pi}{2}\right).$$

Hence, the velocity $\dot{x}$ is a sine wave of amplitude $a\omega$ which is out of *phase* with the position x by $\pi/2$ radians.

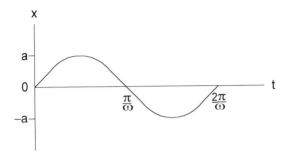

Figure 10.3. One period of simple harmonic motion.

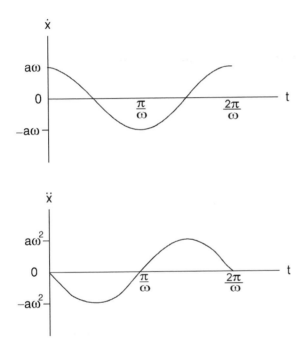

Figure 10.4. Velocity and acceleration of simple harmonic motion.

Differentiating again with respect to t gives the acceleration:

$$\ddot{x} = -a\omega^2 \sin \omega t = a\omega^2 \sin(\omega t + \pi).$$

This is sine wave with amplitude $a\omega^2$ which is out of phase with the velocity $\dot{x}$ by $\pi/2$ radians and with the position x by π radians. The velocity $\dot{x}$ and acceleration $\ddot{x}$ are shown plotted against time t in Figure 10.4.

Notice that, since $x = a \sin \omega t$, the equation for $\ddot{x}$ can be written as:

$$\ddot{x} = -\omega^2 x.$$

This is the basic differential equation defining *simple harmonic motion* with frequency $\omega/2\pi$. Frequency is measured in cycles per second (c/s) or, in metric notation, hertz (Hz). The units Hz and c/s are identical.

Although the basic differential equation defines the motion as being simple harmonic with frequency $\omega/2\pi$, it does not specify either its amplitude or its phase. To find these, we need to know two initial conditions, i.e. that $x = x_0$ and $\dot{x} = v_0$, say, when time $t = 0$. Earlier, we wrote $x = a \sin \omega t$ but, more generally, we should have $x = a \sin(\omega t + \phi)$, where ϕ is the phase in radians when $t = 0$. Hence, $x_0 = a \sin \phi$. Furthermore, $\dot{x} = a\omega \cos(\omega t + \phi)$ and $v_0 = a\omega \cos \phi$.

It follows that the phase ϕ is given by:

$$\frac{x_0 \omega}{v_0} = \tan \phi, \text{ i.e. } \phi = \tan^{-1}\left(\frac{x_0 \omega}{v_0}\right).$$

Then, the amplitude a is given by:

$$x_0^2 \omega^2 + v_0^2 = a^2 \omega^2 (\sin^2 \phi + \cos^2 \phi) = a^2 \omega^2, \text{ i.e. } a = \frac{\sqrt{x_0^2 \omega^2 + v_0^2}}{\omega}.$$

EXERCISE 2

Draw a graph of $y = v/\omega = \dot{x}/\omega$ against x, when x is describing the simple harmonic motion $x = a \sin(\omega t + \phi)$.

Problems 80 and 81.

10.3 Circular motion

Referring to Figure 10.5, let P be a point which moves along a circle of radius r about a fixed point O. If A is a fixed point on the circumference, we can denote the position of P by the angle $\theta = \angle AOP$. If θ is measured in radians, the curvilinear distance AP is $s = r\theta$.

Differentiating s with respect to time t gives the velocity of P along the circle as:

$$v = \dot{s} = r\dot{\theta} = r\omega$$

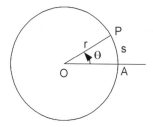

Figure 10.5. Point P moving around a circle of radius r.

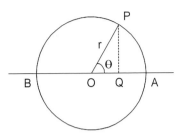

Figure 10.6. Point P moving round a circle at constant speed.

where ω is the *angular velocity* of P about O and is measured in radians per second (rad/s).

Differentiating again gives the acceleration along the circle:

$$\dot{v} = \ddot{s} = r\ddot{\theta} = r\dot{\omega}.$$

Here, $\dot{\omega}$ is the *angular acceleration*, measured in radians per second per second (rad/s^2).

EXERCISE 3

Referring to Figure 10.6, let point P move along the circle just described at constant speed $v = r\omega$, so that the angular velocity ω is also constant. Drop a perpendicular PQ onto the diameter AOB. Show that the point Q moves along AOB with simple harmonic motion about O.

Problems 82 and 83.

10.4 Velocity vectors

Like force, linear velocity is a vector quantity since it has both magnitude and direction. The resultant of two forces acting at a point is the vector sum of the two separate forces. Similarly, if an object has two components of velocity, its resultant velocity is the vector sum of the components. The following are easily visualized examples of objects with two components of velocity.

1. During flight of an aeroplane, when a wind is blowing, the velocity of the aeroplane is a combination of the velocity of the wind and the velocity of the aeroplane relative to the air.

2. A boat sailing across a river has a component of velocity downstream due to the flow of the river together with a component due to the movement of the boat relative to the water.

Take the last example and suppose that the boat is pointing directly across the river, i.e. perpendicular to the river bank. If the boat is moving at 2 m/s relative to the water and the water is flowing at 1 m/s, find the magnitude and direction of the actual velocity **v** of the boat. The latter velocity has two components, as shown in Figure 10.7. The resultant velocity **v** has magnitude v and θ is the angle between its direction and the downstream

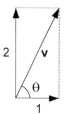

Figure 10.7. Finding the resultant velocity **v** of a boat crossing a river.

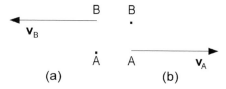

(a)   (b)

Figure 10.8. People at A and B on different trains on parallel tracks.

direction. This is a particularly simple example since the two components of velocity are at right angles to each other. Hence, $v^2 = 2^2 + 1^2 = 5$, $v = \sqrt{5} = 2.236$ m/s. Also, $\tan \theta = 2$, so $\theta = 63°26'$.

EXERCISE 4

For the same boat and river as above, in what direction should the boat be pointed for the resultant velocity to be perpendicular to the river bank? Also, find the magnitude of the resultant velocity in that case.

Problems 84 and 85.

10.5 Relative velocity

Imagine two trains, which are stationary on adjacent tracks in a railway station. Suppose you are sitting at A in one train looking across at another person at B in the other train. If the other train starts to move, as indicated in Figure 10.8a, B will appear to move with the actual speed $\mathbf{v}_B$. On the other hand, if your train started to move as indicated in Figure 10.8b, you could imagine that you were still stationary and that B were moving in the opposite direction. Consequently, in this situation the velocity of B relative to A is: $\mathbf{v}_{BrA} = -\mathbf{v}_A$.

Combining the two situations, i.e. where both trains move, the velocity of B relative to A is: $\mathbf{v}_{BrA} = \mathbf{v}_B - \mathbf{v}_A$.

In fact this is a general formula for relative velocity. For instance, if you are in an aeroplane A flying with velocity $\mathbf{v}_A$ and you look out at another aeroplane B, which has velocity $\mathbf{v}_B$ and a completely different flight path from A, then the velocity of B

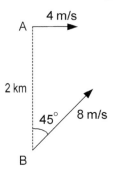

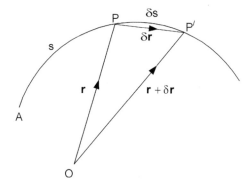

Figure 10.9. Ships A and B moving with different velocities.

Figure 10.10. Point P moving along a curved path.

relative to A is:

$$\mathbf{v}_{BrA} = \mathbf{v}_B - \mathbf{v}_A.$$

EXERCISE 5

A ship A is travelling due east at a speed of 4 m/s. Another ship B is 2 km south of A and is travelling in a north-easterly direction at 8 m/s, as indicated in Figure 10.9. Find the nearest distance of approach as the two ships pass each other.

Problems 86 and 87.

10.6 Motion along a curved path

Referring to Figure 10.10, consider a point P moving along a curve from A with the distance along the curve from A being s at a particular instant. Let the position vector of P at that instant be $\mathbf{r}$ with reference to a fixed point O. Moving on in time by a small interval δt, P moves to P′ a distance δs along the curved path and the position vector changes by $\delta \mathbf{r}$, so that the position vector of P′ is $\mathbf{r} + \delta \mathbf{r}$.

The velocity of P has a direction which is tangential to the path. It is useful to introduce the unit vector $\hat{\mathbf{t}}$ in the direction of the velocity. The average magnitude of the velocity between P and P′ is $\delta s / \delta t$ and so the magnitude of the velocity at P is:

$$v = \lim_{\delta t \to 0} \frac{\delta s}{\delta t} = \frac{ds}{dt} = \dot{s}.$$

We can then introduce the direction of the velocity with the unit vector $\hat{\mathbf{t}}$, so that the *velocity vector* is:

$$\mathbf{v} = v\hat{\mathbf{t}} = \dot{\mathbf{r}} = \frac{d\mathbf{r}}{dt} = \lim_{\delta t \to 0} \frac{\delta \mathbf{r}}{\delta t}.$$

To find the acceleration of the point as it moves along the curve, we must differentiate the velocity $\mathbf{v}$ with respect to time t. Now, when $\mathbf{v}$ is written as $v\hat{\mathbf{t}}$, we see that it is the product of two functions of time. v is a scalar function of time and $\hat{\mathbf{t}}$ is a vector function of time since, although its magnitude is constant, its direction varies with time. We can then use the usual formula for differentiating the product of two functions:

$$\ddot{\mathbf{r}} = \dot{\mathbf{v}} = \frac{dv}{dt}\hat{\mathbf{t}} + v\frac{d\hat{\mathbf{t}}}{dt}.$$

The next problem is to try to interpret $d\hat{\mathbf{t}}/dt$. Referring to the path in Figure 10.11a, we let the unit tangential vector at P be $\hat{\mathbf{t}}$ and the unit tangential vector at P′ be $\hat{\mathbf{t}} + \delta\hat{\mathbf{t}}$. If $\delta\theta$ is the change in tangential direction between P and P′, we can form an isosceles triangle, as shown in Figure 10.11b, with sides corresponding to the vectors: $\hat{\mathbf{t}}$, $\delta\hat{\mathbf{t}}$ and $\hat{\mathbf{t}} + \delta\hat{\mathbf{t}}$. Since $\hat{\mathbf{t}}$ and $\hat{\mathbf{t}} + \delta\hat{\mathbf{t}}$ are both unit vectors, if $\delta\theta$ is small, $\delta\hat{\mathbf{t}}$ is almost perpendicular to $\hat{\mathbf{t}}$. Furthermore, if $\delta\theta$ is measured in radians, the magnitude of $\delta\hat{\mathbf{t}}$ is approximately: $|\hat{\mathbf{t}}| \cdot \delta\theta = \delta\theta$.

By examining Figures 10.11a and b, we see that the direction of $\delta\hat{\mathbf{t}}$ is along the inward drawn normal to the curve. If $\hat{\mathbf{n}}$ is a unit vector with this direction, then $\delta\hat{\mathbf{t}} = \delta\theta \cdot \hat{\mathbf{n}}$. It follows that:

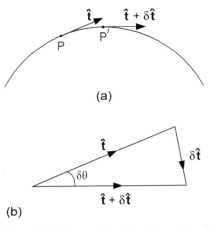

(a)

(b)

Figure 10.11. Tangential unit vectors at P and P′.

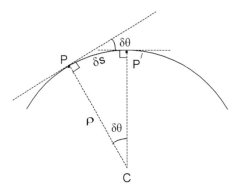

Figure 10.12. Finding the radius of curvature.

$$\frac{d\hat{\mathbf{t}}}{dt} = \lim_{\delta t \to 0} \frac{\delta \hat{\mathbf{t}}}{\delta t} = \lim_{\delta t \to 0} \frac{\delta \theta}{\delta t} \hat{\mathbf{n}} = \frac{d\theta}{dt} \hat{\mathbf{n}} = \dot{\theta} \hat{\mathbf{n}} = \omega \hat{\mathbf{n}},$$

where ω is the rate of change of direction measured in radians per second.

Hence, the *acceleration vector* may be written as:

$$\ddot{\mathbf{r}} = \dot{\mathbf{v}} = \frac{dv}{dt} \hat{\mathbf{t}} + v\omega \hat{\mathbf{n}}.$$

Two other forms of this equation arise if we use the *radius of curvature*. Let the inward-drawn normals at P and P′ intersect at C, as shown in Figure 10.12. Then, the *centre of curvature* of the curve at P is the limiting position of C as P′ moves back along the curve towards P. The corresponding radius of curvature ρ is the limiting length PC. Now:

$$\delta s \approx \rho \delta \theta \quad \text{and} \quad \rho = \lim_{\delta \theta \to 0} \frac{\delta s}{\delta \theta} = \frac{ds}{d\theta}.$$

Finally,

$$\omega = \frac{d\theta}{dt} = \frac{d\theta}{ds}\frac{ds}{dt} = \frac{1}{\rho} v \,.$$

Hence, we can re-write the acceleration of the point as it moves along the curve as:

$$\dot{\mathbf{v}} = \frac{dv}{dt} \hat{\mathbf{t}} + \rho \omega^2 \hat{\mathbf{n}}$$

$$\text{or } \dot{\mathbf{v}} = \frac{dv}{dt} \hat{\mathbf{t}} + \frac{v^2}{\rho} \hat{\mathbf{n}}.$$

EXERCISE 6

Taking into account only the rotation of the earth, find the acceleration of a point on the surface at latitude 30°N. Assume the radius of the earth to be 6370 km.

Problems 88 and 89.

10.7 Answers to exercises

1.   Since, $\ddot{x} = g = constant$, the graph of $\ddot{x}$ against t is a straight line parallel to the t-axis as in Figure 10.13a.

$$v = \int \dot{v}\, dt = \int \ddot{x}\, dt = \int g\, dt = gt + C_1.$$

$v = 0$ when $t = 0$, so $C_1 = 0$ and $v = gt$, so the graph of $\dot{x}$ against t is a straight line from the origin with slope g (see Figure 10.13b).

$$x = \int v\, dt = \int gt\, dt = \frac{1}{2}gt^2 + C_2.$$

$x = 0$ when $t = 0$, so $C_2 = 0$ and $x = \frac{1}{2}gt^2$; the graph of x against t has the shape of a parabola, as shown in Figure 10.14a.

$$\dot{v} = v\frac{dv}{dx} = g, \quad \int v\, dv = \int g\, dx \quad \text{and} \quad \tfrac{1}{2}v^2 = gx + C_3.$$

$v = 0$ when $x = 0$, so $C_3 = 0$ and $v^2 = 2gx$. Since x is measured positively downwards, $v = \dot{x}$ will be positive and the graph is part of a parabola, as shown in Figure 10.14b.

2. $x = a\sin(\omega t + \phi)$, so $v = \dot{x} = a\omega\cos(\omega t + \phi)$ and $\dfrac{v}{\omega} = a\cos(\omega t + \phi)$.

Thus, if $y = \dfrac{v}{\omega}$, $x^2 + y^2 = a^2\sin^2(\omega t + \phi) + a^2\cos^2(\omega t + \phi) = a^2$.

This is the equation for a circle of radius a and centre at the origin, as shown in Figure 10.15.

3.   Referring to Figure 10.6, let the distance of Q to the right of O be x (positive to the right, negative to the left). Since ω is constant, $\dot{\theta} = \omega$ and $\theta = \int \dot{\theta}\, dt = \omega t + \phi$, where ϕ is a constant of integration. Then, $x = r\cos\theta = r\cos(\omega t + \phi)$.

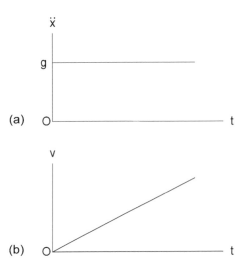

Figure 10.13. Graphs of (a) acceleration and (b) velocity against time.

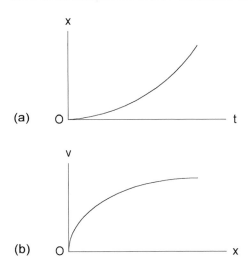

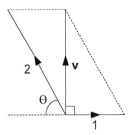

Figure 10.14. Graphs of (a) distance against time and (b) velocity against distance.

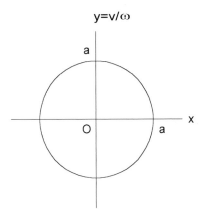

Figure 10.15. A graph of velocity against displacement for simple harmonic motion.

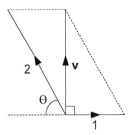

Figure 10.16. Steering a boat to travel at right angles to the bank of a river.

Differentiating with respect to t, we find that $\dot{x} = -r\omega\sin(\omega t + \phi)$ and differentiating again gives $\ddot{x} = -r\omega^2\cos(\omega t + \phi) = -\omega^2 x$. This is the basic differential equation for simple harmonic motion, as shown in Section 10.2.

4.   Figure 10.16 illustrates the solution by the parallelogram rule for vector addition. $\cos\theta = 1/2 = 0.5$. Therefore, $\theta = 60°$ to the upstream direction and this is the direction in which the boat must be pointed.

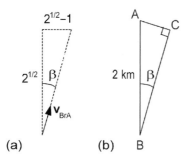

(a)   **(b)**   B

Figure 10.17. Calculating the nearest distance of approach of ship B to ship A.

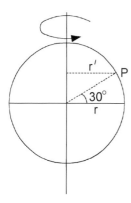

Figure 10.18. Acceleration of P due to rotation of the earth.

Then, the actual velocity is v, where:

$$v^2 = 2^2 - 1^2 = 3, \text{ so } v = \sqrt{3} = 1.732 \, \text{m/s}.$$

5.   In order to solve this problem, we need to examine the movement of B relative to A. Let us write the velocity vectors in terms of their easterly and northerly components:

$$\mathbf{v}_A = (4, 0) \quad \text{and} \quad \mathbf{v}_B = (4\sqrt{2}, 4\sqrt{2}).$$

Then, the relative velocity is:

$$\mathbf{v}_{BrA} = \mathbf{v}_B - \mathbf{v}_A = (4(\sqrt{2} - 1), 4\sqrt{2}),$$

which is in a direction east of north at angle (see Figure 10.17a):

$$\beta = \tan^{-1} \frac{\sqrt{2} - 1}{\sqrt{2}}.$$

If you imagine yourself on ship A looking out at ship B, the nearest distance of approach will be AC in Figure 10.17b.

$$AC = 2 \sin \beta = \frac{2(\sqrt{2} - 1)}{\sqrt{(\sqrt{2})^2 + (\sqrt{2} - 1)^2}} = \frac{2(\sqrt{2} - 1)}{\sqrt{5 - 2\sqrt{2}}} = 0.562 \, \text{km} = 562 \, \text{m}.$$

6.   Referring to Figure 10.18, point P is moving in a circle of radius $r' = r \cos 30°$, where r is the radius of the earth. Its speed around the circle is constant, so the acceleration is $r'\omega^2$ towards the centre of the circle. One rotation per day gives an angular velocity of $\omega = 2\pi/(24 \times 3600)$ rad/s. Thus, the acceleration of P is:

$$r'\omega^2 = \frac{6.37 \times 10^6}{1.155} \times \left(\frac{2\pi}{24 \times 3600}\right)^2 = 0.0292 \, \text{m/s}^2.$$

11 Kinetics of a particle

11.1 Newton's laws of motion

Kinetics is concerned with the relation between force and motion. The basis of this science is the laws of motion formulated by Isaac Newton (1642–1727) in his *Principia*, published in 1687. The essence of these laws is as follows.

First law. If the resultant force acting on a particle is zero, the acceleration of the particle will also be zero, i.e. it will continue in its state of rest or of constant velocity in a straight line.

Second law. If the resultant force acting on a particle is non-zero, the particle will accelerate in the direction of the force with a magnitude proportional to the magnitude of the force and inversely proportional to the mass of the particle.

Third law. To every action there is an equal and opposite reaction. For instance, if particle A collides with particle B, the reaction from B on A will be equal and opposite to the collision force from A on B.

The laws have been stated here with regard to the motion of a particle. This is convenient for the later development of the equations of motion of a body involving rotation as well as translation. However, in many cases the laws apply equally well to the motion of a body as to a particle. Indeed, Newton introduced the laws with regard to bodies rather than particles.

In order to perform calculations using Newton's laws of motion, it is necessary to specify units of measurement. We shall use only SI units (Système International d'Unités) of which the three basic units for dynamics are: the metre (m) for length, the second (s) for time and the kilogramme (kg) for mass. As already stated in Section 1.4, the kilogramme is the mass of a particular piece of platinum–iridium. The latter is the international prototype in the custody of the Bureau International des Poids et Mesures (BIPM), Sèvres, near Paris. At any given point on the earth's surface, the gravitational force on a body, i.e. its weight, is proportional to its mass. Hence, the masses of bodies may be compared by comparing their weights.

The next most important unit in dynamics is that of *force*. This is derived from the three basic units by applying Newton's second law of motion. The SI unit of force is the newton (N), defined as that force which, when applied to a mass of one kilogramme, gives it an acceleration of one metre per second per second ($1 \, \text{m/s}^2$ or $1 \, \text{m s}^{-2}$).

If a body of mass m is allowed to fall freely from a point near to the surface of the earth, it will be subject to a downward gravitational force F proportional to its mass, say $F = mg$, where g is the constant of proportionality. Let x be the distance down from the point of release, then the body's acceleration is $\ddot{x} = d^2x/dt^2$, the second derivative of x with respect to t. Then, from Newton's second law with SI units:

$$m\ddot{x} = F = mg.$$

Thus, neglecting air resistance, as we have done here, g is the acceleration due to gravity.

The value of g varies according to altitude and latitude, and for accuracies better than 1%, this variation has to be taken into account. The international standard value for g is $9.80665 \, \text{m/s}^2$. For all our examples, we shall use the approximate value $g = 9.8 \, \text{m/s}^2$.

EXERCISE 1

A lift, including its passengers, has a total weight of $10\,000 \, \text{N}$ ($10 \, \text{kN}$). It moves upwards from rest with constant acceleration, reaching a speed of $v = 6 \, \text{m/s}$ after travelling a distance $s = 6 \, \text{m}$. Find the force exerted by the lifting cable during this motion.

Subsequently, the lift comes to rest with a constant deceleration from its speed of $6 \, \text{m/s}$ in a time of $2 \, \text{s}$. Find the force transmitted from the floor of the lift to the feet of a passenger of mass $80 \, \text{kg}$ during this part of the motion.

Problems 90 and 91.

11.2 Sliding down a plane

Let us start by considering the hypothetical example of a block sliding down a perfectly smooth plane inclined at angle θ to the horizontal. Since the plane is perfectly smooth, there is no frictional force to oppose sliding and the reaction force acting on the block must be perpendicular to the plane. It is denoted by N in the force diagram of Figure 11.1a. The other force acting on the block is its weight mg. The mass $\times$ acceleration, i.e. $m\dot{v}$, is indicated in Figure 11.1b.

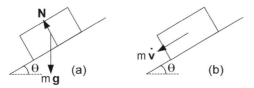

Figure 11.1. A block sliding down a smooth plane.

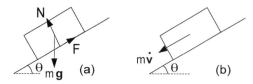

Figure 11.2. A block sliding down a plane while opposed by friction.

Since force and acceleration have direction as well as magnitude, Newton's second law of motion specifies a vector equation. Hence, in a two-dimensional problem like this one, we can form two scalar equations by resolving in two different directions. Resolving perpendicular to the plane gives:

$N - mg \cos \theta = 0.$

Resolving parallel to and down the plane gives:

$mg \sin \theta = m\dot{v}.$

Hence, the block slides down the plane with acceleration $\dot{v} = g \sin \theta$.

Next, we consider the more realistic situation in which the plane is not perfectly smooth. We assume that the block slides rather than topples (see Section 7.2). Let the block be in motion so that the movement down the plane is opposed by a frictional force $F = \mu N$, where μ is the coefficient of kinetic friction (see Section 7.1). Figures 11.2a and b show the force and mass × acceleration diagrams, respectively.

Resolving the vector equation of motion as before gives the equations:

$N - mg \cos \theta = 0$

$mg \sin \theta - F = mg \sin \theta - \mu N = mg \sin \theta - \mu mg \cos \theta$

$= mg(\sin \theta - \mu \cos \theta) = m\dot{v}$

Hence, the acceleration down the plane is:

$\dot{v} = g(\sin \theta - \mu \cos \theta).$

This of course assumes that the slope is steep enough to make $\sin \theta > \mu \cos \theta$.

EXERCISE 2

Let a block slide down a perfectly smooth plane that makes an angle θ to the horizontal. Suppose that it starts from rest at point A and reaches point B after t seconds. Imagine that this is repeated for different angles θ keeping the duration t constant.

If A is on the y-axis, as in Figure 11.3, with y-coordinate $gt^2/4$, show that the locus of B will be a semi-circle about the origin, starting from A, as θ varies from 0 to $\pi/2$. (This question was posed by Galileo (1564–1642)).

EXERCISE 3

A block is released from rest on a plane inclined at $\theta = \pi/4$ to the horizontal. If the length to height ratio of the block is $b/a > 1$, the static friction coefficient $\mu_s = 0.8$ and the kinetic friction coefficient $\mu_k = 0.4$, then: (a) find the acceleration of the block and (b) how far it must slide before it attains a speed of $v = 2$ m/s.

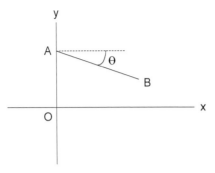

Figure 11.3. A block slides down a smooth plane from A to B in t seconds.

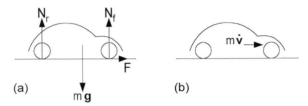

(a)

(b)

Figure 11.4. A vehicle on a horizontal surface.

Problems 92 and 93.

11.3 Traction and braking

Let us consider a four-wheeled vehicle on a horizontal surface. To simplify our problems so that they may be solved without introducing moments, the vehicle must have sophisticated traction and braking systems. It must have four-wheel drive and traction control to prevent the wheels from slipping when the vehicle is accelerating. Also, it must have an antilock braking system (ABS) to prevent the wheels from skidding when the brakes are applied to decelerate the vehicle. In this case, provided sufficient power is available, maximum static friction may be applied by all four wheels to either accelerate or decelerate the vehicle.

Assume now that the coefficient of static friction between the vehicle tyres and the horizontal surface is $\mu = 0.8$. If sufficient power is available, let us find the minimum time and distance travelled when the vehicle accelerates from rest to a speed of 5 m/s and then immediately brakes to decelerate back to rest again.

Referring to Figures 11.4a and b, we assume that the maximum frictional force is applied at all four wheels to give a total thrust $F = \mu(N_r + N_f)$. Resolving vertically and horizontally gives the equations:

$$N_r + N_f - mg = 0 \quad \text{and} \quad F = m\dot{v}.$$

Hence, $\dot{v} = \mu g$. Integrating with respect to t gives $v = \mu g t$, with no integrating constant since $v = 0$ when $t = 0$. To reach $v = 5$ m/s, the time:

$$t = \frac{v}{\mu g} = \frac{5}{0.8 \times 9.8} = 0.638 \text{ s}.$$

To find the distance x travelled during this acceleration phase, we write:

$$\dot{v} = v\frac{dv}{dx} = \mu g \quad \text{and} \quad \int v\, dv = \int \mu g\, dx, \quad \tfrac{1}{2}v^2 = \mu g x \; (v = 0 \text{ when } x = 0).$$

Thus, $x = \dfrac{v^2}{2\mu g} = \dfrac{25}{2 \times 0.8 \times 9.8} = 1.59$ m.

For the deceleration phase, the maximum frictional force F is applied in the opposite direction so that $\dot{v} = -\mu g$. Integrating with respect to time t gives $v = v_0 - \mu g t$, $v_0 = 5$ and v becomes zero when:

$$t = \frac{v_0}{\mu g} = \frac{5}{0.8 \times 9.8} = 0.638 \text{ s}.$$

Also, $v\dfrac{dv}{dx} = -\mu g, \; \int v\, dv = -\int \mu g\, dx, \; \tfrac{1}{2}v^2 = \tfrac{1}{2}v_0^2 - \mu g x, \; v_0 = 5.$

Thus, v returns to zero over a distance:

$$x = \frac{v_0^2}{2\mu g} = \frac{25}{2 \times 0.8 \times 9.8} = 1.59 \text{ m}.$$

It follows that the total time for both acceleration and deceleration is 1.276 s and the corresponding distance travelled is 3.18 m.

EXERCISE 4

Repeat the last example with the vehicle moving up an incline of 15° to the horizontal.

Problems 94 and 95.

11.4 Simple harmonic motion

In Section 10.2, we found that simple harmonic motion was described by the second order differential equation $\ddot{x} = -\omega^2 x$. In this case, x varied sinusoidally with frequency $\omega/2\pi$ Hz. We shall now consider a dynamical system that exhibits simple harmonic motion.

If a weight of mass m is suspended by a spring, as shown in Figure 11.5, it will be in statical equilibrium with the spring stretched from its natural length by x_0, where $kx_0 = mg$. k is the spring stiffness measured in N/m of displacement in either compression or extension from its natural length.

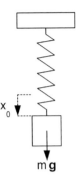

Figure 11.5. A mass suspended from a spring in equilibrium.

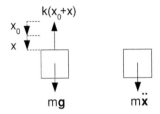

Figure 11.6. Force and mass times acceleration during vertical motion.

Pulling the weight further down and then releasing it will cause it to oscillate up and down. Referring to Figure 11.6, we can equate force to mass times acceleration during the motion to give:

$$m\ddot{x} = mg - k(x_0 + x) = -kx \quad \text{or} \quad \ddot{x} = -\frac{k}{m}x.$$

This is the differential equation of simple harmonic motion, which shows that the weight will oscillate about its statical equilibrium position with frequency $\sqrt{k/m}/2\pi$ Hz, m being given in kilogrammes.

If the weight were released at time $t = 0$ with $x = a$, x would vary subsequently according to the equation $x = a\cos(\omega t)$, where $\omega = \sqrt{k/m}$. The corresponding velocity would be $\dot{x} = -a\omega \sin(\omega t)$, which has maximum magnitude $a\omega$ when $\omega t = n\pi/2$, where n is any odd integer. Notice that this coincides with $x = 0$, i.e. passage through the statical equilibrium position.

EXERCISE 5

If a weight is suspended by a string and released from a position slightly to one side of its statical equilibrium position, show that its subsequent motion will be simple harmonic. Also, derive the relation between its period of oscillation and the length of the string.

Problems 96 and 97.

11.5 Uniform circular motion

We have uniform circular motion when a body is constrained to travel round a circle at constant speed. The general equation for the acceleration of a point in curvilinear motion is (see Section 10.6):

$$\dot{v} = \frac{dv}{dt}\hat{\mathbf{t}} + \rho\omega^2\hat{\mathbf{n}}.$$

In uniform circular motion (see Figure 11.7), the speed v is constant and thus there is no tangential component of acceleration. The only acceleration in this case is the normal component directed towards the centre of the circle. Its magnitude is $\rho\omega^2 = r\dot{\theta}^2$, where r is the radius of the circle and θ is the angular position round the circle measured in radians.

As a dynamical example, let us consider a conical pendulum in which a weight is suspended by a string and travels round a horizontal circle at constant speed. The centre of the circle is vertically below the point of suspension. Referring to Figure 11.8, we resolve force vertically and horizontally, and using Newton's second and third laws of motion, we obtain the following two equations:

$$T \cos\alpha - mg = 0$$
$$T \sin\alpha = mr\dot{\theta}^2$$

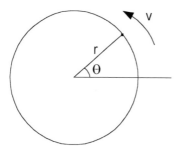

Figure 11.7. Circular motion.

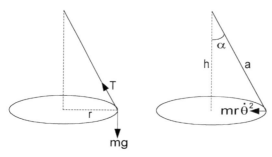

Figure 11.8. Conical pendulum.

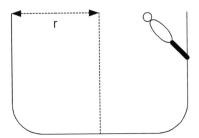

Figure 11.9. A motorcyclist riding round a vertical circular wall.

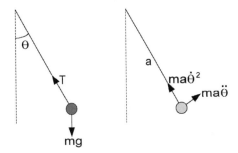

Figure 11.10. A vertical pendulum.

Since $r = a \sin \alpha$, $T = ma\dot{\theta}^2 = ma\omega^2$.

Therefore, $\omega^2 = \dfrac{T}{ma} = \dfrac{mg}{ma \cos \alpha} = \dfrac{g}{h}$ and $\omega = \sqrt{g/h}$.

The period is the time taken for one revolution $(2\pi \, \text{rad}) = 2\pi/\omega = 2\pi \sqrt{h/g}$.

EXERCISE 6

A motorcyclist wishes to ride the bike around the inside of a vertical circular wall, as shown diagrammatically in Figure 11.9. How fast must the bike travel in order not to slide down the wall if the radius of the circular wall is $r = 10 \, \text{m}$ and the coefficient of static friction between the tyres and the wall is $\mu = 0.7$?

Problems 98 and 99.

11.6 Non-uniform circular motion

As an example of non-uniform circular motion, let us consider the pendulum in Exercise 5, Section 11.4, when θ is not restricted to small angles. The weight of mass m is suspended by a string of length a and swings in a vertical plane, which in Figure 11.10 corresponds to the plane of the paper. θ is the angle, in radians, between the string and the vertical.

Resolving tangentially and along the length of the string, we have the following two equations for the motion of the weight:

$$ma\ddot{\theta} = -mg \sin\theta$$

$$ma\dot{\theta}^2 = T - mg \cos\theta.$$

Rewrite the first equation as:

$$a\dot{\theta}\frac{d\dot{\theta}}{d\theta} = -g \sin\theta \quad \text{and then } a \int \dot{\theta}\,d\dot{\theta} = -g \int \sin\theta\,d\theta.$$

Hence, $\frac{1}{2}a\dot{\theta}^2 = g \cos\theta + C,$

where C is the constant of integration.

If the weight were released from rest with the string taut and $\theta = \theta_0 \le \pi/2$, then $0 = g \cos\theta_0 + C$ and the equation relating $\dot{\theta}$ to θ becomes:

$$\frac{1}{2}a\dot{\theta}^2 = g(\cos\theta - \cos\theta_0).$$

It would then swing to and fro between θ_0 and $-\theta_0$, with angular velocity through the lowest point given by:

$$\dot{\theta}^2 = \frac{2g}{a}(1 - \cos\theta_0).$$

Unfortunately, there is no analytical way of finding the period of swing of the pendulum.

EXERCISE 7

Suppose the vertical pendulum is mounted in such a way that if given sufficient velocity, it will swing round and round in a vertical circle. Remembering that the attachment of the weight to its support is not rigid but is a flexible string, find how fast it must pass through its lowest point in order to continue round in a vertical circle.

Problems 100 and 101.

11.7 Projectiles

In studying the flight of a projectile, we will neglect air resistance and assume that our missile does not go very far or very high so that earth curvature and changes in gravity may be neglected. The only force acting on the projectile of mass m is then the gravitational force $\mathbf{F} = m\mathbf{g}$, where the direction of this vector is vertically downwards. If we represent the position of the projectile P by the vector $\mathbf{r}$ from a reference point O, as shown in Figure 11.11, Newton's second law of motion tells us that $m\ddot{\mathbf{r}} = m\mathbf{g}$. Hence, the kinematics of the projectile's motion is prescribed by the second order vector differential equation:

$$\ddot{\mathbf{r}} = \mathbf{g}.$$

Figure 11.11. Position vector **r** of projectile P.

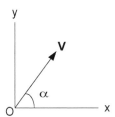

Figure 11.12. Cartesian coordinates for a projectile's trajectory.

If we now integrate once with respect to time, we obtain the velocity $\dot{\mathbf{r}} = \mathbf{g}t + \dot{\mathbf{r}}_0$, where $\dot{\mathbf{r}}_0$ is the initial velocity when $t = 0$. Integrating again gives the position:

$$\mathbf{r} = \tfrac{1}{2}\mathbf{g}t^2 + \dot{\mathbf{r}}_0 t + \mathbf{r}_0,$$

where $\mathbf{r}_0$ is the initial position vector when $t = 0$.

To study the geometry of the projectile's trajectory, we introduce a Cartesian system of coordinates with x-axis horizontal and y-axis vertically upwards, as indicated in Figure 11.12. We take the origin O as the position of the projectile when $t = 0$ and the initial velocity of magnitude V at angle α above the horizontal.

Consider now the x- and y-components of the vector equations of motion. $\ddot{\mathbf{r}} = \mathbf{g}$ is replaced by the two equations:

$$\ddot{x} = 0 \quad \text{and} \quad \ddot{y} = -g.$$

The velocity equation $\dot{\mathbf{r}} = \mathbf{g}t + \dot{\mathbf{r}}_0$ is replaced by the two equations:

$$\dot{x} = V\cos\alpha \quad \text{and} \quad \dot{y} = V\sin\alpha - gt.$$

Finally, the position equation $\mathbf{r} = \tfrac{1}{2}\mathbf{g}t^2 + \dot{\mathbf{r}}_0 t + \mathbf{r}_0$ is replaced by the two equations:

$$x = Vt\cos\alpha \quad \text{and} \quad y = Vt\sin\alpha - \tfrac{1}{2}gt^2.$$

To investigate the geometry of the projectile's trajectory, we must eliminate t between these two equations. From the first, $t = x/(V\cos\alpha)$, which when substituted into the second gives:

$$y = \left(\frac{x}{V\cos\alpha}\right)V\sin\alpha - \frac{1}{2}g\left(\frac{x}{V\cos\alpha}\right)^2 = x\tan\alpha - \frac{g}{2V^2\cos^2\alpha}x^2.$$

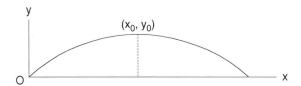

Figure 11.13. Projectile's parabolic trajectory.

Now multiply by $-(2V^2 \cos^2\alpha)/g$ and complete the square on x as follows:

$$x^2 - \frac{2V^2 \sin\alpha \cos\alpha}{g} x + \frac{V^4 \sin^2\alpha \cos^2\alpha}{g^2} = \left(x - \frac{V^2 \sin\alpha \cos\alpha}{g} \right)^2$$

$$= -\frac{2V^2 \cos^2\alpha}{g} y + \frac{V^4 \sin^2\alpha \cos^2\alpha}{g^2} = -\frac{2V^2 \cos^2\alpha}{g} \left(y - \frac{V^2 \sin^2\alpha}{2g} \right).$$

This is the equation of a parabola (see Figure 11.13):

$$(x - x_0)^2 = -4a(y - y_0),$$

with axis $x = x_0 = (V^2 \sin\alpha \cos\alpha)/g$ and apex at (x_0, y_0), where $y_0 = (V^2 \sin^2\alpha)/(2g)$.

The latter, i.e. $y_0 = (V^2 \sin^2\alpha)/(2g)$, is the height of the trajectory. Also, by the symmetry of the parabola about its axis, we see that the range is:

$$R = 2x_0 = \frac{2V^2 \sin\alpha \cos\alpha}{g} = \frac{V^2 \sin 2\alpha}{g}.$$

From this, we see that $R = V^2/g$ is the maximum range which is achieved when $\alpha = 45°$.

Since the x-component of velocity is constant, the time of flight is:

$$\frac{R}{\dot{x}} = \frac{2V^2 \sin\alpha \cos\alpha}{gV \cos\alpha} = \frac{2V \sin\alpha}{g}.$$

To achieve a given range $R < V^2/g$, the angle of projection is determined by the equation:

$$2\alpha = \sin^{-1}\left(\frac{gR}{V^2} \right).$$

Since $\sin 2\alpha = \sin(180° - 2\alpha)$ and $\alpha < 45°$, the desired range is obtained by angles of elevation of either α or $90° - \alpha$, i.e. a low trajectory or a high trajectory.

EXERCISE 8

Show that the range R up a plane, inclined at angle β to the horizontal ($\beta < \alpha$), is given by the formula:

$$R = \frac{V^2}{g} \frac{\sin(2\alpha - \beta) - \sin\beta}{\cos^2\beta}.$$

Problems 102 and 103.

11.8 Motion of connected weights

So far we have been applying Newton's laws of motion to the motion of a single body. Before leaving Chapter 11, it is worthwhile considering cases in which bodies are connected together. Let two weights of masses M and $(M + m)$ be connected by a string which passes over a light frictionless pulley, as indicated in Figure 11.14. When released, the larger mass will move downwards with an acceleration $\dot{v}$ and the lighter one will obviously move upwards with the same acceleration. With a light frictionless pulley, we can assume that the string tension T is the same on each side. Applying Newton's second law of motion to each mass in turn gives the following two equations:

$(M + m)\dot{v} = (M + m)g - T$

$M\dot{v} = T - Mg.$

Adding, to eliminate T, gives:

$(2M + m)\dot{v} = mg$

and the acceleration is thus:

$$\dot{v} = \frac{mg}{2M + m}.$$

Next, consider the more complicated example of a chain sliding off a smooth horizontal table. Referring to Figure 11.15, let the mass per unit length of the chain be σ, the total length be a, the overhang be x and the tensile force in the chain at the corner be T.

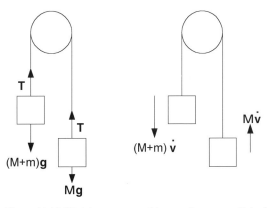

Figure 11.14. Weights connected by a string over a light frictionless pulley.

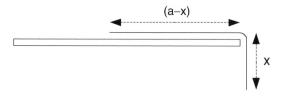

Figure 11.15. Chain sliding off a smooth table.

Now apply Newton's second law of motion firstly to the part of the chain on the table and secondly to the part which overhangs to obtain the following two equations:

$$T = \sigma(a - x)\ddot{x}$$

$$\sigma x g - T = \sigma x \ddot{x}$$

Adding these two equations to eliminate T, we obtain the second order differential equation:

$$\sigma a \ddot{x} = \sigma g x, \text{ i.e. } \ddot{x} = gx/a.$$

The exponential functions $\exp \sqrt{g/a}t$ and $\exp -\sqrt{g/a}t$, each satisfy this differential equation, so the general solution is:

$$x = A \exp \sqrt{g/a}t + B \exp -\sqrt{g/a}t,$$

where A and B are constants.

To find A and B, we need two initial conditions, so let us suppose that the chain starts from rest with an initial overhang x_0 and $\dot{x} = 0$ when $t = 0$. Differentiating x with respect to t gives:

$$\dot{x} = A\sqrt{g/a} \exp \sqrt{g/a}t - B\sqrt{g/a} \exp -\sqrt{g/a}t.$$

Then at $t = 0$:

$$x_0 = A + B \quad \text{and} \quad 0 = A\sqrt{g/a} - B\sqrt{g/a} \quad \text{or} \quad 0 = A - B.$$

Therefore, $A = B = x_0/2$.

Hence, $x = (\exp \sqrt{g/a}t + \exp -\sqrt{g/a}t)x_0/2 = x_0 \cosh(\sqrt{g/a}t)$,

(see Section 22.2).

EXERCISE 9

Consider the system shown in Figure 11.16. A block of mass m_1 rests on a smooth plane inclined at angle α to the horizontal. It is attached to a string which passes round two light frictionless pulleys as shown with one pulley supporting another weight of mass m_2. Find the acceleration $\dot{v}$ of the latter.

Problems 104 and 105.

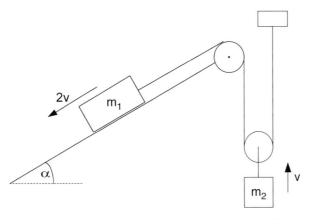

Figure 11.16. System of weights, pulleys and a smooth inclined plane.

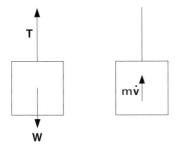

Figure 11.17. Force and mass × acceleration diagrams for a lift moving upwards with constant acceleration.

11.9 Answers to exercises

1. In the first part of the question, we consider the motion of the lift during the acceleration mode. There are two vertical forces: the cable tension T and the weight W of the lift and passengers. Since Newton's second law equates resultant force to mass times acceleration, it is helpful to draw two diagrams as shown in Figure 11.17: the force diagram and the mass × acceleration diagram.

The resultant upward force is $(T - W)$ and the mass × acceleration is $m\dot{v}$, where m is the mass of the lift and passengers and $\dot{v}$ is the acceleration, written as the derivative of the velocity v with respect to time. Hence, from Newton's second law:

$$T - W = m\dot{v}.$$

The weight W is given as $W = 10\,000\,\text{N}$ but the mass m must be derived from this using the equation $W = mg$, where g is the acceleration due to gravity, i.e. $g = 9.8\,\text{m/s}^2$.

Before we can find T from the equation of motion, we must firstly evaluate $\dot{v}$, the acceleration. Apart from being told that $\dot{v}$ is constant, the only information is that $v = 6\,\text{m/s}$ when the distance travelled, $s = 6\,\text{m}$. This suggests that we should write the acceleration $\dot{v}$ in terms of v and s. This is possible since $v = ds/dt$ and therefore:

$$\dot{v} = \frac{dv}{dt} = \frac{dv}{ds}\frac{ds}{dt} = v\frac{dv}{ds} = C,$$

where C is the constant acceleration. It then follows that:

$$\int v\,dv = \int C\,ds \quad \text{and} \quad \tfrac{1}{2}v^2 = Cs + B,$$

where B is the integrating constant. Since the lift starts from rest, $v = 0$ when $s = 0$. Substitution shows that $B = 0$ and therefore, $v^2 = 2Cs$. Next, $v = 6$ when $s = 6$ and therefore, $C = 3\,\text{m/s}^2$.

Returning to the equation of motion,

$$T = W + m\dot{v} = W\left(1 + \frac{C}{g}\right) = 10\,000\left(1 + \frac{3}{9.8}\right) = 13\,060\,\text{N}.$$

In the second part of the question, we are concerned with a passenger of mass $m_1 = 80\,\text{kg}$ as the lift decelerates. Let P be the upward force from the lift floor on the passenger's feet. Apply Newton's second law to the motion of the passenger to give the equation $P - m_1 g = m_1 \dot{v}_1$, where v_1 is the velocity of the passenger during the deceleration phase.

$\dot{v}_1 = C_1 = constant$, so $v_1 = C_1 t + A_1$.

Let $t = 0$ at the start of the deceleration, at which time $v_1 = 6\,\text{m/s}$. Hence, $A_1 = 6$.
Now, $t = 2$ when the lift comes to rest, i.e. when $v_1 = 0$.

Therefore, $0 = 2C_1 + 6$ and $C_1 = -3\,\text{m/s}^2$.

Returning to the equation of motion,

$P = m_1(g + \dot{v}_1) = 80(9.8 - 3) = 544\,\text{N}$.

2. Let s be the distance down the slope from A, with $s = 0$ and $\dot{s} = 0$ when $t = 0$. The acceleration of the block as it slides down the slope is $\ddot{s} = g\sin\theta$, where, as in figure 11.18, θ is the angle of the slope to the horizontal. Integrating twice with respect to t gives:

firstly $\dot{s} = gt\sin\theta$ and secondly $s = \tfrac{1}{2}gt^2\sin\theta$.

The integrating constants are zero since the initial conditions are zero.
We now let t be the time when the block reaches B and hence, $AB = \tfrac{1}{2}gt^2\sin\theta$. Given that A is the point with coordinates $(0, gt^2/4)$, the coordinates of B are:

$x_b = \tfrac{1}{2}gt^2\sin\theta\cos\theta = \tfrac{1}{4}gt^2\sin 2\theta$,

$y_b = \tfrac{1}{4}gt^2 - \tfrac{1}{2}gt^2\sin\theta\sin\theta = \tfrac{1}{4}gt^2(1 - 2\sin^2\theta) = \tfrac{1}{4}gt^2\cos 2\theta$.

Therefore, $x_b^2 + y_b^2 = (gt^2/4)^2(\sin^2 2\theta + \cos^2 2\theta) = (gt^2/4)^2$.

This is the equation of a circle about the origin with radius $gt^2/4$.

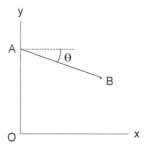

Figure 11.18. AB represents a smooth slope.

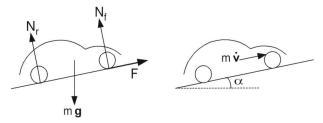

Figure 11.19. A vehicle on an inclined plane.

3. Since $b/a > 1 = \tan(\pi/4) = \tan\theta$, we see from Section 7.2 that the block will slide rather than topple and also, since $\tan\theta = 1 > 0.8 = \mu_s$, the block will start to slide when it is released from rest. Its acceleration down the slope will be:

$$\dot{v} = (\sin\theta - \mu_k \cos\theta)g = 9.8(1 - 0.4)/\sqrt{2} = 4.16\,\text{m/s}^2.$$

Since we want to find the distance travelled by the block when it reaches a speed of $v = 2$ m/s, we write the acceleration as:

$$\dot{v} = v\frac{dv}{ds}. \text{ Then, } \int v\,dv = \int \dot{v}\,ds \quad \text{so} \quad v^2/2 = \dot{v}s + C.$$

$v = 0$ when $s = 0$, so $C = 0$ and:

$$s = \frac{v^2}{2\dot{v}} = \frac{4}{2 \times 4.16} = 0.481\,\text{m}.$$

4. Referring to Figure 11.19, $F = \mu(N_r + N_f)$. Resolve perpendicular to and along the inclined plane to obtain the equations:

$$N_r + N_f - mg\cos\alpha = 0$$

$$F - mg\sin\alpha = m\dot{v}.$$

Combining these with the equation for F, we see that:

$$\dot{v} = (\mu\cos\alpha - \sin\alpha)g = 5.036\,\text{m/s}^2.$$

Therefore, $v = 5.036t$, so $t = 5/5.036 = 0.993$ s.

Also, $v\dfrac{dv}{dx} = 5.036$, $v^2/2 = 5.036x$, $x = \dfrac{25}{2 \times 5.036} = 2.482\,\text{m}$.

For the deceleration phase, the direction of F is reversed. Hence,

$$\dot{v} = -(\mu\cos\alpha + \sin\alpha)g = -10.109\,\text{m/s}^2.$$

$v = v_0 - 10.109t$ and $v = 0$ when $t = \dfrac{5}{10.109} = 0.495$ s.

Also, $v^2/2 = v_0^2/2 - 10.109x$ and $v = 0$ when $x = \dfrac{25}{2 \times 10.109}$, i.e. $x = 1.236\,\text{m}$.

Then, for both acceleration and deceleration phases, the total time elapsed is 1.49 s and the distance travelled is 3.72 m.

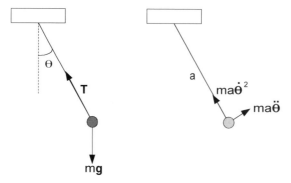

Figure 11.20. A simple pendulum, i.e. a small-oscillation vertical pendulum.

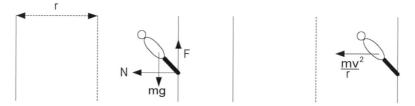

Figure 11.21. Motorcycling on a vertical cylindrical wall.

5.   In Section 10.6, we found that a point moving along a curve has two components of acceleration, i.e.

$$\dot{\mathbf{v}} = \frac{dv}{dt}\hat{\mathbf{t}} + \rho\omega^2\hat{\mathbf{n}},$$

where $\mathbf{v}$ is the velocity vector, ρ is the radius of curvature, ω is the rate of change of direction (rad/s), $\hat{\mathbf{t}}$ is the unit vector in the direction of travel and $\hat{\mathbf{n}}$ is the unit vector along the inward drawn normal to the curve. Referring to Figure 11.20, $\rho = a$ (the length of the string), $v = a\dot{\theta}$ (θ in radians), $dv/dt = a\ddot{\theta}$ and $\omega = \dot{\theta}$. The two components of mass × acceleration are indicated in the second diagram of Figure 11.20.

Now apply Newton's second law of motion to the tangential components:

$$ma\ddot{\theta} = -mg\sin\theta \approx -mg\theta,$$

since θ is small. This corresponds to:

$$\ddot{\theta} = -\frac{g}{a}\theta,$$

which is the differential equation of simple harmonic motion. The period T of oscillation is the reciprocal of the frequency $\sqrt{g/a}/2\pi$, i.e. $T = 2\pi\sqrt{a/g}$. Measuring a in metres (m) and g in m/s^2 (≈ 9.8) gives T in seconds (s). Notice that T is proportional to the square root of the length of the string.

A weight suspended by a string and swinging through a small angle θ in this way is referred to as a simple pendulum.

6.   Referring to Figure 11.21, since we are concerned with the speed v, it is convenient to write the acceleration towards the centre of the circle (of radius r) as v^2/r. Also, the reaction from the wall has a normal component N and a frictional component F. If the coefficient of static friction is μ, then $F \le \mu N$.

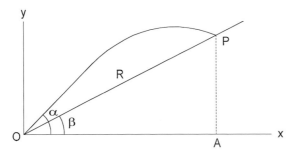

Figure 11.22. Projectile's trajectory above an inclined plane.

Resolving vertically and horizontally:

$$F - mg = 0 \quad \text{and} \quad N = mv^2/r.$$

Combining these with the restriction on F gives $mg \leq \mu mv^2/r$.

Hence, $v^2 \geq rg/\mu = \dfrac{10 \times 9.8}{0.7} = 140$, so $v \geq 11.8\,\text{m/s}$.

7.   To solve this problem, we must make use of the inward normal equation of motion, i.e. $ma\dot{\theta}^2 = T - mg\cos\theta$. Now, for the weight to continue round in a complete vertical circle, the tension T in the string must always be greater than or equal to zero. T will be least in the uppermost position, i.e. when $\theta = \pi$. Thus, in that position, $T = ma\dot{\theta}^2 - mg \geq 0$, i.e. $a\dot{\theta}^2 \geq g$. Let us take the minimum value, i.e. $a\dot{\theta}^2 = g$ when $\theta = \pi$. Referring back to the equation $a\dot{\theta}^2/2 = g\cos\theta + C$ in Section 11.6, we have $a\dot{\theta}^2/2 = g/2 = -g + C$ and $C = 3g/2$. Thus, when $\theta = 0$, $a\dot{\theta}^2/2 = g + 3g/2 = 5g/2$. Then, if the speed through the lowest point is $v = a\dot{\theta}$, we have $v^2 = 5ag$. Hence, to continue round in a vertical circle, we must have $v \geq \sqrt{5ag}$.

8.   Let P be the point of intersection of the trajectory with the inclined plane, as shown in Figure 11.22. If P has coordinates (x, y), from the trajectory equation:

$$y = x\tan\alpha - \frac{g}{2V^2\cos^2\alpha}x^2$$

and from the plane: $y = x\tan\beta$.

Therefore, $x\left(\tan\alpha - \tan\beta - \dfrac{gx}{2V^2\cos^2\alpha}\right) = 0$

and $x = \dfrac{2V^2\cos^2\alpha}{g}(\tan\alpha - \tan\beta)$.

Now, the x-coordinate of P corresponds to OA, so the range:

$$R = \frac{OA}{\cos\beta} = \frac{2V^2\cos^2\alpha}{g\cos\beta}(\tan\alpha - \tan\beta) = \frac{2V^2\cos^2\alpha}{g\cos\beta}\frac{\sin(\alpha - \beta)}{\cos\alpha\cos\beta}$$

$$= \frac{2V^2\cos\alpha\sin(\alpha - \beta)}{g\cos^2\beta} = \frac{V^2}{g}\frac{\sin(2\alpha - \beta) - \sin\beta}{\cos^2\beta},$$

(see Section 18.4). In this case the range will be maximized if $2\alpha - \beta = 90°$, i.e. $\alpha = 45° + \beta/2$.

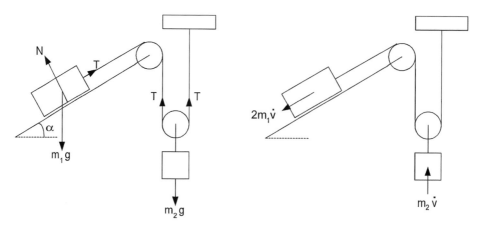

Figure 11.23. A weight and pulley system.

9.   Referring to Figure 11.23, since the pulleys are light and frictionless, the tension T is constant throughout the length of the string. Also, since the inclined plane is smooth, the reaction from the plane on mass m_1 is normal to the plane and is denoted by N in the force diagram of Figure 11.23.

Resolve the forces acting on m_1 normal to and parallel to the plane, so that Newton's second law gives the following two equations:

$$N - m_1 g \cos \alpha = 0$$

$$m_1 g \sin \alpha - T = 2m_1 \dot{v}.$$

For the second mass m_2, all forces and motion are vertical, so we have just one equation:

$$2T - m_2 g = m_2 \dot{v}.$$

(Note that if m_2 moves up by x, then m_1 moves down the plane by $2x$. Hence, the corresponding accelerations are $\dot{v}$ and $2\dot{v}$, respectively.)

We are not interested in N, so we can ignore the first equation. Then, adding twice the second equation to the third eliminates T and gives:

$$2m_1 g \sin \alpha - m_2 g = (4m_1 + m_2)\dot{v}.$$

Hence, $\dot{v} = \dfrac{2m_1 \sin \alpha - m_2}{4m_1 + m_2} g.$

12 Plane motion of a rigid body

12.1 Introduction

In Chapter 11, we were able to study the dynamics of a body provided no rotation was involved or if it were, then its effect could be neglected. Thus we were only concerned with the direct relation between force and acceleration. In this chapter, we shall also consider the turning effect of a force. We can no longer regard the rigid body like a single particle but rather like a whole collection of particles fixed in position relative to each other.

To simplify the analysis, we will restrict ourselves to the study of the plane motion of a rigid body. If the body is a plane lamina, then movement will take place in the plane of the lamina, which will be fixed. If it is not a plane lamina, the movement of any small part of the body will be restricted to a plane which is parallel to a given fixed plane. Thus, any rotation of the body will be about an axis perpendicular to this fixed plane.

An important concept in studying dynamics is that of *momentum*. The momentum of a particle of mass m and velocity $\mathbf{v}$ is simply $m\mathbf{v}$. Notice that this is a vector quantity with the direction of the velocity $\mathbf{v}$. Assuming the mass m to be constant, then the *rate of change of momentum* is $m\dot{\mathbf{v}}$, which is the mass times the acceleration of the particle. Thus, another way of stating Newton's second law of motion is to say that the force applied to a particle equates to its rate of change of momentum.

12.2 Moment

Let a force $\mathbf{F}$ act on a particle P, which has position vector $\mathbf{r}$ from a reference point O, as shown in Figure 12.1. In Section 2.1, we found that the moment of $\mathbf{F}$ about O is given by the vector product $\mathbf{r} \times \mathbf{F}$. The direction of this vector is perpendicular to both $\mathbf{r}$ and $\mathbf{F}$, using the right-hand thread rule. Hence, with $\mathbf{r}$ and $\mathbf{F}$ in the plane of the paper, as shown in Figure 12.1, $\mathbf{r} \times \mathbf{F}$ is perpendicularly upwards out of the paper. The magnitude, which is the turning effect of $\mathbf{F}$ about O, is $r F \sin\theta$.

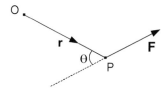

Figure 12.1. The moment of a force.

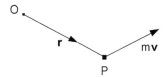

Figure 12.2. The moment of momentum of a particle.

Referring to Figure 12.2, this concept of moment can be applied to any localized vector and in particular to the momentum vector $m\mathbf{v}$ of the particle P, where m is the mass and $\mathbf{v}$ is the velocity of the particle. Hence, the *moment of momentum* of particle P about O is $\mathbf{r} \times m\mathbf{v}$.

If we now differentiate with respect to time t, we can find the rate of change of moment of momentum. Differentiating a vector product follows the usual product rule for differentiation, giving:

$$\dot{\mathbf{r}} \times m\mathbf{v} + \mathbf{r} \times m\dot{\mathbf{v}}.$$

Now, $\dot{\mathbf{r}} = \mathbf{v}$ and $\mathbf{v} \times \mathbf{v} = \mathbf{0}$. Hence, the first term is zero, leaving $\mathbf{r} \times m\dot{\mathbf{v}}$ as the rate of change of moment of momentum.

We notice also that:

$$\mathbf{r} \times m\dot{\mathbf{v}} = \mathbf{r} \times \mathbf{F},$$

from Newton's second law of motion. Hence, the rate of change of moment of momentum of P about O equals the moment about O of the force $\mathbf{F}$ applied to the particle P.

12.3 Instantaneous centre of rotation

In this section, we wish to show that, at a given instant, the plane motion of a body may be represented by a rotation about a point I. The point I may itself be moving and so it is referred to as the instantaneous centre of rotation.

We start by showing that any displacement of the body in the plane of motion may be achieved by a rotation about a point in the plane. Referring to Figure 12.3, let A and B be two points of the body in its plane of motion; their position determines the position of the body. Let the body move so that A moves to A′ and B to B′. Draw the perpendicular

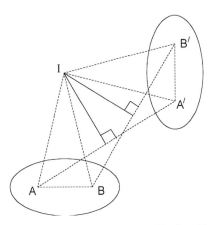

Figure 12.3. Change in position of body achieved by a rotation about I.

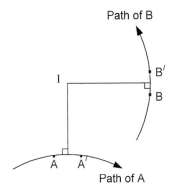

Figure 12.4. Centre of rotation I for small movement of the body.

bisectors of AA′ and BB′, and let I be their point of intersection. Comparing triangles ABI and A′B′I, we see that:

$$AB = A'B', \quad AI = A'I \quad \text{and} \quad BI = B'I.$$

Hence, the triangles ABI and A′B′I are equal and the second is obtained from the first by a rotation about I. Correspondingly, the new position of the body may have been achieved by a rotation about I from its original position. If the movement had been a pure translation, then I would have been positioned at infinity.

In order to find the instantaneous centre of rotation, we need to take the limit of the construction to find I as the amount of movement tends to zero. Figure 12.4 shows the position of I for short distances AA′ and BB′. From this it becomes obvious how to find the instantaneous centre of rotation: draw the perpendicular to the path of A at A and the perpendicular to the path of B at B, as in Figure 12.5. The point of intersection of these two perpendiculars is then the instantaneous centre of rotation.

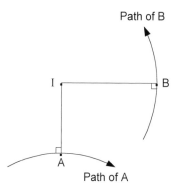

Figure 12.5. Instantaneous centre of rotation.

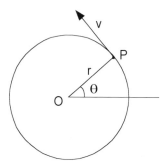

Figure 12.6. Point P moving along a circular path.

EXERCISE 1

The ends of a straight rod of length a are constrained to move in straight tracks. One track is horizontal and the other is vertical with its bottom end connected to the horizontal track. Using coordinate geometry, find the path followed by the instantaneous centre of rotation of the rod as the rod moves from the vertical to the horizontal.

EXERCISE 2

Find the instantaneous centre of rotation as a circular disc rolls along a horizontal plane.

Problem 106.

12.4 Angular velocity

Suppose a point P is moving with velocity v in a circle of radius r about a fixed point O, as in Figure 12.6. Let the position of P be given by the angle θ in radians, which is the angle between OP and a given radial direction. Since θ is measured in radians, its rate of change $\dot{\theta}$ is related to the velocity of P by the equation $v = r\dot{\theta}$. $\dot{\theta}$ is the angular velocity of P about O and since θ is measured in radians, $\dot{\theta}$ is given in radians per second (rad/s).

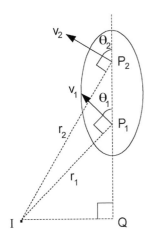

Figure 12.7. Plane lamina moving in its own plane.

Next consider a plane lamina moving in a general manner, i.e. with both translation and rotation, in its own plane. Its motion at any given instant is described by a rotational velocity ω (rad/s) about its instantaneous centre of rotation I. Referring to Figure 12.7, we pick any two points P_1 and P_2 in the lamina and examine the angular velocity of P_2 relative to P_1.

If the distances $I P_1$ and $I P_2$ are r_1 and r_2, respectively, then the corresponding velocities of P_1 and P_2 must be $v_1 = r_1 \omega$ and $v_2 = r_2 \omega$ as indicated in Figure 12.7. Their directions are perpendicular to $I P_1$ and $I P_2$, respectively, and let θ_1 and θ_2 be the angles between these directions and the line $P_1 P_2$. Complete the construction of the diagram, as shown in Figure 12.7, by extending $P_2 P_1$ down to Q, where $I Q \perp Q P_1$.

Since P_1 and P_2 are points in a rigid body, the distance $P_1 P_2$ is constant. Hence, the velocity of P_2 relative to P_1 is perpendicular to $P_1 P_2$ and must have magnitude

$$v_{2r1} = v_2 \sin \theta_2 - v_1 \sin \theta_1.$$

Dividing this by the length $P_1 P_2$ gives the angular velocity of P_2 relative to P_1,

i.e. $\dfrac{v_2 \sin \theta_2 - v_1 \sin \theta_1}{P_1 P_2} = \dfrac{(r_2 \sin \theta_2 - r_1 \sin \theta_1)\omega}{P_1 P_2} = \dfrac{(Q P_2 - Q P_1)\omega}{P_1 P_2} = \omega.$

Since the points P_1 and P_2 were picked arbitrarily, we draw the conclusion that the angular velocity of the lamina is the angular velocity of any one point relative to any other, where the two points are points of the lamina itself, which is moving in its own plane.

EXERCISE 3

A circular disc of radius r is rolling along a horizontal surface with velocity v. Find the magnitude and direction of the velocity **u** of a point P on the rim of the disc ahead of and at the same height as the centre of the disc.

Problems 107 and 108.

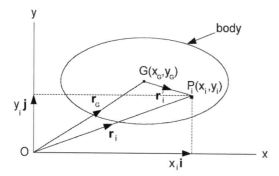

Figure 12.8. Body viewed relative to the x, y plane of motion.

12.5 Centre of gravity

We studied the centre of gravity and its role in statics in Chapter 3. It is also very important in dynamics, so we shall give it further consideration here.

DEFINITION.

The centre of gravity is the unique point in the body through which the resultant gravitational force always acts for any orientation of the body.

Since we are only studying plane motion, the body will be regarded as a plane lamina. For a general body points become lines perpendicular to the plane of motion.

Imagine a body in an x, y plane, as shown in Figure 12.8. The centre of gravity G has coordinates (x_G, y_G) and the ith particle P_i has coordinates (x_i, y_i). We shall find that position vectors are also useful. $\mathbf{r}_G$ and $\mathbf{r}_i$ are the position vectors relative to O of G and P_i, respectively, and $\mathbf{r}'_i$ is the position vector of P_i relative to G.

Let particle P_i have mass m_i so that the total mass of the body is $\sum_i m_i$, where we sum over all the particles of the body. Now, think of gravity acting perpendicular to the x, y plane and take moments of the gravitational force about the y-axis. From the definition of the position G, it follows that:

$$\left(\sum_i m_i\right) g x_G = \sum_i m_i g x_i = g \sum_i m_i (x_G + x'_i) = g\left(\sum_i m_i\right) x_G + g\sum_i m_i x'_i,$$

where x'_i is the x-component of $\mathbf{r}'_i$. From this we see that:

$$g \sum_i m_i x'_i = 0, \text{ i.e. } \sum_i m_i x'_i = 0.$$

Similarly, by taking moments about the x-axis, we find that $\sum_i m_i y'_i = 0$.

If we now multiply by the unit vector $\mathbf{i}$ in one case and by the unit vector $\mathbf{j}$ in the other, it follows that:

$$\sum_i m_i x_i' \mathbf{i} = 0 \quad \text{and} \quad \sum_i m_i y_i' \mathbf{j} = 0.$$

Consequently, $\sum_i m_i x_i' \mathbf{i} + \sum_i m_i y_i' \mathbf{j} = \sum_i m_i (x_i' \mathbf{i} + y_i' \mathbf{j}) = \sum_i m_i \mathbf{r}_i' = 0.$

This result is of fundamental importance in developing the general dynamic equations for the plane motion of a rigid body. Notice also that, if we differentiate once and twice with respect to time, we obtain:

$$\sum_i m_i \dot{\mathbf{r}}_i' = 0 \quad \text{and} \quad \sum_i m_i \ddot{\mathbf{r}}_i' = 0.$$

12.6 Acceleration of the centre of gravity

Staying with our concept of a rigid body as developed in Section 12.5, we look firstly at the linear momentum of the body. For the ith particle, the linear momentum is $m_i \dot{\mathbf{r}}_i$. Hence, the linear momentum for the whole body is:

$$\sum_i m_i \dot{\mathbf{r}}_i = \sum_i m_i (\dot{\mathbf{r}}_G + \dot{\mathbf{r}}_i') = \sum_i m_i \dot{\mathbf{r}}_G + \sum_i m_i \dot{\mathbf{r}}_i' = \left(\sum_i m_i \right) \dot{\mathbf{r}}_G,$$

which is simply the total mass times the velocity of the centre of gravity. Furthermore, by differentiating with respect to time, we find that the rate of change of linear momentum is:

$$\left(\sum_i m_i \right) \ddot{\mathbf{r}}_G,$$

i.e. the total mass times the acceleration of the centre of gravity.

Next we consider the vector sum of the external forces acting on the body. This sum equals the vector sum of all the forces acting on all the particles of the body, since the vector sum of the internal forces (intermolecular forces) is zero. The sum of all the forces equals $\sum_i \mathbf{F}_i$, where $\mathbf{F}_i$ is the resultant force acting on the ith paticle. We then have:

$$\sum_i \mathbf{F}_i = \sum_i m_i \ddot{\mathbf{r}}_i = \sum_i m_i (\ddot{\mathbf{r}}_G + \ddot{\mathbf{r}}_i') = \left(\sum_i m_i \right) \ddot{\mathbf{r}}_G.$$

Hence, the vector sum of the external forces acting on the body is equal to its total mass times the acceleration of its centre of gravity.

EXERCISE 4

Figure 12.9 shows a circular drum on a steep slope of angle α to the horizontal. A rope is wrapped round the drum and the rope is pulled upwards parallel to the slope with a force of magnitude P. The

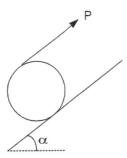

Figure 12.9. A circular drum on a steep slope.

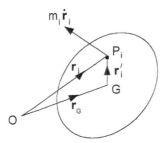

Figure 12.10. Finding the moment of momentum about an axis through O perpendicular to the plane of motion of the body.

drum is assumed to be slipping down the slope with a coefficient of kinetic friction μ. If m is the mass of the drum, find the acceleration of its centre of gravity.

12.7 General dynamic equations

We see from Section 12.2 (and referring here to Figure 12.10) that the moment of momentum of particle P_i about an axis through O perpendicular to the plane of motion is given in vector form as:

$$\mathbf{r}_i \times m_i \mathbf{v}_i = \mathbf{r}_i \times m_i \dot{\mathbf{r}}_i.$$

Hence, for the whole body, its *moment of momentum* about the same axis through O is given by taking the vector sum over all the particles of the body to obtain:

$$\mathbf{H}_\mathrm{O} = \sum_i \mathbf{r}_i \times m_i \dot{\mathbf{r}}_i = \sum_i (\mathbf{r}_\mathrm{G} + \mathbf{r}'_i) \times m_i (\dot{\mathbf{r}}_\mathrm{G} + \dot{\mathbf{r}}'_i)$$

$$= \sum_i \mathbf{r}_\mathrm{G} \times m_i \dot{\mathbf{r}}_\mathrm{G} + \sum_i \mathbf{r}_\mathrm{G} \times m_i \dot{\mathbf{r}}'_i + \sum_i \mathbf{r}'_i \times m_i \dot{\mathbf{r}}_\mathrm{G} + \sum_i \mathbf{r}'_i \times m_i \dot{\mathbf{r}}'_i$$

$$= \mathbf{r}_\mathrm{G} \times \left(\sum_i m_i \right) \dot{\mathbf{r}}_\mathrm{G} + \mathbf{r}_\mathrm{G} \times \left(\sum_i m_i \dot{\mathbf{r}}'_i \right) + \left(\sum_i m_i \mathbf{r}'_i \right) \times \dot{\mathbf{r}}_\mathrm{G} + \sum_i m_i \mathbf{r}'_i \times \dot{\mathbf{r}}'_i.$$

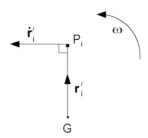

Figure 12.11. Motion of particle P_i relative to axis through G.

From Section 12.5, we see that the second and third terms are both zero. Hence,

$$\mathbf{H}_O = \mathbf{r}_G \times \left(\sum_i m_i \right) \dot{\mathbf{r}}_G + \sum_i m_i \mathbf{r}'_i \times \dot{\mathbf{r}}'_i.$$

Before commenting on this equation, let us examine the second term further. Referring now to Figure 12.11, the distance GP_i is constant and therefore the velocity of P_i relative to G must be perpendicular to $\mathbf{r}'_i$, i.e. $\dot{\mathbf{r}}'_i \perp \mathbf{r}'_i$. If the body is turning with angular velocity ω, the angular velocity of P_i relative to G is ω and therefore the magnitude of its relative velocity is $\dot{r}'_i = \omega r'_i$, remembering that ω is measured in rad/s. Since they are perpendicular, the magnitude of the vector product of the vectors $\mathbf{r}'_i$ and $\dot{\mathbf{r}}'_i$ is $|\mathbf{r}'_i \times \dot{\mathbf{r}}'_i| = r'^2_i \omega$. The direction of ω, the vector form of the angular velocity, is the same as that of $\mathbf{r}'_i \times \dot{\mathbf{r}}'_i$. Therefore,

$$\mathbf{r}'_i \times \dot{\mathbf{r}}'_i = r'^2_i \omega.$$

If we substitute the latter into our equation for moment of momentum, it becomes:

$$\mathbf{H}_O = \mathbf{r}_G \times \left(\sum_i m_i \right) \dot{\mathbf{r}}_G + \left(\sum_i m_i r'^2_i \right) \omega = \mathbf{r}_G \times M \dot{\mathbf{r}}_G + I_G \omega.$$

Here, $M = \sum_i m_i$, which is the total mass and

$$I_G = \sum_i m_i r'^2_i,$$

which is called the *moment of inertia* of the body about an axis through G perpendicular to the plane of motion.

If we now differentiate the moment of momentum with respect to time, we obtain the rate of change of moment of momentum:

$$\dot{\mathbf{H}}_O = \dot{\mathbf{r}}_G \times M \dot{\mathbf{r}}_G + \mathbf{r}_G \times M \ddot{\mathbf{r}}_G + I_G \dot{\omega} = \mathbf{r}_G \times M \ddot{\mathbf{r}}_G + I_G \dot{\omega},$$

since $\dot{\mathbf{r}}_G \times \dot{\mathbf{r}}_G = 0$.

To find the dynamic equation concerned with moments, we must consider the moments of forces about the axis through O perpendicular to the plane of motion. Let us indicate the sum of the moments of the external forces by the expression $\sum \mathbf{r}_e \times \mathbf{F}_e$, where $\mathbf{r}_e$ is the position vector from O of the point of application of $\mathbf{F}_e$. Since, the effects of internal forces cancel,

$$\sum \mathbf{r}_e \times \mathbf{F}_e = \sum_i \mathbf{r}_i \times \mathbf{F}_i,$$

where $\mathbf{F}_i$ is the resultant of all the forces acting on particle P_i,

$$= \sum_i \mathbf{r}_i \times m_i \ddot{\mathbf{r}}_i, \text{ from Newton's second law of motion,}$$

$$= \sum_i \dot{\mathbf{r}}_i \times m_i \dot{\mathbf{r}}_i + \sum_i \mathbf{r}_i \times m_i \ddot{\mathbf{r}}_i, \text{ since the first term is zero,}$$

$$= \frac{d}{dt}\left(\sum_i \mathbf{r}_i \times m_i \dot{\mathbf{r}}_i\right) = \frac{d}{dt}\mathbf{H}_O = \dot{\mathbf{H}}_O.$$

Summarizing the results of Sections 12.6 and 12.7, we have the following. The vector sum of the external forces,

$$\sum \mathbf{F}_e = M\ddot{\mathbf{r}}_G,$$

which is the total mass times the acceleration of the centre of gravity.

Then, the sum of the moments about O of the external forces,

$$\sum \mathbf{r}_e \times \mathbf{F}_e = \mathbf{r}_G \times M\ddot{\mathbf{r}}_G + I_G\dot{\omega},$$

which is the moment about O of the total mass times the acceleration of the centre of gravity plus the moment of inertia about G times the angular acceleration.

Since we are considering plane motion, the force equation may be resolved into two different directions, usually at right angles but not necessarily so, to give two scalar equations; the moment equation also corresponds to one scalar equation. These three equations are the general dynamic equations of a rigid body in plane motion.

EXERCISE 5

A uniform circular hoop of radius r is supported at a point O of its rim and can turn freely in its own plane about the point O. If the hoop is released from rest with its centre at the same height as O, find the angular velocity of the hoop as it swings through its lowest position.

Problems 109 and 110.

12.8 Moments of inertia

In Exercise 5 of Section 12.7, we saw that the moment of inertia of a circular hoop about an axis through its centre and perpendicular to its plane is simply its total mass times the square of its radius. We can use this as the starting point for finding the corresponding moment of inertia of a uniform circular disc. The circular strip shown in Figure 12.12

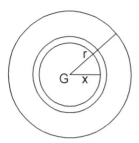

Figure 12.12. Finding the moment of inertia of a uniform circular disc.

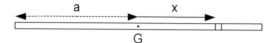

Figure 12.13. Finding the moment of inertia of a uniform rod.

is like a circular hoop of radius x. If the width of the strip is dx, its mass is:

$$\frac{2\pi x dx}{\pi r^2} M = \frac{2M}{r^2} x dx,$$

where M is the total mass of the disc. Hence, the moment of inertia of the strip about its centre G is:

$$\left(\frac{2M}{r^2} x dx\right) x^2 = \frac{2M}{r^2} x^3 dx.$$

The total moment of inertia for the disc is the sum over all such concentric discs as x goes from 0 to r, which is the integral:

$$I_G = \int_0^r \frac{2M}{r^2} x^3\, dx = \frac{2M}{r^2}\left[\frac{x^4}{4}\right]_0^r = Mr^2/2.$$

The moment of inertia is sometimes written as $I = Mk^2$, in which case k is called the *radius of gyration*. Thus, for a uniform circular disc, the radius of gyration is $k = r/\sqrt{2}$.

A uniform solid right circular cylinder can be considered as a whole lot of uniform circular discs stuck together. Hence, the moment of inertia about its axis has the same form, i.e. it is $I_G = Mr^2/2$, where M is now the mass of the solid cylinder.

Next, let us find the moment of inertia of a uniform straight rod about an axis through its centre and perpendicular to its length. For a small segment of length dx, where x is its distance from the centre G, as indicated in Figure 12.13, its mass is $(M/2a)dx$, where $2a$ is the length of the rod and M is its mass. The moment of inertia about G of the small segment is:

$$\left(\frac{dx}{2a} M\right) x^2 = \frac{M}{2a} x^2 dx.$$

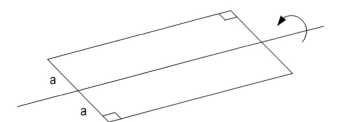

Figure 12.14. Finding the moment of inertia of a flat plate about an axis along its middle.

Summing over all the segments gives the total moment of inertia as the integral:

$$\int_{-a}^{a} \frac{M}{2a} x^2 \, dx = \frac{M}{2a} \left[\frac{x^3}{3} \right]_{-a}^{a} = \frac{Ma^2}{3} = Mk^2,$$

where the radius of gyration is $k = a/\sqrt{3}$.

Since a rectangular plate, as shown in Figure 12.14, is like a whole lot of rods stuck together, its moment of inertia about the axis shown in the diagram has exactly the same formula as the rod, i.e. $I_G = Ma^2/3$.

EXERCISE 6

Given that a sphere of radius r has surface area $4\pi r^2$, use the formula for the moment of inertia of a circular hoop to show that the moment of inertia of a uniform spherical shell about an axis through its centre is $I_G = 2Mr^2/3$, where M is its mass and r its radius.

EXERCISE 7

Given that a sphere of radius r has volume $4\pi r^3/3$, use the result of Exercise 6 to show that the moment of inertia of a uniform solid sphere about an axis through its centre is $I_G = 2Mr^2/5$. This time, M and r are the mass and radius of the solid sphere.

Problem 111.

12.9 Perpendicular axis theorem

The perpendicular axis theorem applies only to a plane lamina. However, it need not have uniform density and the axes need not pass through the centre of gravity.

Draw Cartesian axes (which of course are perpendicular to each other) in the plane of a lamina as in Figure 12.15. Let P_i be a particle of mass m_i at the point (x_i, y_i). Summing over all particles, the moment of inertia about the y-axis is $I_y = \sum_i m_i x_i^2$. Similarly, the moment of inertia about the x-axis is $I_x = \sum_i m_i y_i^2$.

Next, we look at the moment of inertia of the lamina about the axis through O perpendicular to its plane which is:

$$I_O = \sum_i m_i r_i^2 = \sum_i m_i \left(x_i^2 + y_i^2 \right) = \sum_i m_i x_i^2 + \sum_i m_i y_i^2 = I_x + I_y.$$

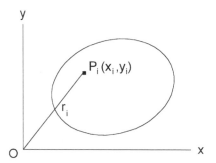

Figure 12.15. A lamina in the x, y plane.

EXERCISE 8

Given that the moment of inertia of a uniform circular lamina about an axis through its centre and perpendicular to its plane is $Mr^2/2$, find its moment of inertia about a diameter.

EXERCISE 9

Use the formula $Ma^2/3$ for the moment of inertia of a straight rod of length $2a$ to find the moment of inertia of a rectangular lamina of length $2a$ and width $2b$ about an axis through its centre and perpendicular to its plane.

Problems 112 and 113.

12.10 Rotation about a fixed axis

Let us continue Exercise 5 of Section 12.7 of a hoop swinging in its own vertical plane about a point O in its rim. By taking moments about O, we found an equation which gives the angular acceleration as:

$$\ddot{\theta} = \frac{g}{2r}\cos\theta.$$

Furthermore, on releasing the hoop from rest with the centre G at the same height as O, the equation for the angular velocity was found to be:

$$\dot{\theta}^2 = \frac{g}{r}\sin\theta.$$

Let us now examine the reaction at the hinge O during this motion. Since the acceleration of the centre of gravity G is given in terms of its radial and tangential components, it is convenient to express the reaction at the hinge in terms of corresponding components $\mathbf{R}_r$ and $\mathbf{R}_t$, as shown in Figure 12.16. The mass times the acceleration of the centre of gravity equals the vector sum of the external forces acting on the hoop. Thus, if we resolve in the direction of $-\mathbf{R}_t$ and then in the direction of $\mathbf{R}_r$, we obtain:

$$Mg\cos\theta - R_t = Mr\ddot{\theta} = \frac{Mg}{2}\cos\theta.$$

Therefore, $R_t = \dfrac{Mg}{2}\cos\theta.$

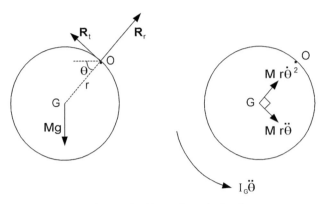

Figure 12.16. Reactions at the hinge of a swinging hoop.

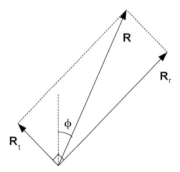

Figure 12.17. Finding the resultant reaction **R** at the hinge of a swinging hoop.

$$R_r - Mg \sin\theta = Mr\dot\theta^2 = Mg \sin\theta.$$

Therefore, $R_r = 2Mg \sin\theta$.

At the moment of release, the total reaction is vertically upwards and given by $R_t = Mg/2$. At the bottom of its swing, the reaction is again vertically upwards and given by $R_r = 2Mg$. In between, the total reaction **R** veers away from the vertical as in Figure 12.17. For instance, when $\theta = \pi/4$:

$$R_t = \frac{Mg}{2\sqrt{2}} \quad \text{and} \quad R_r = Mg\sqrt{2}.$$

Then if ϕ is the angle between the direction of **R** and the vertical,

$$\tan(\phi + 45°) = \frac{R_r}{R_t} = 4, \ \phi + 45° = 76° \quad \text{and} \quad \phi = 31°.$$

For a second example of rotation about a fixed axis, consider a uniform straight rod of length $2a$, hinged at one end to swing freely in a vertical plane, as illustrated in Figure 12.18. In Section 12.8, we found that the moment of inertia of the rod about its centre is $I_G = Ma^2/3$, where M is its mass.

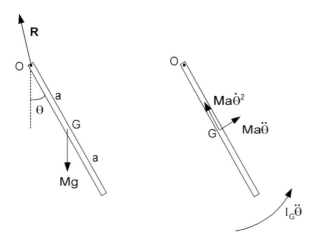

Figure 12.18. A rod swinging freely in a vertical plane about one end.

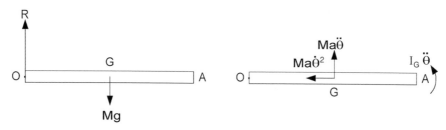

Figure 12.19. A swinging rod in a horizontal position.

If we are not interested in the hinge reaction **R**, we can investigate the motion of the rod by taking moments about the hinge O. The moment equation states that the sum of the moments of the external forces equals the moment of the mass times the acceleration of the centre of gravity plus the moment of inertia about the centre of gravity times the angular acceleration. Thus, in this case, the moment about O is:

$$-Mga\sin\theta = Ma\ddot{\theta} \cdot a + \tfrac{1}{3}Ma^2\ddot{\theta} = \tfrac{4}{3}Ma^2\ddot{\theta}.$$

Therefore, $\ddot{\theta} = -\dfrac{3g}{4a}\sin\theta$.

For small swings, $\sin\theta \approx \theta$ and $\ddot{\theta} = -\omega^2\theta$ with $\omega = \sqrt{3g/4a}$. Notice the difference from the simple pendulum in Exercise 5 of Section 11.4 in which $\omega = \sqrt{g/a}$, where a was the length of the pendulum. A pendulum, like the swinging rod, which involves moment of inertia, is referred to as a *compound pendulum*. In this particular case, the frequency of oscillation is $\omega/2\pi$, where $\omega = 0.866\sqrt{g/a}$.

Next, suppose that the free end A of the rod is released from rest with A at the same height as O, as in Figure 12.19. Let us find the acceleration of A and the hinge reaction immediately after the rod has been released.

The acceleration of A is $2a\ddot{\theta}$ and:

$$\ddot{\theta} = -\frac{3g}{4a} \sin\theta \text{ with } \theta = \frac{\pi}{2}.$$

Therefore, the acceleration of A is: $-3g/2$, where the negative sign just means that the acceleration is downwards. At the instant of release $\dot{\theta} = 0$, so the mass times the acceleration of G is vertical. The gravitational force Mg is also vertical, so the hinge reaction **R** must be vertical.

Hence, $R - Mg = Ma\ddot{\theta} = -\frac{3}{4}Mg$ and $R = \frac{1}{4}Mg$.

Now, immediately before the release, R would be half the weight, i.e. $Mg/2$. Thus, R immediately reduces by a half when A is released.

EXERCISE 10

Re-examine the uniform straight rod as a compound pendulum but this time with the point of suspension O a distance $b < a$ from the centre G. Find the value of b to maximize the frequency of small-amplitude oscillation and compare this frequency with that when O was at the end of the rod.

Problems 114 and 115.

12.11 General plane motion

In motion of a rigid body about a fixed axis, we were able to find the angular acceleration by taking moments about the axis. This involved only one equation but in general plane motion, we not only need a moment equation but also a vector force equation resolved into two scalar equations. The basic equations were summarized at the end of Section 12.7. To illustrate the application of these results, we shall study the motion under gravity of a uniform solid sphere on an inclined plane.

We shall start by considering the case where the sphere rolls down the plane without slipping. Let the mass of the sphere be M and the radius be r, so from Exercise 7 of Section 12.8, we see that its moment of inertia about an axis through its centre G is $2Mr^2/5$.

Referring to Figure 12.20, the forces acting on the sphere are its weight Mg and the reaction from the inclined plane at the point of contact P. The reaction has been indicated on the force diagram by two components, N normal to the plane and F, the friction force, along the plane. On the acceleration diagram, we show $M\dot{v}$, the mass times the acceleration of the centre of gravity, and $2Mr^2\dot{\omega}/5$, the moment of inertia times the angular acceleration (rad/s^2).

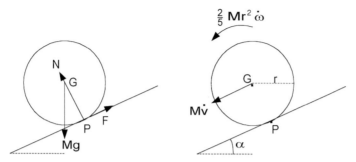

Figure 12.20. A sphere rolling down an inclined plane.

Resolving perpendicular to and down the plane and taking moments about G give the equations:

$$N - Mg \cos \alpha = 0$$
$$Mg \sin \alpha - F = M\dot{v}$$
$$Fr = \tfrac{2}{5} Mr^2 \dot{\omega}.$$

The moments have been taken about G but the same result would be obtained by taking moments about P. In the latter case we would have to include the moment of the mass times the acceleration of G.

In these three equations, we have the following four unknowns: N, F, $\dot{v}$ and $\dot{\omega}$. To find a fourth equation, we make use of the fact that, since the sphere is rolling without slipping, the point of contact P is an instantaneous centre of rotation. Hence, the velocity of G is $v = r\omega$ and differentiating with respect to time gives the required fourth equation:

$$\dot{v} = r\dot{\omega}.$$

Substituting the latter into the moment equation and dividing by r gives: $F = 2M\dot{v}/5$. Sustituting this into the second equation gives:

$$Mg \sin \alpha - \tfrac{2}{5} M\dot{v} = M\dot{v}.$$

Hence, the acceleration of the sphere down the slope is:

$$\dot{v} = \tfrac{5}{7} g \sin \alpha.$$

So far in this example, we have assumed that the sphere rolls down the plane without slipping. Let us now examine the relationship between the coefficient of static friction μ_s and the angle of elevation of the plane α for this to be true. In the non-slip case:

$$F \le \mu_s N = \mu_s Mg \cos \alpha$$
$$\text{and } F = \tfrac{2}{5} M\dot{v} = \tfrac{2}{7} Mg \sin \alpha.$$

Hence, it follows that $\tan\alpha \le 7\mu_s/2$. Alternatively, the sphere will slip as well as roll if:

$$\mu_s < \tfrac{2}{7}\tan\alpha.$$

Let us now consider the slip-and-roll situation, i.e with the coefficient of kinetic friction $\mu_k < \mu_s < \tfrac{2}{7}\tan\alpha$. The three dynamic equations:

$$N - Mg\cos\alpha = 0$$
$$Mg\sin\alpha - F = M\dot{v}$$
$$Fr = 2Mr^2\dot{\omega}/5$$

are the same as before but now:

$$F = \mu_k N.$$

Substituting the latter into the second equation and using the first equation for N gives:

$$M\dot{v} = Mg\sin\alpha - \mu_k Mg\cos\alpha \text{ or } \dot{v} = g(\sin\alpha - \mu_k\cos\alpha).$$

For the angular acceleration, we use the third equation and substitute for F as before:

$$2Mr\dot{\omega}/5 = \mu_k Mg\cos\alpha \text{ so } \dot{\omega} = \frac{5\mu_k g}{2r}\cos\alpha.$$

EXERCISE 11

Referring to Figure 12.21, a yo-yo has a light string wound round the inner core, which has radius r. Find the acceleration of the yo-yo if the end of the string is held fixed and the yo-yo is allowed to fall and unwind under its own weight, given that the radius of gyration of the yo-yo about its axis is k.

EXERCISE 12

A uniform circular disc of radius r is spinning about a horizontal axis through its centre and perpendicular to its plane when it is placed on a horizontal plane (see Figure 12.22). If the initial angular

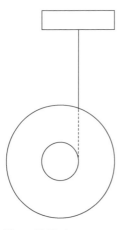

Figure 12.21. A yo-yo.

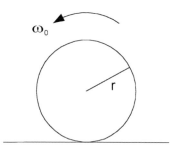

Figure 12.22. A skidding disc.

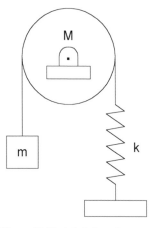

Figure 12.23. A belt hanging over a drum and attached to a weight at one end and a spring at the other.

velocity is ω_0 and the coefficient of kinetic friction with the surface is μ_k, find what horizontal velocity the disc will eventually develop and how far it will travel before skidding ceases.

Problems 116 and 117.

12.12 More exercises

EXERCISE 13

A flywheel with moment of inertia about its axis corresponding to that of a uniform circular disc has radius 1 m and mass 120 kg. Find the magnitude of a constant torque which will accelerate the flywheel from rest to 300 rpm in 20 s.

EXERCISE 14

A uniform solid circular drum can rotate freely about its horizontal axis. A light belt hangs over the drum carrying a mass m at one end and attached to a spring of constant k at the other, as shown in Figure 12.23. Let $m = 30$ kg, the mass of the drum $M = 120$ kg and $k = 1$ N/mm. If m is pulled

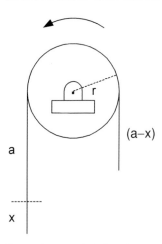

Figure 12.24. A chain hanging over a pulley.

down from its equilibrium position and then released, it will oscillate up and down. Find the period of oscillation, assuming that the belt does not slip on the drum.

EXERCISE 15

A pulley wheel of mass M and radius r has moment of inertia corresponding to that of a uniform circular disc. It carries a uniform chain (see Figure 12.24) of length $2a + \pi r$ and mass per unit length of σ. Assuming no friction in the pulley axle and no slip of the chain on the pulley, find x (see diagram) as a function of time for $x < a$ if the system is released from rest with a small $x = x_0 > 0$.

EXERCISE 16

A bicycle is ridden round a corner with radius of curvature of 3 m at a speed of 3 m/s. Find the angle by which the cyclist must lean over from the vertical towards the centre of the curve.

EXERCISE 17

A uniform solid sphere and a uniform solid right circular cylinder are released from rest at the same time and same height and both roll down the same slope without slipping. When the sphere reaches the bottom, the cylinder has still 1 m to go. Find the length of the slope.

EXERCISE 18

A uniform solid sphere rolls to and fro in the bottom of a hollow circular cylinder. The axis of the cylinder is horizontal and the sphere rolls in a vertical plane perpendicular to the axis. Find the period of small-amplitude oscillation given that the radii of the sphere and cylinder are 0.1 m and 0.4 m, respectively.

EXERCISE 19

A uniform solid right circular cylinder has its axis horizontal. A string is wound round it and then attached to a fixed point A as shown in Figure 12.25. The string unwinds as the cylinder turns and

Figure 12.25. A cylinder slipping down an inclined plane.

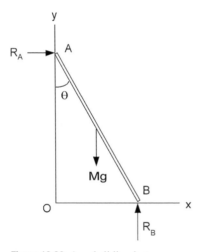

Figure 12.26. A rod sliding between a smooth wall and a smooth floor.

slips down a plane inclined at 60° to the horizontal. Find the acceleration of the centre of the cylinder, given that the coefficient of kinetic friction between the cylinder and the plane is $\mu_k = 1/3$, assuming no slipping between string and cylinder.

EXERCISE 20

A uniform straight rod is constrained to stay in a vertical plane perpendicular to a vertical wall and horizontal floor, as shown in Figure 12.26. Let the wall and floor be perfectly smooth so that the rod will slip down the wall and along the floor. If the motion starts from rest with the rod almost vertical, show that the reactions R_A and R_B (see the diagram) are related to the angle θ by the equations:

$$R_A = \tfrac{3}{4}Mg(3\cos\theta - 2)\sin\theta \quad \text{and} \quad R_B = \tfrac{1}{4}Mg(1 - 3\cos\theta)^2.$$

12.13 Answers to exercises

1. Let the horizontal and vertical tracks coincide with the x- and y-axes. Drawing the perpendiculars to the horizontal and vertical paths at the ends of the rod gives I as the instantaneous centre of rotation, as shown in Figure 12.27.

    The coordinates of I are $x = a \sin \theta$ and $y = a \cos \theta$. Hence, the equation for the path followed by I is: $x^2 + y^2 = a^2(\sin^2 \theta + \cos^2 \theta) = a^2$, i.e. it is a quarter circle of radius a and centre of origin O.

2. Referring to Figure 12.28, the centre C of the disc moves horizontally and therefore the instantaneous centre of rotation must lie on the vertical line through C. P is the point of the disc in contact with the horizontal surface at that instant. Hence, that point of the disc has zero velocity at that instant. That point P must therefore be the instantaneous centre of rotation.

    Note: If any point of a moving body is instantaneously at rest, that point must be the instantaneous centre of rotation.

3. Referring to Figure 12.29, since the disc is rolling (without slipping), I, the point of contact with the horizontal surface, must be stationary at that instant. Hence, I is the instantaneous centre of rotation and every point of the disc is rotating about I with angular velocity ω at that instant.

    Since the velocity of the centre is v, the angular velocity of the disc is $\omega = v/r$. P is also rotating about I at that instant, so its velocity **u** is perpendicular to the line IP, i.e. it is directed downwards from the horizontal at an angle of 45°. Its magnitude is:

$$u = IP \cdot \omega = \sqrt{2}r \cdot v/r = \sqrt{2}v.$$

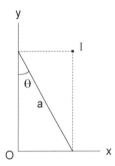

Figure 12.27. Instantaneous centre of rotation I for a rod with its ends on horizontal and vertical tracks.

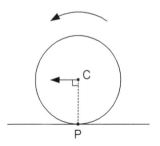

Figure 12.28. A rolling disc.

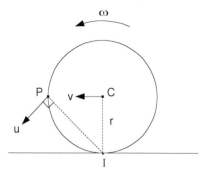

Figure 12.29. Finding the velocity of point P of a rolling disc.

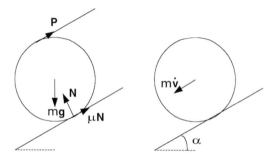

Figure 12.30. A drum sliding down an inclined plane.

4. The left-hand diagram of Figure 12.30 shows the external forces acting on the drum. Their vector sum equals the mass times the acceleration of the centre of gravity, which is shown as $m\dot{v}$ in the right-hand diagram.

By resolving forces perpendicular to and parallel to the plane, we obtain the following two equations:

$$N - mg\cos\alpha = 0$$
$$mg\sin\alpha - \mu N - P = m\dot{v}.$$

Substituting for N from the first equation and then dividing by m gives:

$$\dot{v} = g(\sin\alpha - \mu\cos\alpha) - P/m.$$

5. In a uniform circular hoop of radius r, all of its particles are at distance r from the centre, which in turn is obviously the centre of gravity position G. The moment of inertia about an axis through G and perpendicular to the plane of the hoop is:

$$I_G = \sum_i m_i r_i^2 = \sum_i m_i r^2 = Mr^2.$$

As the hoop swings as shown in Figure 12.31, G moves in a circular arc of radius r about O. If θ is the angle in radians of the line GO down from the horizontal, G will have a component of acceleration $r\omega^2 = r\dot{\theta}^2$ towards O and a tangential component of acceleration $r\dot{\omega} = r\ddot{\theta}$. In order to write down the scalar version of $\mathbf{r}_G \times M\ddot{\mathbf{r}}_G$, all we need to do is to take the sum of the moments of these two

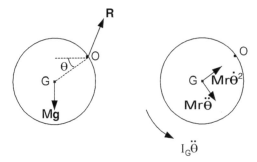

Figure 12.31. A hoop swinging in its own vertical plane about a point on its rim.

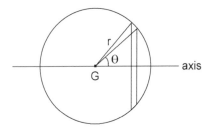

Figure 12.32. Finding the moment of inertia of a spherical shell.

orthogonal components of acceleration about O and multiply by M. Only the tangential component has a moment so:

$$|\mathbf{r}_G \times M\ddot{\mathbf{r}}_G| = r \cdot Mr\ddot{\theta} = Mr^2\ddot{\theta}.$$

Also, $|I_G\dot{\omega}| = Mr^2\ddot{\theta}.$

In the force diagram, **R** represents the hinge reaction at O. Thus, if we take moments about O, we have:

$$Mgr\cos\theta = Mr^2\ddot{\theta} + Mr^2\ddot{\theta} \text{ or } 2r\ddot{\theta} = 2r\dot{\theta}\frac{d\dot{\theta}}{d\theta} = g\cos\theta.$$

Therefore, $\displaystyle\int 2r\dot{\theta}\,d\dot{\theta} = \int g\cos\theta\,d\theta$, i.e. $r\dot{\theta}^2 = g\sin\theta + C.$

Since $\dot{\theta} = 0$ when $\theta = 0$, $C = 0$ and $\dot{\theta}^2 = \dfrac{g}{r}\sin\theta.$

At the lowest position, $\theta = \pi/2$ and $\dot{\theta} = \sqrt{g/r}$.
    If $r = 0.2$ m, say, then $\dot{\theta} = \sqrt{9.8/0.2} = \sqrt{49} = 7$ rad/s.

6.   Start by taking a thin slice through the spherical shell and perpendicular to the axis about which the moment of inertia is to be found, as shown in Figure 12.32. If θ is the angle subtended at the centre G as shown in the diagram, the width of the strip may be denoted by $rd\theta$. The radius of the strip is $r\sin\theta$, so its area is: $2\pi r\sin\theta \cdot rd\theta = 2\pi r^2 \sin\theta d\theta$. Regarding the strip like a circular hoop, its moment of inertia about the axis is:

$$\frac{2\pi r^2 \sin\theta d\theta}{4\pi r^2} \cdot M \cdot (r\sin\theta)^2 = \tfrac{1}{2}Mr^2\sin^3\theta d\theta.$$

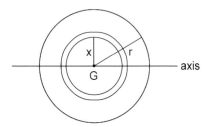

Figure 12.33. Finding the moment of inertia of a uniform solid sphere.

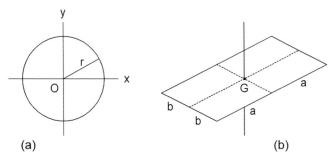

(a)   (b)

Figure 12.34. Using the perpendicular axis theorem for moments of inertia.

Now, we sum over all such slices as θ changes from 0 to π to obtain the total moment of inertia as the integral:

$$\int_0^\pi \frac{1}{2} Mr^2 \sin^3\theta \, d\theta = \frac{Mr^2}{2} \int_0^\pi (1 - \cos^2\theta) \sin\theta \, d\theta$$

$$= \frac{Mr^2}{2} \left[-\cos\theta + \frac{1}{3}\cos^3\theta \right]_0^\pi = \frac{Mr^2}{2} \frac{4}{3} = \frac{2}{3} Mr^2.$$

In this case, the radius of gyration is $k = \sqrt{2/3}\,r$.

7.   To find the moment of inertia of a uniform solid sphere, we regard it as a collection of concentric spherical shells. Referring to Figure 12.33, the mass of a shell of radius x and thickness dx is:

$$\frac{4\pi x^2 dx}{4\pi r^3/3} M = \frac{3M}{r^3} x^2 dx.$$

Summing over all such shells as x changes from 0 to r gives the total moment of inertia as the integral:

$$\int_0^r \frac{2}{3} x^2 \frac{3M}{r^3} x^2 \, dx = \frac{2M}{r^3} \int_0^r x^4 \, dx = \frac{2M}{r^3} \frac{r^5}{5} = \frac{2}{5} Mr^2.$$

In this case the radius of gyration is $\sqrt{2/5}\,r$.

8.   Referring to Figure 12.34a and using the perpendicular axis theorem:

$$I_O = I_x + I_y.$$

But $I_x = I_y$ and therefore, $I_O = 2I_x$.

Hence, $I_x = \frac{1}{2} I_O = \frac{1}{2}\frac{1}{2} Mr^2 = \frac{1}{4} Mr^2.$

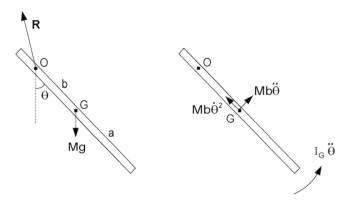

Figure 12.35. A rod swinging as a pendulum about point O.

9.   Referring to Figure 12.34b, the moment of inertia about an axis parallel to side of length $2a$ is $Mb^2/3$. The moment of inertia about an axis parallel to side of length $2b$ is $Ma^2/3$. Therefore, the moment of inertia about a perpendicular axis through G is:

$$I_G = Ma^2/3 + Mb^2/3 = M(a^2 + b^2)/3.$$

10.   We are not interested in the hinge reaction **R** so, referring to Figure 12.35, we just take moments about O:

$$-Mgb\sin\theta = Mb\ddot\theta \cdot b + \tfrac{1}{3}Ma^2\ddot\theta = \tfrac{1}{3}M(a^2 + 3b^2)\ddot\theta.$$

Hence, $\ddot\theta = -\dfrac{3gb}{a^2 + 3b^2}\sin\theta \approx -\omega^2\theta,$

for small oscillations with:

$$\omega^2 = \frac{3gb}{a^2 + 3b^2}.$$

To maximize the frequency with respect to b, we find the stationary value for ω^2 as follows:

$$\frac{d(\omega^2)}{db} = \frac{3g(a^2 + 3b^2) - 18gb^2}{(a^2 + 3b^2)^2} = 0$$

if $a^2 + 3b^2 - 6b^2 = a^2 - 3b^2 = 0$, i.e. $b^2 = a^2/3$, so $b = 0.577a$.

Substituting back into the expression for ω^2 gives:

$$\omega^2 = \frac{3ga/\sqrt3}{a^2 + a^2} = \frac{\sqrt3}{2}\frac{g}{a} \quad \text{and} \quad \omega = 0.931\sqrt{g/a}.$$

Therefore, the ratio of the new frequency to the old one is $0.931/0.866 = 1.075$.

It is worth noting that since $b = 0.577a$ gives a stationary value of ω^2 with respect to b, small changes in b from that value will produce negligible changes in ω. Hence, high accuracy in the time keeping of a clock is obtained by choosing the suspension point of the pendulum in this way. Such a pendulum is referred to as a *Schuler pendulum* after the name of its inventor.

11.   The force and acceleration diagrams for the yo-yo are shown in Figure 12.36. The vertical force and moment about G equations are, respectively:

$$Mg - T = M\dot v$$

$$Tr = Mk^2\dot\omega.$$

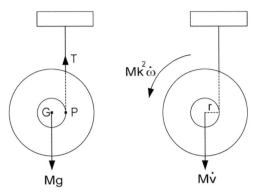

Figure 12.36. A yo-yo accelerating downwards.

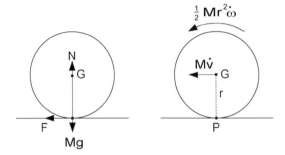

Figure 12.37. Force and acceleration diagrams for a skidding disc.

P is an instantaneous centre of rotation, so $v = r\omega$ and $\dot{v} = r\dot{\omega}$.

Therefore, $Mg - T = Mg - M\dfrac{k^2}{r}\dot{\omega} = Mg - M\dfrac{k^2}{r^2}\dot{v} = M\dot{v}$.

Hence, $\left(1 + \dfrac{k^2}{r^2}\right)\dot{v} = \dfrac{r^2 + k^2}{r^2}\dot{v} = g$ and $\dot{v} = \dfrac{r^2}{r^2 + k^2}g$.

12.   Referring to Figure 12.37, we see that the vertical and horizontal force equations and the moment about G equation are, respectively:

$N - Mg = 0$

$F = M\dot{v}$

$-Fr = \tfrac{1}{2}Mr^2\dot{\omega}$.

Add to these the friction equation: $F = \mu_k N$ and we deduce the following:

$F = \mu_k Mg = M\dot{v}$ so $\dot{v} = \mu_k g$.

Also, $-F = -\mu_k Mg = Mr\dot{\omega}/2$ so $r\dot{\omega} = -2\mu_k g$.

Then, integrating with respect to time gives:

$v = \mu_k gt$ and $r\omega = r\omega_0 - 2\mu_k gt$.

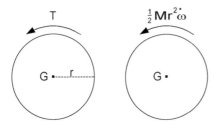

Figure 12.38. Accelerating a flywheel.

Skidding will cease when P becomes an instantaneous centre of rotation, which is when $v = r\omega$. This happens after time t when:

$$\mu_{\mathrm{k}} g t = r\omega_0 - 2\mu_{\mathrm{k}} g t, \text{ i.e. } t = \frac{r\omega_0}{3\mu_{\mathrm{k}} g}.$$

Then, the velocity developed is:

$$v = \mu_{\mathrm{k}} g \frac{r\omega_0}{3\mu_{\mathrm{k}} g} = \frac{r\omega_0}{3}.$$

The distance travelled before skidding ceases is:

$$x = \int_0^{r\omega_0/3\mu_{\mathrm{k}} g} \mu_{\mathrm{k}} g t \, dt = \mu_{\mathrm{k}} g \left[\frac{t^2}{2}\right]_0^{r\omega_0/3\mu_{\mathrm{k}} g} = \frac{r^2\omega_0^2}{18\mu_{\mathrm{k}} g}.$$

13. Referring to Figure 12.38, the moment equation is:

$$T = \frac{Mr^2}{2}\dot{\omega}.$$

Then, $\dot{\omega}$ is required in rad/s^2, hence,

$$\dot{\omega} = \frac{300 \times 2\pi}{60 \times 20} = \frac{\pi}{2}.$$

Finally, the torque $T = \dfrac{120}{2}\dfrac{\pi}{2} = 30\pi \text{ N m}.$

14. Referring to Figure 12.39, in static equilibrium: $kx_0 = mg$.
  For the mass m: $mg - T_1 = m\ddot{x}$.
  For the drum, taking moments about G: $(T_1 - T_2)r = Mr^2\dot{\omega}/2$.
  For the spring: $T_2 = k(x_0 + x)$.
  Also, $\dot{x} = r\omega$ and $\ddot{x} = r\dot{\omega}$. Hence, $T_1 - T_2 = M\ddot{x}/2$.
  From the other two equations:

$$T_1 - T_2 = mg - m\ddot{x} - kx_0 - kx = -m\ddot{x} - kx.$$

Consequently, $M\ddot{x}/2 = -m\ddot{x} - kx$ and $\ddot{x} = -\omega^2 x$,

where $\omega^2 = \dfrac{k}{m + M/2} = \dfrac{10^3}{30 + 60} = \dfrac{100}{9}, \quad \omega = \dfrac{10}{3}.$

Since ω is in rad/s, the period is: $2\pi/\omega = 6\pi/10 = 1.88 \text{ s}.$

15. Referring to Figure 12.40, for the section of chain on the left, the force holding it up is the tension T_1 as it leaves the wheel. Hence, for that section of chain:

$$(a + x)\sigma g - T_1 = (a + x)\sigma r\dot{\omega}.$$

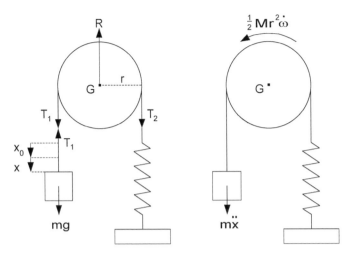

Figure 12.39. A weight, drum and spring system.

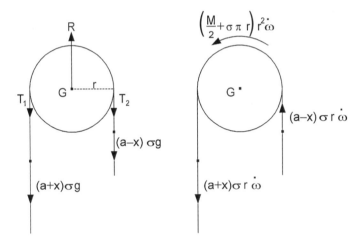

Figure 12.40. A chain hanging over a pulley wheel.

For the section of chain on the right:

$$T_2 - (a - x)\sigma g = (a - x)\sigma r\dot{\omega}.$$

Finally, for the pulley and the section of chain on top of it, taking moments about G:

$$(T_1 - T_2)r = (M/2 + \sigma \pi r)r^2\dot{\omega}.$$

From the first two equations:

$$T_1 - T_2 = (a + x)\sigma g - (a + x)\sigma r\dot{\omega} - (a - x)\sigma g - (a - x)\sigma r\dot{\omega} = 2x\sigma g - 2a\sigma r\dot{\omega}$$
$$= (M/2 + \sigma \pi r)r\dot{\omega}, \text{ from the moment equation.}$$

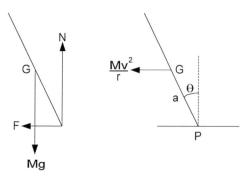

Figure 12.41. Forces on a cyclist rounding a corner.

Now, $\dot{x} = r\omega$ and $\ddot{x} = r\dot{\omega}$, so we can write the previous equation as:

$$(M/2 + \sigma\pi r + 2a\sigma)\ddot{x} = 2\sigma g x$$

or simply $\ddot{x} - k^2 x = 0$,

where $k^2 = \dfrac{2\sigma g}{M/2 + \sigma(\pi r + 2a)}$.

Using the D operator for d/dt, the second order differential equation becomes: $(D^2 - k^2)x = (D + k)(D - k)x = 0$. The solution is: $x = A\exp kt + B\exp -kt$, where A and B are constants, and $\dot{x} = Ak\exp kt - Bk\exp -kt$. At $t = 0$, $x = x_0$ and $\dot{x} = 0$. Hence, $A + B = x_0$ and $A - B = 0$. Therefore, $A = B = x_0/2$ and the solution is: $x = x_0(\exp kt + \exp -kt)/2 = x_0\cosh(kt)$ (see Section 22.2).

16.   Denote the bicycle speed by v and the radius of curvature of the corner by r, so that the acceleration inwards from the curve is v^2/r. With N and F denoting the normal and frictional components of the reaction force from the ground see (Figure 12.41), we have the following three dynamical equations found by resolving forces upwards and then to the left, and taking moments about G:

$$N - Mg = 0$$
$$F = Mv^2/r$$
$$Na\sin\theta - Fa\cos\theta = 0.$$

Hence, $\tan\theta = \dfrac{F}{N} = \dfrac{Mv^2}{r}\dfrac{1}{Mg} = \dfrac{v^2}{rg} = \dfrac{9}{3 \times 9.8}$ and $\theta = 17°$.

17.   Refer to Figure 12.42 to obtain the following three dynamical equations for the sphere, by resolving forces perpendicular to and then down the plane, and taking moments about G:

$$N - Mg\cos\alpha = 0$$
$$Mg\sin\alpha - F = M\dot{v}$$
$$Fr = \tfrac{2}{5}Mr^2\dot{\omega}.$$

Also, since P is the instantaneous centre of rotation: $v = r\omega$ and $\dot{v} = r\dot{\omega}$. It follows that:

$$F = \tfrac{2}{5}M\dot{v} = Mg\sin\alpha - M\dot{v}, \quad \dot{v} = \tfrac{5}{7}g\sin\alpha.$$

Let time $t = 0$ when it starts to roll, then $v = 0$ and $s = 0$ when $t = 0$, where s is the distance travelled. Integrating $\dot{v}$ with respect to t gives $v = (5/7)gt\sin\alpha$ and integrating again gives

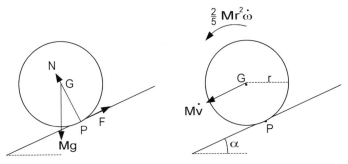

Figure 12.42. A sphere rolling down a plane.

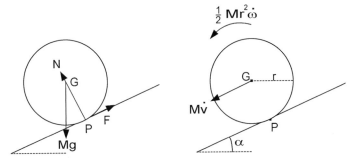

Figure 12.43. A cylinder rolling down a plane.

$s = (5/14)gt^2 \sin\alpha$. Notice that this result is independent of M and r. Hence, there is no loss of generality if we use the same symbols to study the motion of the cylinder.

Referring to Figure 12.43, the corresponding three dynamical equations for the cylinder are:

$$N - Mg\cos\alpha = 0$$
$$Mg\sin\alpha - F = M\dot{v}$$
$$Fr = \tfrac{1}{2}Mr^2\dot{\omega}.$$

Also, as before: $\dot{v} = r\dot{\omega}$.

Therefore, $F = \tfrac{1}{2}M\dot{v} = Mg\sin\alpha - M\dot{v}$, $\dot{v} = \tfrac{2}{3}g\sin\alpha$.

Integrating: $v = \tfrac{2}{3}gt\sin\alpha$ and $s = \tfrac{1}{3}gt^2\sin\alpha$.

Now, let $t = T$ be the time when the sphere reaches the bottom of the slope after travelling a distance $s = S$. It follows that:

$$S = \frac{5}{14}gT^2\sin\alpha \quad \text{and} \quad S - 1 = \frac{1}{3}gT^2\sin\alpha.$$

Hence, $\dfrac{S}{S-1} = \dfrac{15}{14}$ and $S = 15\,m$.

18.   Although the question is only concerned with small-amplitude oscillations, we start as shown in Figure 12.44 by formulating the dynamical equations for larger θ and make the small-angle approximations after that. Notice also that G follows a circular path of radius $d = R - r$. Furthermore, since P is the instantaneous centre of rotation of the sphere, the velocity of G is: $d\dot{\theta} = -r\omega$. Also, $d\ddot{\theta} = -r\dot{\omega}$.

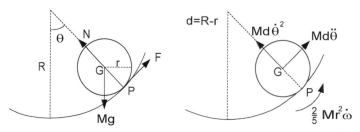

Figure 12.44. A sphere rolling in the bottom of a hollow circular cylinder.

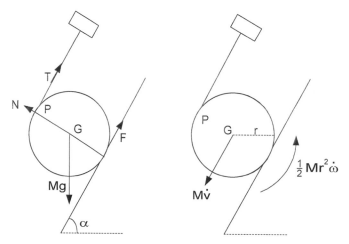

Figure 12.45. A cylinder sliding down an inclined plane.

Resolving forces normal to and tangential to the surface of the cylinder and taking moments about G give:

$$N - Mg \cos \theta = M d \dot{\theta}^2$$

$$F - Mg \sin \theta = M d \ddot{\theta}$$

$$Fr = \frac{2}{5} Mr^2 \left(-\frac{d}{r} \ddot{\theta} \right).$$

Therefore, $F = Mg \sin \theta + M d \ddot{\theta} = -\frac{2}{5} M d \ddot{\theta}$ and $\ddot{\theta} = -\frac{5g}{7d} \sin \theta$.

For small-amplitude oscillations, $\sin \theta = \theta$ and $\ddot{\theta} = -\omega^2 \theta$, where $\omega^2 = 5g/7d$. The period of oscillation is:

$$\frac{2\pi}{\omega} = 2\pi \sqrt{\frac{7 \times 0.3}{5 \times 9.8}} = 1.3 \text{ s}.$$

19.   Referring to Figure 12.45, P is an instantaneous centre of rotation so $v = -r\omega$ and $\dot{v} = -r\dot{\omega}$. Also, $F = \mu_k N$.

Resolving forces perpendicular to and down the plane, and taking moments about G:

$$N - Mg \cos \alpha = 0$$

$$Mg \sin \alpha - F - T = M \dot{v}$$

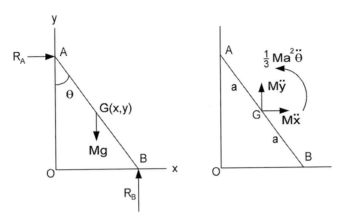

Figure 12.46. A rod sliding down and along a smooth wall and smooth floor.

$$(F - T)r = \frac{1}{2}Mr^2\left(-\frac{\dot{v}}{r}\right)$$

Therefore, $M\dot{v} = 2T - 2F = 2T - 2\mu_k N = 2T - 2\mu_k Mg\cos\alpha$.

Also, $2M\dot{v} = -2T - 2F + 2Mg\sin\alpha = -2T - 2\mu_k Mg\cos\alpha + 2Mg\sin\alpha$.

Adding gives $3M\dot{v} = 2Mg(\sin\alpha - 2\mu_k\cos\alpha)$.

If we now substitute $\alpha = 60°$ and $\mu_k = 1/3$, we find that:

$$\dot{v} = \frac{2}{3}\left(\frac{\sqrt{3}}{2} - \frac{1}{3}\right)g = 0.355\,g.$$

20.   Let G have coordinates (x, y), as shown in Figure 12.46. Resolving forces vertically and horizontally, and taking moments about G:

$$R_B - Mg = M\ddot{y}$$

$$R_A = M\ddot{x}$$

$$R_B a\sin\theta - R_A a\cos\theta = \frac{1}{3}Ma^2\ddot{\theta}.$$

Now, $x = a\sin\theta$, $\dot{x} = a\cos\theta \cdot \dot{\theta}$, $\ddot{x} = -a\sin\theta \cdot \dot{\theta}^2 + a\cos\theta \cdot \ddot{\theta}$

and $y = a\cos\theta$, $\dot{y} = -a\sin\theta \cdot \dot{\theta}$, $\ddot{y} = -a\cos\theta \cdot \dot{\theta}^2 - a\sin\theta \cdot \ddot{\theta} \cdot$

Thus, $R_B = Mg - Ma(\cos\theta \cdot \dot{\theta}^2 + \sin\theta \cdot \ddot{\theta})$,

$$R_A = -Ma(\sin\theta \cdot \dot{\theta}^2 - \cos\theta \cdot \ddot{\theta})$$

and the third dynamical equation becomes:

$$Mga\sin\theta - Ma(\cos\theta \cdot \dot{\theta}^2 + \sin\theta \cdot \ddot{\theta})a\sin\theta + Ma(\sin\theta \cdot \dot{\theta}^2 - \cos\theta \cdot \ddot{\theta})a\cos\theta = \frac{1}{3}Ma^2\ddot{\theta}.$$

Ma and the $\dot{\theta}^2$ terms cancel leaving:

$$\tfrac{1}{3}a\ddot{\theta} + a\ddot{\theta} = \tfrac{4}{3}a\ddot{\theta} = g\sin\theta \ or \ \ddot{\theta} = \frac{3g}{4a}\sin\theta.$$

Now, $\ddot{\theta} = \dot{\theta}\dfrac{d\dot{\theta}}{d\theta}$, so $\displaystyle\int\dot{\theta}\,d\dot{\theta} = \frac{3g}{4a}\int\sin\theta\,d\theta$ and $\ \dfrac{1}{2}\dot{\theta}^2 = -\dfrac{3g}{4a}\cos\theta + C.$

$\dot{\theta} = 0$ when $\theta = 0$, so $C = \dfrac{3g}{4a}$. Therefore, $\dot{\theta}^2 = \dfrac{3g}{2a}(1 - \cos\theta)$.

We can now substitute for $\dot{\theta}^2$ and $\ddot{\theta}$ in the equations for R_A and R_B to give:

$$R_A = -Ma\sin\theta \cdot \frac{3g}{2a}(1-\cos\theta) + Ma\cos\theta \cdot \frac{3g}{4a}\sin\theta$$
$$= \tfrac{3}{4}Mg\sin\theta \cdot (3\cos\theta - 2) \quad \text{and}$$
$$R_B = Mg - Ma\cos\theta \cdot \frac{3g}{2a}(1-\cos\theta) - Ma\sin\theta \cdot \frac{3g}{4a}\sin\theta$$
$$= Mg(1 - \tfrac{3}{2}\cos\theta + \tfrac{3}{2}\cos^2\theta - \tfrac{3}{4} + \tfrac{3}{4}\cos^2\theta)$$
$$= \tfrac{1}{4}Mg(1 - 6\cos\theta + 9\cos^2\theta) = \tfrac{1}{4}Mg(1 - 3\cos\theta)^2.$$

13 Impulse and momentum

13.1 Definition of impulse and simple applications

If a force $\mathbf{F}$ is constant in both magnitude and direction, then its *impulse* over a period of time from t_1 to t_2 is defined as:

$$\mathbf{I} = (t_2 - t_1)\mathbf{F}.$$

If $\mathbf{F}$ varies with time in magnitude or direction or both, then the impulse:

$$\mathbf{I} = \int_{t_1}^{t_2} \mathbf{F}\, dt.$$

The important measure of an impulse is its effect. To appreciate this, let us start by considering the force $\mathbf{F}$ applied to a particle of mass m. By Newton's second law of motion:

$$\mathbf{F} = m\dot{\mathbf{v}} = m\frac{d\mathbf{v}}{dt}$$

and $$\mathbf{I} = \int_{t_1}^{t_2} m\frac{d\mathbf{v}}{dt}\, dt = \int_{\mathbf{v}_1}^{\mathbf{v}_2} m\, d\mathbf{v} = m\mathbf{v}_2 - m\mathbf{v}_1,$$

where $\mathbf{v}_1$ and $\mathbf{v}_2$ are the velocities of the particle at times t_1 and t_2, respectively. Hence, in this case, the measure of the impulse is the change in momentum of the particle over the period of duration of the impulse.

Since an impulse is basically force multiplied by time, its unit of magnitude is newton second (N s).

Usually an impulse is provided by a very large force acting over a very short time. Such impulses occur in an explosion or a collision. Let us consider examples to illustrate these.

If a shell of mass m is fired with a velocity of v from a gun of mass M, find the initial velocity of recoil V of the gun. Although the explosive force F will vary with time as the shell is projected along the barrel of the gun as shown in Figure 13.1, it will be applied equally in opposite directions to the shell and the gun. Hence, the two

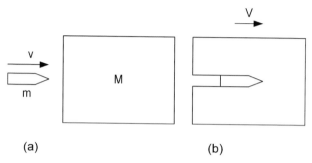

Figure 13.1. A shell being fired from a gun.

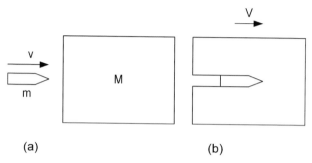

(a)   (b)

Figure 13.2. A bullet fired into a block of wood.

impulses have the same magnitude and impart the same changes in momentum, albeit in opposite directions. Thus:

$$MV = mv \quad \text{and} \quad V = \frac{m}{M}v.$$

As another example, suppose we have a block of wood of mass 2 kg resting on a smooth horizontal surface. A bullet of mass 10 g is fired horizontally into the block with a velocity v, as shown in Figure 13.2a. If the bullet embeds itself in the block, as shown in Figure 13.2b, and imparts a velocity of 3 m/s to it, find the initial velocity v of the bullet.

Since action and reaction are equal and opposite (Newton's third law of motion), the impulse supplied by the bullet to the block is equal and opposite to that received by the bullet from the block. Thus, if the magnitude of the impulse is I, then:

$$I = MV \quad \text{and} \quad I = mv - mV.$$

Therefore, $v = \dfrac{M+m}{m}V = \dfrac{2.01}{0.01} \times 3 = 603 \text{ m/s}.$

Note: When there is a collision between two particles (or two bodies which may be treated like particles as in the last example), since the change in momentum of one particle is equal and opposite to the change in momentum of the other, the total momentum of the two particles remains unchanged. This is sometimes referred to as the *principle of conservation of momentum.*

EXERCISE 1

Suppose that two particles travelling at right angles to each other collide and coalesce, as indicated in Figure 13.3. Let one particle have mass m and velocity $2v$, and the other have mass $3m$ and velocity v.

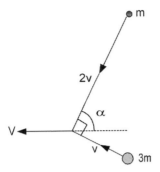

Figure 13.3. Two particles collide and coalesce.

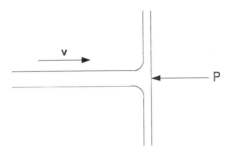

Figure 13.4. Water jet playing on a wall.

Find the magnitude V and direction α (see diagram) of the velocity of the composite particle after the impact.

   Problems 118, 119 and 120.

13.2 Pressure of a water jet

Let us examine the force on a wall exerted by a water jet with cross-sectional area A, velocity $\mathbf{v}$ and density ρ. The force on the wall is equal and opposite to the reaction force $\mathbf{P}$ from the wall (see Figure 13.4). The latter force destroys the momentum of the water in the direction perpendicular to the wall.

   Over a period of one second, the impulse provided by $\mathbf{P}$ is:

$$\mathbf{I} = \int_0^1 \mathbf{P}\,dt = \mathbf{P}.$$

This equates to the change in momentum of the water in this direction during this time, i.e.

$$\mathbf{P} = Av\rho(\mathbf{0} - \mathbf{v}) = -A\rho v\mathbf{v},$$

since the mass of water striking the wall in one second is $Av\rho$. Hence, the magnitude of the force exerted on the wall by the jet is: $A\rho v^2$.

EXERCISE 2

Find the force exerted on a wall by a water jet with cross-sectional area $3 \times 10^{-4}\,\mathrm{m}^2$, velocity $30\,\mathrm{m/s}$ and density $10^3\,\mathrm{kg/m^3}$.

13.3 Elastic collisions

If a ball bounces off a fixed surface as in Figure 13.5a, its velocity v after the impact is a constant e times its velocity u before the impact, with $0 < e < 1$. Similarly, if two balls collide directly as in Figure 13.5b, their relative velocity of departure $(v_2 - v_1)$ is a constant e times their relative velocity of approach $(u_1 - u_2)$.

The constant e is referred to as the *coefficient of restitution* and its value depends on the nature of the materials of the bodies which collide. Newton established the relationship experimentally and formulated the equation:

$$v_1 - v_2 = -e(u_1 - u_2),$$

which is sometimes referred to as *Newton's rule*.

The rule (Newton's) also applies to the components of velocities perpendicular to the surfaces of contact when two smooth spheres collide obliquely. Referring to Figure 13.6, Newton's rule becomes:

$$v_1 \cos \phi_1 - v_2 \cos \phi_2 = -e(u_1 \cos \theta_1 - u_2 \cos \theta_2).$$

Since the surfaces of contact are smooth, velocities parallel to the surfaces of contact are unaffected, i.e.

$$v_1 \sin \phi_1 = u_1 \sin \theta_1 \quad \text{and} \quad v_2 \sin \phi_2 = u_2 \sin \theta_2.$$

Along the line of centres, total momentum is conserved, i.e.

$$m_1 v_1 \cos \phi_1 + m_2 v_2 \cos \phi_2 = m_1 u_1 \cos \theta_1 + m_2 u_2 \cos \theta_2.$$

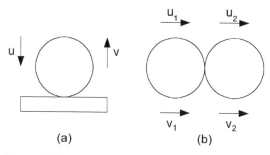

(a)   (b)

Figure 13.5. Bouncing balls.

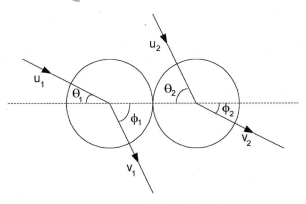

Figure 13.6. Smooth spheres colliding obliquely.

EXERCISE 3

A smooth sphere is at rest on a horizontal surface when it is struck by an identical sphere travelling along the surface with velocity **u** at an angle of 45° to the line of centres when the collision occurs. Find the velocities (both magnitude and direction) of the two spheres after the impact, if the coefficient of restitution is $e = 0.94$.

Problems 121 and 122.

13.4 Moments of impulse and momentum

So far, we have considered the effect of impulses on particles or on bodies which may be regarded like particles. Now, we wish to study the effect of impulses on the general two-dimensional motion of rigid bodies, including rotational as well as translational responses. We will, however, restrict the impulses to large forces applied over short times, as occur in collisions.

We start from the general dynamic equations of a rigid body, summarized at the end of Section 12.7. Firstly,

$$\sum \mathbf{F}_e = M\ddot{\mathbf{r}}_G,$$

which says that the vector sum of the external forces applied to the body equals the total mass times the acceleration of its centre of gravity. Secondly,

$$\sum \mathbf{r}_e \times \mathbf{F}_e = \mathbf{r}_G \times M\ddot{\mathbf{r}}_G + I_G\dot{\omega},$$

which says that the sum of the moments of the external forces about an axis perpendicular to the plane of motion equals the moment of the mass times the acceleration of the centre of gravity plus the moment of inertia about the axis through the centre of gravity and perpendicular to the plane of motion times the angular acceleration of the body.

By considering only impulses consisting of large forces acting over short periods of time, we can ignore the other forces and any movement will be assumed to be negligible over the short duration of the impulses. The latter word is plural since more than one impulse may occur simultaneously.

Let $\mathbf{F}_{\mathrm{ei}}$ be large impulsive forces applied to a rigid body during a short duration from t_1 to t_2. Then the vector sum of these external impulses will be:

$$\sum \int_{t_1}^{t_2} \mathbf{F}_{\mathrm{ei}}\, dt = \int_{t_1}^{t_2} \sum \mathbf{F}_{\mathrm{ei}}\, dt = \int_{t_1}^{t_2} M\ddot{\mathbf{r}}_{\mathrm{G}}\, dt = M\dot{\mathbf{r}}_{\mathrm{G2}} - M\dot{\mathbf{r}}_{\mathrm{G1}}.$$

This is the change in momentum of the body defined as the total mass times the change in velocity of the centre of gravity.

Next we consider the sum of the moments of the external impulses about an axis through a point O and perpendicular to the plane of motion. Since we are only concerned with the plane motion of a rigid body, a moment equation will be a scalar equation. However, it is convenient to use vectors to derive the equation starting from the dynamic moment equation:

$$\sum \mathbf{r}_{\mathrm{e}} \times \mathbf{F}_{\mathrm{e}} = \mathbf{r}_{\mathrm{G}} \times M\ddot{\mathbf{r}}_{\mathrm{G}} + I_{\mathrm{G}}\dot{\omega}.$$

Again, we only include the large impulsive forces $\mathbf{F}_{\mathrm{ei}}$ applied over the short time from t_1 to t_2. One of them is indicated in Figure 13.7 acting at a point with position vector $\mathbf{r}_{\mathrm{ei}}$ from O. The vector sum of the moments of the impulses is:

$$\sum \mathbf{r}_{\mathrm{ei}} \times \int_{t_1}^{t_2} \mathbf{F}_{\mathrm{ei}}\, dt = \int_{t_1}^{t_2} \sum \mathbf{r}_{\mathrm{ei}} \times \mathbf{F}_{\mathrm{ei}}\, dt = \int_{t_1}^{t_2} \dot{\mathbf{H}}_{\mathrm{O}}\, dt = \mathbf{H}_{\mathrm{O}}(t_2) - \mathbf{H}_{\mathrm{O}}(t_1),$$

where $\mathbf{H}_{\mathrm{O}}$ is the moment of momentum of the body about the axis through O.

We also know from Section 12.7 that:

$$\mathbf{H}_{\mathrm{O}} = \mathbf{r}_{\mathrm{G}} \times M\dot{\mathbf{r}}_{\mathrm{G}} + I_{\mathrm{G}}\omega.$$

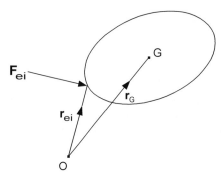

Figure 13.7. Body acted on by large impulsive force $\mathbf{F}_{\mathrm{ei}}$.

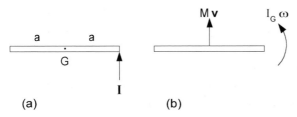

Figure 13.8. Effect of an impulse applied to a rod on a smooth surface.

Substituting this into our equation for the vector sum of the moments of the impulses, we obtain:

$$\sum \mathbf{r}_{ei} \times \int_{t_1}^{t_2} \mathbf{F}_{ei}\, dt = \mathbf{r}_G \times M(\dot{\mathbf{r}}_{G2} - \dot{\mathbf{r}}_{G1}) + I_G(\boldsymbol{\omega}_2 - \boldsymbol{\omega}_1),$$

which is the moment of the mass times the change in velocity of the centre of gravity plus the moment of inertia about the centre of gravity (axis perpendicular to the plane of motion) times the change in angular velocity.

Suppose we have a uniform straight rod resting on a smooth horizontal surface. Let the rod have mass M and length $2a$. Subject the rod to a horizontal impulse $\mathbf{I}$ at one end and perpendicular to its length. Let us investigate the initial motion of the rod.

Figure 13.8a shows the impulse applied to the rod and Figure 13.8b shows the resulting linear and angular momentum.

We can equate the impulse to the mass times the change in velocity of the centre of gravity G. If the velocity of G changes from zero to $\mathbf{v}$, then $\mathbf{v}$ must have the same direction as $\mathbf{I}$ and magnitude given by:

$I = Mv$, i.e. $v = I/M$.

Next take moments about G to give: $Ia = I_G\omega$, the moment of inertia about G times the change in angular velocity.

Now, $I_G = \dfrac{1}{3}Ma^2$, so $Ia = \dfrac{1}{3}Ma^2\omega$ and $\omega = \dfrac{3I}{Ma}$.

EXERCISE 4

Let us extend the last example by linking another identical rod end-to-end with the first one with a smooth hinge. Suppose they are resting in a straight line on a smooth horizontal surface when the free end of one is struck a blow with impulse $\mathbf{I}$ at right-angles to the line of the rods. Find the linear and angular velocities imparted to the rods by $\mathbf{I}$ and also find the reactive impulses $\pm\mathbf{I}_r$ that occur at the hinge.

Problems 123 and 124.

13.5 Centre of percussion

Referring back to the example in Section 13.4 (see Figure 13.9), it was found that the impulse **I** gave G a velocity **v** with $v = I/M$ and an angular velocity $\omega = 3I/Ma$. Now, if we label the ends of the rod A and B, with the impulse applied at A, the velocity imparted to A is $v + a\omega = 4I/M$. However, the velocity imparted to B is $v - a\omega = -2I/M$, where the negative sign means that it is in the opposite direction from **v**. Hence, there is a point P at distance $2a/3$ from B which has zero velocity imparted to it and is therefore the instantaneous centre of rotation.

Next, let us fasten A to a fixed point with a smooth hinge (see Figure 13.10). Then apply an impulse at a point P of the rod and perpendicular to the rod, such that it incurs no impulsive reaction at the hinge and let $GP = b$. The change in linear momentum is $I = Mv$ and the change in moment of momentum about G is $Ib = I_G\omega$. Since the rod can only rotate about A, $v = a\omega$.

Also, $I_G = \frac{1}{3}Ma^2$, so $Ib = Mab\omega = \frac{1}{3}Ma^2\omega$ and $b = a/3$.

This point P, which is distance $2a/3$ from B, is called the *centre of percussion*. Notice that its position is the same as that of the instantaneous centre of rotation in the previous

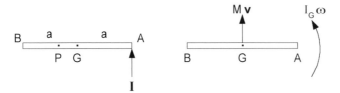

Figure 13.9. Finding P, the centre of percussion.

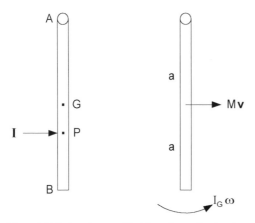

Figure 13.10. A rod smoothly hinged to a fixed point at A.

example. For a two- or three-dimensional rigid body, its centre of percussion P lies on the straight line which passes through the hinge point A and its centre of gravity G.

EXERCISE 5

Find the position P of the centre of percussion for a uniform solid sphere which is smoothly hinged to a fixed point at a point A on its surface.

Problems 125 and 126.

13.6 Conservation of moment of momentum

If a moving rigid body receives just one impulse **I**, the latter will have no moment about the point O at which it is applied. Consequently, the moment of momentum of the body, about an axis through O perpendicular to the plane of motion, will not be affected by **I** and the moment of momentum will be conserved.

Let us consider an example to illustrate this phenomenon. Suppose that a uniform circular disc is rotating with angular velocity ω_1 about its own axis through G which is stationary on a smooth horizontal surface. Find the new angular velocity ω_2 if a point O on the rim is suddenly fixed so that the disc starts to rotate about O. Figure 13.11 shows the before and after situations.

At the instant the point O is fixed, the disc will receive an impulse through O. This will not affect the moment of momentum about O, which will be conserved.

Before the fixing of O, the point G is stationary and the moment of momentum about O is simply $I_G\omega_1$. If the angular velocity becomes ω_2 after O is fixed, the moment of momentum about O is then $I_G\omega_2$ plus the moment about O of the mass M times the velocity $a\omega_2$ of G. Since $I_G = Ma^2/2$, we have the equation:

$$\tfrac{1}{2}Ma^2\omega_2 + a \cdot Ma\omega_2 = \tfrac{3}{2}Ma^2\omega_2 = \tfrac{1}{2}Ma^2\omega_1. \text{ Hence, } \omega_2 = \omega_1/3.$$

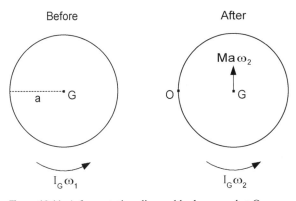

Figure 13.11. A free rotating disc suddenly pegged at O.

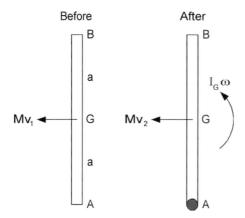

Figure 13.12. Finding the subsequent motion when a sliding rod is caught at one end.

EXERCISE 6

A rod of length 1 m is travelling sideways at 8 m/s along a smooth horizontal surface when one end is suddenly fixed, so that the rod starts to rotate about that end. Find the angular velocity of this new rotation. The before and after situations are illustrated in Figure 13.12.

Problems 127 and 128.

13.7 Impacts

When two bodies collide, the impulses that occur between the two bodies are equal and opposite. Consequently, provided no other simultaneous impulses occur, the total linear momentum will be conserved over the impact. However, if one of the bodies is constrained to move about a hinge, the collision with a free body would usually incur a reactive impulse from the hinge. Nevertheless, the total moment of momentum of the two bodies about the hinge would be conserved over the impact. We shall illustrate this with an example.

A rod AB of length $2a$ and mass M is suspended by a smooth hinge at A, as shown in Figure 13.13. The end B is struck by a small body of mass m, which is travelling horizontally with velocity $\mathbf{v}$. If the small body sticks to the rod, find the angular velocity ω of the rod about A immediately after the impact.

The total moment of momentum about A immediately before the impact is simply that of the small body, i.e. $2amv$. After the impact, it is that of the small body plus that of the rod, i.e. $2a \cdot m \cdot 2a\omega + (a \cdot M \cdot a\omega + \frac{1}{3}Ma^2 \cdot \omega)$. Equating the total moment of momentum about A after impact to before gives:

$$4a^2(m + M/3)\omega = 2amv \quad \text{and} \quad \text{therefore} \quad \omega = \frac{mv}{2a(m + M/3)}.$$

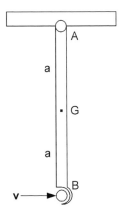

Figure 13.13. Vertically suspended rod struck by small body which adheres to it.

EXERCISE 7

A uniform rectangular metal lamina is supported by a smooth horizontal hinge along its upper edge. The plate is hanging vertically at rest when it is struck at its mid-point by a steel ball travelling horizontally at 20 m/s and perpendicular to the plane of the lamina. The length of the lamina from top to bottom is 0.2 m and its mass is three times that of the steel ball. If the coefficient of restitution between the ball and the plate is $e = 0.95$, find the velocity of the ball and the angular velocity of the plate immediately after the impact.

Problems 129 and 130.

13.8 Answers to exercises

1. The equation of conservation of momentum, i.e. total momentum before impact equals total momentum after impact, is a vector equation. By resolving in two different directions, we obtain two scalar equations. Let us resolve perpendicular to and parallel to the direction of the velocity after impact. Referring to Figure 13.3, we see that:

   $$3mv \cos \alpha - 2mv \sin \alpha = 0$$
   $$3mv \sin \alpha + 2mv \cos \alpha = 4mV.$$

   From the first equation, $\tan \alpha = 3/2$ and therefore, $\alpha = 56.31°$. Then, from the second equation,

   $$V = \tfrac{3}{4}v \sin \alpha + \tfrac{1}{2}v \cos \alpha = 0.9014v.$$

2. As we saw in Section 13.2, the force exerted by a jet of water playing on a wall is $A\rho v^2$, where A is the cross-sectional area of the jet, ρ is the density of water in the jet and v is its velocity. In this exercise, $A = 3 \times 10^{-4}$, $\rho = 10^3$ and $v = 30$, all in basic SI units. Hence, the force exerted is:

   $$P = 3 \times 10^{-4} \times 10^3 \times 9 \times 10^2 = 270 \, \text{N}.$$

3. Since the spheres are smooth, the impulse supplied to the initially stationary sphere is perpendicular to the surface at the point of contact. Hence, as shown in Figure 13.14, the velocity $\mathbf{v}_2$ given to the

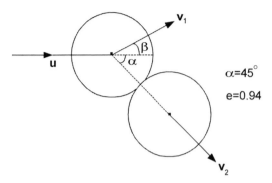

Figure 13.14. Collision of two smooth spheres.

initially stationary sphere is at $45°$ to the direction of the initial velocity $\mathbf{u}$ of the other sphere. Let the velocity of the latter sphere change to $\mathbf{v}_1$ due to the impact with a change in direction of angle β.

Since the masses of the spheres are the same, we can omit the mass from the conservation of momentum equation and write it as:

$$v_1 \cos(\alpha + \beta) + v_2 = u \cos \alpha.$$

Then, Newton's rule gives us the equation:

$$v_1 \cos(\alpha + \beta) - v_2 = -eu \cos \alpha.$$

Finally, the velocity parallel to the surface of contact is unaffected, so:

$$v_1 \sin(\alpha + \beta) = u \sin \alpha.$$

Subtracting the second equation from the first equation gives:

$$2v_2 = (1 + e)u \cos \alpha \quad \text{and} \quad v_2 = \frac{1.94}{2\sqrt{2}}u = 0.686u.$$

Dividing the third equation by the first equation with v_2 on the other side gives:

$$\tan(\alpha + \beta) = \frac{u \sin \alpha}{u \cos \alpha - v_2} = \frac{1}{1 - 0.686\sqrt{2}} = 33.5.$$

Therefore, $\alpha + \beta = 88.3°$ and $\beta = 43.3°$.

Then, from the third equation:

$$v_1 = \frac{u \sin \alpha}{\sin(\alpha + \beta)} = \frac{u}{\sqrt{2} \sin 88.3°} = 0.707u.$$

4.   We deal with this problem by treating each rod separately as with the example of Section 13.4. However, one rod now has two impulses, $\mathbf{I}$ at one end and the reactive impulse $\mathbf{I}_r$ at the other (see Figure 13.15). For each rod, we have two equations, one for translation and one for rotation. Hence:

$$I + I_r = Mv_1$$
$$(I - I_r)a = \tfrac{1}{3}Ma^2\omega_1$$
$$-I_r = Mv_2$$
$$-I_r a = \tfrac{1}{3}Ma^2\omega_2.$$

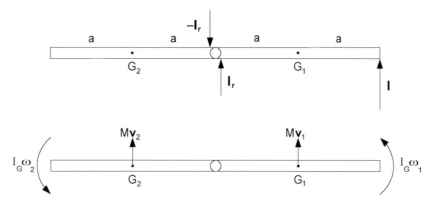

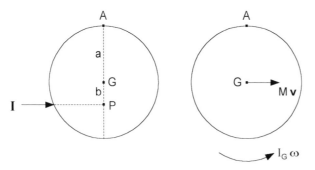

Figure 13.15. Impulse applied to an outer end of two linked rods.

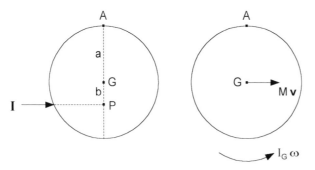

Figure 13.16. A suspended sphere subjected to an impulse through its centre of percussion.

Since we have five unknowns: v_1, ω_1, v_2, ω_2 and I_r, we need another equation in order to solve the problem. This we obtain from the velocity of the hinge after the impulse, since this must be the same for each rod. Hence:

$$v_1 - a\omega_1 = v_2 + a\omega_2.$$

Substituting into this last equation from the previous four equations gives:

$$I + I_r - 3(I - I_r) = -I_r - 3I_r, \text{ therefore } I_r = I/4.$$

Substituting this into the first four equations gives:

$$v_1 = \frac{5I}{4M}, \quad \omega_1 = \frac{9I}{4aM}, \quad v_2 = -\frac{I}{4M}, \quad \omega_2 = -\frac{3I}{4aM}.$$

The negative signs for v_2 and ω_2 imply that the directions are opposite to those indicated by the arrows in the diagram.

If a numerical problem had been set using SI units with I in N s, M in kg and a in m, then the v's would be in m/s and the ω's in rad/s.

5. The moment of inertia of the sphere about an axis through its centre is $I_G = 2Ma^2/5$. Referring to Figure 13.16, the position of P must be such that there is no reactive impulse at A. Hence, we have the following translational and rotational equations:

$$I = Mv = Ma\omega \quad \text{and} \quad Ib = I_G\omega = 2Ma^2\omega/5.$$

Therefore, $b = 2a/5$.

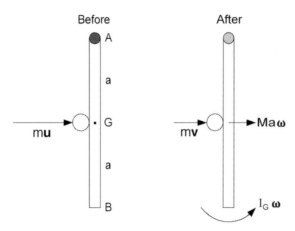

Figure 13.17. A suspended metal plate hit by a steel ball.

6. Referring to the before and after diagrams of Figure 13.12, we can see that the rod will receive an impulse at the end A when A is suddenly fixed. However, the rod receives no other impulse, so the moment of momentum about A will be conserved. Equating the moment of momentum about A after to that before gives the equation:

$$a M v_2 + I_G \omega = M a^2 \omega + \tfrac{1}{3} M a^2 \omega = \tfrac{4}{3} M a^2 \omega = a M v_1.$$

Therefore, $\omega = \dfrac{3 v_1}{4a} = \dfrac{3 \times 8}{4 \times 0.5} = 12 \, \text{rad/s}.$

7. Referring to the before and after diagrams of Figure 13.17, conservation of total moment of momentum about A gives:

$$a(m v + M a \omega) + \tfrac{1}{3} M a^2 \omega = \tfrac{4}{3} M a^2 \omega + a m v = a m u.$$

By Newton's rule: $a \omega - v = e u.$
    Eliminating v: $(4M/3 + m) a \omega = (1 + e) m u.$

Therefore, $\omega = \dfrac{(1 + e) m u}{(4M/3 + m) a} = \dfrac{1.95 \times 20}{5 \times 0.1} = 78 \, \text{rad/s}.$

Then, $v = a \omega - e u = 0.1 \times 78 - 0.95 \times 20 = -11.2 \, \text{m/s}.$

The negative sign indicates that the direction of the velocity of the ball is reversed by the impact.

14 Work, power and energy

14.1 Work done by force on a particle

In Chapter 9, we defined the work done by a force as the force multiplied by the distance moved by its point of application in the direction of the force. This is fine if the force is constant, but what happens if the force varies in magnitude or direction or both?

Let us suppose that the point of application of force $\mathbf{F}$ moves from A to B along the path Γ, as indicated in Figure 14.1. For a small change $\delta\mathbf{r}$ in the position $\mathbf{r}$ of the point of application, the work done is approximately the scalar product $\mathbf{F}.\delta\mathbf{r}$, where $\mathbf{F}$ is the force at position $\mathbf{r}$. If we imagine the path split into little bits $\delta\mathbf{r}$, then the work done is approximately the sum $\sum \mathbf{F}.\delta\mathbf{r}$ over all the little bits $\delta\mathbf{r}$ as we move along the curve Γ followed by the point of application of $\mathbf{F}$ from A to B. The accuracy increases with the number of sub-divisions and the work done is:

$$W = \lim_{\delta r \to 0} \sum \mathbf{F}.\delta\mathbf{r} = \int_{\Gamma} \mathbf{F}.d\mathbf{r}.$$

This is called a *line integral* along the path Γ from A to B.

Next, let us see what happens if $\mathbf{F}$ is the resultant force acting on a particle of mass m and Γ is the consequential path followed by the particle. (Since the particle may have an initial velocity, the direction of motion need not coincide with the direction of $\mathbf{F}$.) In this case, $\mathbf{F} = m\ddot{\mathbf{r}} = m\dot{\mathbf{v}}$, where $\mathbf{v} = \dot{\mathbf{r}}$ is the velocity of the particle and $\dot{\mathbf{v}} = \ddot{\mathbf{r}}$ is its acceleration.

On substituting for $\mathbf{F}$ in the line integral, the work done by $\mathbf{F}$ becomes:

$$W = \int_{\Gamma} m\ddot{\mathbf{r}}.d\mathbf{r} = m\int_{\Gamma} \frac{d\dot{\mathbf{r}}}{dt}.d\mathbf{r} = m\int_{t_1}^{t_2} \frac{d\dot{\mathbf{r}}}{dt}.\frac{d\mathbf{r}}{dt}\,dt,$$

where t_1 and t_2 represent the times when the particle is at the start and the end, respectively, of its path Γ.

On substituting $\mathbf{v}$ for $\dot{\mathbf{r}}$ and $d\mathbf{r}/dt$, we have:

$$W = m\int_{t_1}^{t_2} \dot{\mathbf{v}}.\mathbf{v}\,dt = \frac{1}{2}m\int_{t_1}^{t_2} \frac{d}{dt}(\mathbf{v}.\mathbf{v})\,dt = \frac{1}{2}m\int_{t_1}^{t_2} \frac{d}{dt}(v^2)\,dt = \frac{1}{2}mv_2^2 - \frac{1}{2}mv_1^2,$$

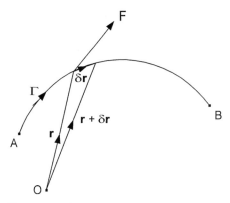

Figure 14.1. Finding the work done by force **F** as its point of application moves along path Γ.

where v_1 and v_2 are the magnitudes of the velocity of the particle at times t_1 and t_2, respectively. The quantity $\frac{1}{2}mv^2$ is the *kinetic energy* of the particle. Hence, the change in the kinetic energy of the particle between times t_1 and t_2 equals the work done by the force **F** during that time.

The unit of kinetic energy is obviously the same as that of work. In SI units, it is the joule (J), which is the same as the newton metre (N m).

The relationship between work done and change in kinetic energy is sometimes called the *principle of work*. Although it has only been developed for the motion of a particle, it may be applied immediately to the motion of a rigid body provided no rotation is involved. The extension to include rotation will be considered later.

In Section 9.5, we introduced the concept of potential energy as being the work done against the field of force in moving a body from a reference position to its present position. Then, provided no heat is generated in the motion, the principle of work corresponds to the *principle of conservation of energy*, i.e. that gain in kinetic energy equals loss in potential energy and vice versa. However, as we shall see from the following example, we may be able to apply the principle of work even when heat is generated, in this case by friction.

Consider a block resting on a plane inclined at an angle θ to the horizontal, as shown in Figure 14.2. Assume that θ is large enough to overcome static friction so that when the block is released from rest, it starts to slide down the plane. Let μ be the coefficient of kinetic friction and use the principle of work to find the velocity v when the block has moved a distance a down the plane from where it was released from rest.

Figure 14.2 shows the forces acting on the block. The normal component of reaction N does no work since there is no movement in that direction. The frictional component $F = \mu N = \mu Mg \cos\theta$ does negative work: $-a\mu Mg \cos\theta$. The gravitational force

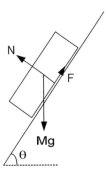

Figure 14.2. A block sliding down an inclined plane.

does positive work: $aMg \sin\theta$. Hence, the increase in kinetic energy is:

$$\tfrac{1}{2}Mv^2 = aMg(\sin\theta - \mu\cos\theta) \quad \text{and} \quad v = \sqrt{2ag(\sin\theta - \mu\cos\theta)}.$$

Note: Use of the principle of work gives an immediate relation between velocity and distance travelled.

EXERCISE 1

The driver of a motor car slams on the brakes when travelling at 80 km/h and the vehicle skids to a standstill. Find the skid distance if the coefficient of kinetic friction between the tyres and the road is $\mu = 0.5$.

Problems 131 and 132.

14.2 Conservation of energy

In both the example and the exercise in Section 14.1, the principle of conservation of energy, i.e. kinetic energy + potential energy = constant, could not be used since in both cases heat would have been generated due to work being done in overcoming friction. However, the conservation of energy principle is very useful when there is no friction or when it is small enough to be neglected.

Consider a weight of mass M attached to a fixed support by a taut string of length a. If the weight is released from rest with the string horizontal, we can use the principle of conservation of energy to find the velocity of the weight through its lowest point.

Referring to Figure 14.3, let the potential energy be zero with the weight at its lowest point. Then it must be Mga (weight × height above lowest point) when the weight is released with zero velocity and with zero kinetic energy. Then, by the principle of conservation of energy, the kinetic energy at its lowest point equals the potential energy when the weight is released. Thus:

$$\tfrac{1}{2}Mv^2 = Mga \quad \text{and} \quad v = \sqrt{2ga}.$$

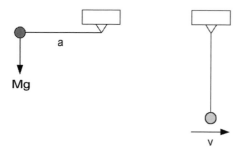

Figure 14.3. A weight on a string released from rest in a horizontal position.

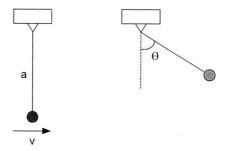

Figure 14.4. Finding maximum θ given v.

EXERCISE 2

Suppose we have a weight on a string, as in the last example, with the length of string $a = 5$ m. If, as in Figure 14.4, θ is the angle which the string makes with the downward vertical, find the maximum value of θ given that the weight passes through its lowest point with a velocity $v = 7$ m/s. (It may be assumed that the maximum $\theta < 90°$.)

Problem 133.

14.3 Spring energy

Referring to Figure 14.5, we assume that the force required to hold a spring in a stretched or compressed condition, with its length differing by x from its natural length, is $F = kx$, where k is the *spring constant*. The work required to increase the length by an infinitessimal amount dx is $kx\,dx$. Then the work required to achieve a displacement in length of a from its natural length is:

$$W = \int_0^a kx\,dx = k\left[\frac{x^2}{2}\right]_0^a = \frac{1}{2}ka^2.$$

Hence, assuming zero potential energy in the spring when it is in its natural state, when stretched or compressed a distance a, the potential energy stored in the spring is: $\frac{1}{2}ka^2$.

Figure 14.5. A spring stretched by x from its natural length.

EXERCISE 3

A weight of 0.5 kg mass rests on a smooth horizontal surface against the end of a coil spring, the other end of which is fixed. The weight is pushed back to compress the spring by 5 cm. If the weight is then released, find the velocity given to it by the spring, assuming a spring constant of 2 N/m.

Problem 134.

14.4 Power

Power is simply the rate of doing work. The SI unit of work is the joule (J) so it follows that the SI unit of power is the joule per second (J/s). This unit is called the watt (W).

If the point of application of a force $\mathbf{F}_i$ undergoes a very small displacement $d\mathbf{r}_i$, the corresponding work done by $\mathbf{F}_i$ is: $dW_i = \mathbf{F}_i.d\mathbf{r}_i$. In this case the power is:

$$P_i = \frac{dW_i}{dt} = \mathbf{F}_i.\frac{d\mathbf{r}_i}{dt} = \mathbf{F}_i.\mathbf{v}_i,$$

where $\mathbf{v}_i$ is the velocity of the point of application of $\mathbf{F}_i$.

If we have a system of n forces $\mathbf{F}_i$, the total power is:

$$P = \sum_{i=1}^{n} P_i = \sum_{i=1}^{n} \mathbf{F}_i.\mathbf{v}_i.$$

EXERCISE 4

Find the power output of a car, which has a total mass (complete with passengers) of 1500 kg, as it is driven at a constant speed of 72 km/h up a slope of 1 in 10 (1 m rise for 10 m along road) against a resistance (friction and air) of 2 kN.

Problems 135 and 136.

14.5 Kinetic energy of translation and rotation

Consider a rigid body moving with both translation and rotation in a fixed plane of motion. Axes through a reference point O and through the centre of gravity G are perpendicular to the plane of motion. In Figure 14.6, P_i represents the ith particle of the body and its mass will be denoted by m_i. The vectors $\mathbf{r}_G$ and $\mathbf{r}_i$ are perpendicular to the axis through O and the vector $\mathbf{r}'_i$ is perpendicular to the axis through G.

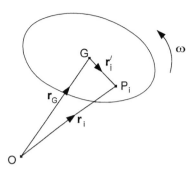

Figure 14.6. Position vectors parallel to the plane of motion of a body.

The kinetic energy of the body is:

$$T = \sum_i \frac{1}{2} m_i (\dot{\mathbf{r}}_i . \dot{\mathbf{r}}_i) = \frac{1}{2} \sum_i m_i (\dot{\mathbf{r}}_G + \dot{\mathbf{r}}'_i) . (\dot{\mathbf{r}}_G + \dot{\mathbf{r}}'_i)$$

$$= \frac{1}{2} \sum_i m_i \dot{r}_G^2 + \frac{1}{2} \left(\sum_i m_i \dot{\mathbf{r}}'_i \right) . \dot{\mathbf{r}}_G + \frac{1}{2} \dot{\mathbf{r}}_G . \left(\sum_i m_i \dot{\mathbf{r}}'_i \right) + \frac{1}{2} \sum_i m_i \dot{r}_i'^2.$$

Now $\sum_i m_i = M$, the total mass and from Section 12.5, $\sum_i m_i \dot{\mathbf{r}}'_i = 0$. Also, since the body is rigid, $\dot{\mathbf{r}}'_i$ is perpendicular to $\mathbf{r}'_i$ and the magnitude of the relative velocity $\dot{\mathbf{r}}'_i$ is $\dot{r}'_i = r'_i \omega$, where ω is the angular velocity of the body (in rad/s). It follows that:

$$\sum_i m_i \dot{r}_i'^2 = \left(\sum_i m_i r_i'^2 \right) \omega^2 = I_G \omega^2,$$

where I_G is the moment of inertia about the axis through G as defined in Section 12.7.
Consequently, the kinetic energy of the body is:

$$T = \tfrac{1}{2} M \dot{r}_G^2 + \tfrac{1}{2} I_G \omega^2.$$

EXERCISE 5

Find the kinetic energy of a uniform solid sphere of mass 2 kg and radius 5 cm which is rolling along a surface without slipping with an angular velocity of 12 rad/s.

Problem 137.

14.6 Energy conservation with both translation and rotation

Provided no heat is generated by the work done in overcoming friction or by an impact, the sum of potential and kinetic energies will be conserved during the motion of a rigid body which involves rotation as well as translation. We just have to include the extra

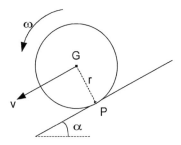

Figure 14.7. A uniform solid sphere rolling down an inclined plane.

rotational term in the expression for kinetic energy. We can then use the principle of conservation of energy to find a direct relationship between velocity and displacement.

EXERCISE 6

A uniform solid sphere of mass M and radius r is allowed to roll from rest down a plane surface at angle α to the horizontal, as indicated in Figure 14.7. Assuming that α is small enough and the coefficient of static friction is large enough for there to be no slipping, use the principle of conservation of energy to derive an expression for the velocity v of the sphere after it has rolled a distance d down the slope.

Problems 138 and 139.

14.7 Energy and moment of momentum

In some situations, it is possible to use the principle of conservation of moment of momentum for one phase of the motion and the principle of conservation of energy for another phase. This can be illustrated by the following example.

Referring to Figure 14.8, a uniform circular disc of radius $r = 0.2$ m is rolling along a horizontal surface with velocity $u = 1$ m/s when it strikes a kerb of height $h = 2.5$ cm. Assuming no slip or rebound, find the velocity v of the disc after it has mounted the kerb.

Firstly, we consider what happens when the disc strikes the edge of the kerb P, as in Figure 14.9. The disc will be given a reactive impulse through P and then start to rotate about P with an angular velocity ω, say. Since the impulse acts through P, the disc's moment of momentum about P will be conserved over the instant of the impact.

Immediately before the impact, the disc's moment of momentum about P is the moment of the mass times the velocity of G, i.e. $(r - h)Mu$, plus the moment of inertia about G times the angular velocity, i.e. $\frac{1}{2}Mr^2 \cdot u/r$. Immediately after the impact the moment of momentum about P is: $r \cdot Mr\omega + \frac{1}{2}Mr^2 \cdot \omega$. Thus,

$$\frac{3}{2}Mr^2\omega = \frac{3}{2}Mru - Mhu \quad \text{and} \quad \omega = \frac{u}{r}\left(1 - \frac{2h}{3r}\right).$$

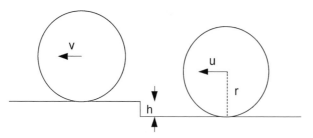

Figure 14.8. A rolling disc before and after mounting kerb.

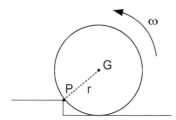

Figure 14.9. A rolling disc striking a kerb.

As the disc rotates about P and subsequently rolls away from the kerb, the energy is conserved. If we let the potential energy be zero at the bottom of the kerb, the total energy there (immediately after the impact) is:

$$\frac{1}{2}M(r\omega)^2 + \frac{1}{2} \cdot \frac{1}{2}Mr^2 \cdot \omega^2 = \frac{3}{4}Mr^2\omega^2.$$

The potential energy becomes Mgh when the disc has climbed the kerb and the kinetic energy is then:

$$\frac{1}{2}Mv^2 + \frac{1}{2} \cdot \frac{1}{2}Mr^2 \cdot \left(\frac{v}{r}\right)^2 = \frac{3}{4}Mv^2.$$

Hence, by conservation of energy:

$$\tfrac{3}{4}Mv^2 + Mgh = \tfrac{3}{4}Mr^2\omega^2, \text{ or } v^2 = r^2\omega^2 - \tfrac{4}{3}gh.$$

On substituting the numerical data:

$$\omega = \frac{1}{0.2}\left(1 - \frac{2 \times 0.025}{3 \times 0.2}\right) = 4.583\,\text{rad/s}$$

and $v^2 = (0.2 \times 4.583)^2 - \tfrac{4}{3} \times 9.8 \times 0.025 = 0.5135, \ v = 0.717\,\text{m/s}.$

EXERCISE 7

A uniform straight rod AB is hinged to a fixed support at A so that it can swing freely in a vertical plane. It is released from rest in the horizontal position, as shown in Figure 14.10a. When it reaches the vertical position, it strikes a small stationary ball with its end B and projects it horizontally with

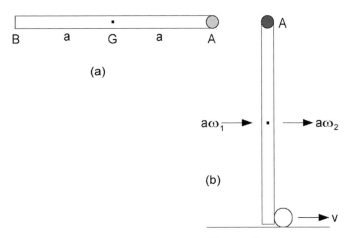

Figure 14.10. A rod swinging down and striking a ball.

velocity v (see Figure 14.10b). Find v given that the length and mass of the rod are 1 m and 4 kg, respectively, the mass of the ball is 0.4 kg and the coefficient of restitution between the rod and the ball is $e = 0.92$.

Problems 140 and 141.

14.8 Answers to exercises

1. Let N be the normal component of the resultant reaction from the road (see Figure 14.11). During the skid the frictional component of reaction will be $F = \mu N$.

   Resolving vertically, $N - Mg = 0$, so $F = \mu Mg$. By the principle of work, the work done by F during the skid equals the change in kinetic energy. Therefore, if the skid distance is a:

   $$-aF = -a\mu Mg = 0 - \tfrac{1}{2}Mv^2,$$

   where v is the initial velocity.

   Therefore, $a = \dfrac{v^2}{2\mu g} = \dfrac{(80/3.6)^2}{2 \times 0.5 \times 9.8} = 50.4\,\mathrm{m},$

   where the factor $1/3.6$ changes km/h to m/s.

2. Let the potential energy be zero when the weight is at its lowest point. At its highest point, the velocity is zero and hence the kinetic energy is zero. Thus:

   $$Mga(1 - \cos\theta) = \tfrac{1}{2}Mv^2, \quad (1 - \cos\theta) = \dfrac{v^2}{2ga}$$

   and $\cos\theta = 1 - \dfrac{v^2}{2ga} = 1 - \dfrac{49}{2 \times 9.8 \times 5} = 1 - \dfrac{1}{2} = \dfrac{1}{2}$, so $\theta = 60°$.

3. Figure 14.12a shows the initial condition with the spring compressed and Figure 14.12b shows the weight being projected by the spring with velociy v. If the weight is released with the spring

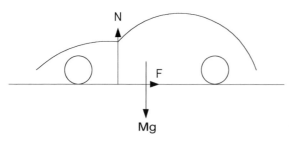

Figure 14.11. Forces on a motor car as it skids to a standstill.

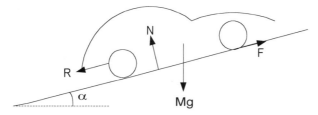

Figure 14.12. Weight being projected by a compressed spring.

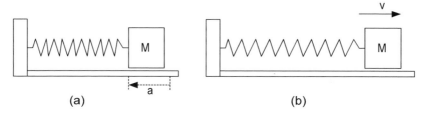

Figure 14.13. A car travelling up an incline at constant speed.

compressed by a, the potential energy stored in the spring is $\frac{1}{2}ka^2$. All of this potential energy is transformed into the kinetic energy $\frac{1}{2}Mv^2$ of the weight.

Thus, $\dfrac{1}{2}Mv^2 = \dfrac{1}{2}ka^2$, $v^2 = \dfrac{ka^2}{M} = \dfrac{2 \times (0.05)^2}{0.5} = 0.01$, $v = 0.1$ m/s $= 10$ cm/s.

4. In Figure 14.13, F represents the traction force, R the resistance and N the total normal reaction from the road.

Since the car is travelling at constant speed, the forces balance, so resolving parallel to the road:

$F - R - Mg \sin \alpha = 0.$

Now, $R = 2$ kN, $M = 1500$ kg and $\sin \alpha = 0.1$.

Therefore $F = 2000 + 1500 \times 9.8 \times 0.1 = 3470$ N.

The speed $v = 72$ km/h $= 72/3.6 = 20$ m/s.

Power output $= Fv = 3470 \times 20 = 69.4$ kW.

Alternatively, 1 hp (horsepower)$= 745.7$ W, so the power output $= 93$ hp.

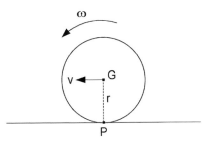

Figure 14.14. A uniform solid sphere rolling along a fixed surface.

5. Referring to Figure 14.14, the point of contact P is an instantaneous centre of rotation. Hence, $\dot{r}_G = v = r\omega$. Also, recall that $I_G = \frac{2}{5}Mr^2$. Therefore, the kinetic energy of the rolling sphere is:

$$T = \tfrac{1}{2}M\dot{r}_G^2 + \tfrac{1}{2}I_G\omega^2 = \tfrac{1}{2}Mr^2\omega^2 + \tfrac{1}{2} \times \tfrac{2}{5}Mr^2\omega^2 = 0.7Mr^2\omega^2$$

$$= 0.7 \times 2 \times (0.05 \times 12)^2 = 0.504\,\text{J}.$$

6. The loss in potential energy as the sphere moves a distance d down the slope equals the work done by gravity, which is $Mgd\sin\alpha$. (Mg is the gravitational force and $d\sin\alpha$ is the distance moved in the direction of the force.) If v and ω are the velocity and angular velocity of the sphere after it has moved a distance d from rest, then the increase in kinetic energy is: $\frac{1}{2}Mv^2 + \frac{1}{2}I_G\omega^2$.

    The point of contact P is the instantaneous centre of rotation and therefore $v = r\omega$. Also, $I_G = \frac{2}{5}Mr^2$. Thus, by conservation of energy, the increase in kinetic energy equals the loss in potential energy, i.e.:

$$\tfrac{1}{2}Mv^2 + \tfrac{1}{2}\cdot\tfrac{2}{5}Mv^2 = \tfrac{7}{10}Mv^2 = Mgd\sin\alpha, \text{ so } v = \sqrt{\tfrac{10}{7}gd\sin\alpha}.$$

7. Referring to Figure 14.10, the angular velocity ω_1 when the rod reaches the vertical is given by equating its kinetic energy there to its loss in potential energy in swinging down from the horizontal, i.e.:

$$\frac{1}{2}M(a\omega_1)^2 + \frac{1}{2}\cdot\frac{1}{3}Ma^2\cdot\omega_1^2 = \frac{2}{3}Ma^2\omega_1^2 = Mga, \quad \omega_1^2 = \frac{3g}{2a}.$$

    When the rod strikes the ball, the only external impulse to the system of rod and ball is applied through the hinge A. Hence, the total moment of momentum about A is conserved over the impact. Using the letter m for the mass of the ball, the total moment of momentum equation is:

$$a\cdot Ma\omega_2 + \tfrac{1}{3}Ma^2\cdot\omega_2 + 2a\cdot mv = \tfrac{4}{3}Ma^2\omega_2 + 2mav = \tfrac{4}{3}Ma^2\omega_1.$$

    We have two unknowns v and ω_2, so we need another equation, which is given by Newton's rule:

$$-2a\omega_2 + v = e\cdot 2a\omega_1.$$

    Re-writing the moment of momentum equation as:

$$2a\omega_2 + \frac{3m}{M}v = 2a\omega_1$$

and adding it to the Newton's rule equation gives:

$$\left(1 + \frac{3m}{M}\right) v = 2(1 + e)a\omega_1.$$

Substituting the numerical data gives:

$$\omega_1^2 = \frac{3 \times 9.8}{2 \times 0.5} = 29.4, \text{ so } \omega_1 = 5.422 \text{ rad/s} \quad \text{and}$$

$$\left(1 + \frac{3 \times 0.4}{4}\right) v = 2 \times 1.92 \times 0.5 \times 5.422, \quad v = 8.01 \text{ m/s}.$$

Part III

Problems

15 Statics

PROBLEM 1

A uniform straight rod AB is suspended by light strings AC and BD as shown in Figure 15.1. If AB is at $30°$ to the horizontal and BD is at $30°$ to the vertical, what angle must the string AC make with the vertical?

PROBLEM 2

Repeat Problem 1 with rod AB replaced by a uniform lamina in the shape of an equilateral triangle ABE. (The centre of gravity is one third of the way up the median from the base.)

PROBLEM 3

Find the magnitude and direction of the resultant $\mathbf{R}$ of the two forces $\mathbf{F}_1$ and $\mathbf{F}_2$ indicated in Figure 15.2, given that $F_1 = 2\,\text{N}$ and $F_2 = 1\,\text{N}$.

PROBLEM 4

Two weights are attached to the ends of a light string which passes over smooth pegs A and B. As shown in Figure 15.3, a third weight is suspended from a point C in the middle of the string. If the masses of the weights are in the ratio $3 : 4 : 5$ as indicated in the diagram, find the angles θ and ϕ which the sections of string AC and BC make with the vertical. (Since $3^2 + 4^2 = 5^2$, a triangle with sides in the ratio $3 : 4 : 5$ is a right-angle triangle.)

PROBLEM 5

Assume the same set-up as in Problem 4 but with the weights 3, 4 and 5 replaced by W_1, W_2 and W_3, respectively. Find the weights W_2 and W_3, if $W_1 = 20\,\text{N}$ and the angles θ and ϕ are $30°$ and $45°$, respectively.

PROBLEM 6

A light rigid strut AB is hinged to a wall so that it can turn freely about A in a vertical plane perpendicular to the wall. The strut is held at $45°$ to the vertical by a horizontal light rope BC, as shown in Figure 15.4. A weight W is suspended from B by another

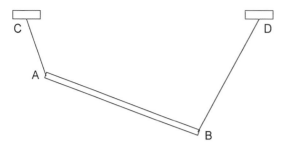

Figure 15.1. A uniform rod suspended by light strings.

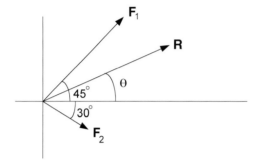

Figure 15.2. Finding the resultant of two forces acting at a point.

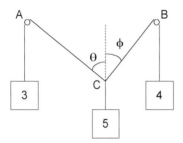

Figure 15.3. Three weights held by a string passing over two smooth pegs.

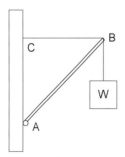

Figure 15.4. A weight supported by a light strut and light ropes.

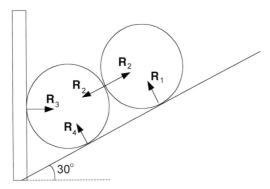

Figure 15.5. Two uniform smooth spheres resting at the bottom of a slope.

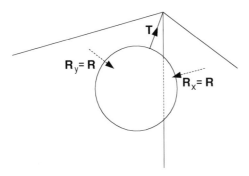

Figure 15.6. A uniform smooth sphere suspended by a light string in the corner between two walls.

light rope. Find the tension T in the rope CB and the compressive force F in the strut AB.

PROBLEM 7

Two uniform smooth spheres rest at the bottom of a 30° slope against a vertical wall as shown in Figure 15.5. Find the reactions at the surfaces of contact given that the weight of each sphere is W.

PROBLEM 8

Referring to Figure 15.6, a uniform smooth sphere of weight W is suspended by a light string in the corner between two vertical walls at right angles to each other. The string is attached to the corner and its angle to the vertical is 45°. Find the tension T in the string and the equal reactions R from the vertical walls.

PROBLEM 9

Referring to Figure 15.7, a light strut AB is hinged to a wall so that it can turn freely about A in a vertical plane perpendicular to the wall. A weight W is suspended from B and the strut is held at an angle α to the vertical by a light horizontal cable DC which

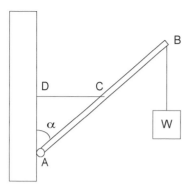

Figure 15.7. A weight supported by a hinged strut and cable.

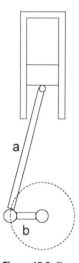

Figure 15.8. Representation of a piston engine.

is attached to the wall at D and to the mid-point of the strut at C. If the tension in the cable is T and the length of the strut is $2a$, find (a) the moment about A applied by the tension T and (b) the moment about A applied by the weight W.

PROBLEM 10

Figure 15.8 represents a *piston engine*. Noting that the sideways reactions from the cylinder and the crank allow the downward force on the piston to be transferred unaltered to the crank, find the moment exerted on the crankshaft given that: the diameter of the cylinder is 0.1 m, the gas pressure is 800 kN/m² above atmospheric and the lengths of the connecting rod and crank are $a = 0.64$ m and $b = 0.16$ m, respectively.

PROBLEM 11

The dimensions of a gravity dam are as shown in Figure 15.9: height 9 m, top and bottom thicknesses 3 m and 6 m, respectively. The weights of the rectangular and triangular sections of the dam are represented by the forces $40k$ and $20k$, respectively.

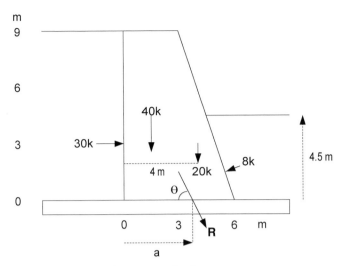

Figure 15.9. Forces acting on a gravity dam.

Figure 15.10. Plan view of a ship with horizontal forces acting on it.

The corresponding water pressure resultant forces on either side correspond to $30k$ and $8k$ acting at one third depth as shown. Find the magnitude R and direction θ of the resultant of these four forces and also the distance a from the bottom corner of the point where the resultant cuts the base of the dam.

PROBLEM 12

A ship (see plan view in Figure 15.10) is being eased into a quayside by two tugs pushing against the side of the ship, each applying a force of 2×10^6 N as indicated in the diagram. Simultaneously, twin propellers are spun in opposite directions providing a push of 1.5×10^6 N and a pull of 1.2×10^6 N as indicated, the propeller shafts being 12 m apart. Find the magnitude R and direction ϕ (measured relative to the axis of the ship) of the resultant of these four forces. Also find the distance a behind the bow of the point where the resultant force cuts the axis of the ship.

PROBLEM 13

A uniform straight plank AB is suspended horizontally by vertical ropes attached to the ends A and B, as shown in Figure 15.11. A weight W rests on the plank one quarter of the

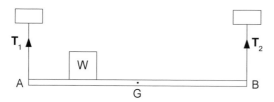

Figure 15.11. A suspended plank carrying an extra weight.

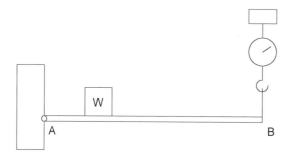

Figure 15.12. A hinged plank supported by a spring balance.

way along it from the end A. If the weight of the plank is also W, find the magnitude and the line of action of the resultant weight. Use this to determine the magnitude and the line of action of the resultant of the two tensions and hence, find the tensions T_1 and T_2.

PROBLEM 14

A uniform straight plank AB is smoothly hinged to a support at A. The plank is held in a horizontal position with a spring balance attached to the end B, as shown in Figure 15.12. With a weight $W = 200\,\text{N}$ resting on the plank one quarter the way along it from A, the spring balance registers a vertical force of $F = 100\,\text{N}$ acting upwards on the plank at B. Find the magnitude, direction and line of action of the resultant of the two forces W and F.

PROBLEM 15

Referring to Figure 15.13, let **F** be a force acting on a rigid body at A. Now find equivalent systems of forces as follows.
(a) Find equivalent forces P and Q acting along aa and cc, respectively.
(b) Find equivalent forces P and Q acting along bb and cc, respectively.
(c) Find forces P and Q when force **F** at A is replaced by P at B together with the couple formed by Q along bb and $-Q$ along cc.

PROBLEM 16

A mast of weight W has a flange which rests on a support at C (see Figure 15.14). The mast is kept vertical by brackets at A and B. Neglecting friction, find the reaction forces R_a, R_b and R_c, as indicated in the diagram, noting also the distances a and b.

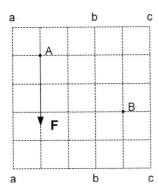

Figure 15.13. Lattice for describing equivalent force systems.

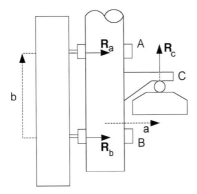

Figure 15.14. Support for a vertical mast.

PROBLEM 17

A light beam AB is hinged to a vertical wall at A. The beam is 3 m long and it is held in a horizontal position by a light cable attached to a point C of the beam, where $AC = 2\,\text{m}$, and to a point on the wall 2 m above A, as shown in Figure 15.15. The beam supports two equal weights $W = 2\,\text{kN}$ attached at B and C as shown. Find the tension T in the cable, the magnitude of the reaction force R at the hinge and the angle θ which it makes with the upward vertical.

PROBLEM 18

Figure 15.16 shows a ladder resting on a smooth horizontal surface and leaning against a smooth vertical wall. It is held in place by a light rope stretched between a point C of the ladder and the bottom of the wall. The force $W = 1\,\text{kN}$ represents the combined weight of the ladder and a man standing on the middle rung. If the rope makes an angle of $30°$ with the horizontal, find its tension T and the reactions R_a and R_b acting on the ends of the ladder A and B, respectively. Also, A and B are 2 m and 4 m from the bottom of the wall as indicated in the diagram.

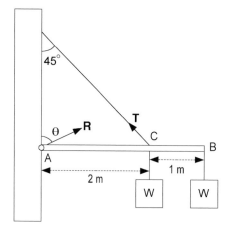

Figure 15.15. A light horizontal beam supporting two weights W and held in position by a light cable.

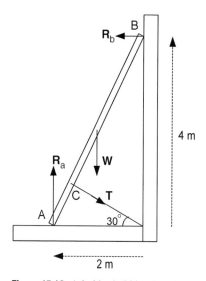

Figure 15.16. A ladder held in place by a rope.

PROBLEM 19

Figure 15.17 shows a mechanism for raising a bridge over a canal. If the weight of the bridge deck is $W = 20\,kN$ and the counterweight is $W_c = 16.5\,kN$, find the extra pull P required to lift the bridge, assuming that any other weights may be neglected. Also, find P when the deck has been raised through an angle of $45°$.

PROBLEM 20

Figure 15.18 is supposed to represent *lifting tongs* suspended at H. The pivot blocks D and F grip the weight W and lift it by friction. The bell cranks CAD and EBF pivot about A and B at the ends of the crossbar AB.

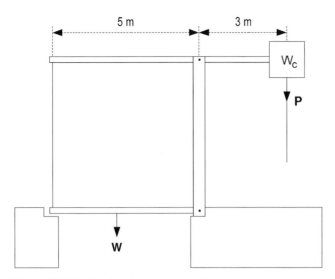

Figure 15.17. Mechanism for raising a bridge.

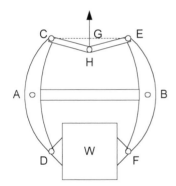

Figure 15.18. Lifting tongs lifting a cube-shaped block.

Find the tension T in the crossbar AB and the gripping force F_g on the load given the following: the bell cranks are symmetrical with both CD and EF vertical, $CD = EF = 1$ m, $CH = HE = 0.5$ m, $GH = 0.1$ m and $W = 10$ kN.

PROBLEM 21

Let four coplanar parallel forces: $F_1 = 3$ N, $F_2 = 2$ N, $F_3 = 5$ N and $F_4 = 6$ N act at the points: $A_1(2, -2)$, $A_2(-3, 0)$, $A_3(0, 2)$ and $A_4(4, -2)$, respectively, where the unit of length for the two-dimensional Cartesian position coordinates is 0.1 m. Find the coordinates x_c and y_c for the centre of action of the parallel forces.

PROBLEM 22

A light horizontal beam AB is suspended by vertical ropes at A and B, as shown in Figure 15.19. Three weights W_1, W_2 and W_3 are attached to the beam at points $1/4$, $1/2$

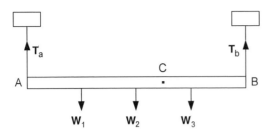

Figure 15.19. A suspended light beam supporting three weights.

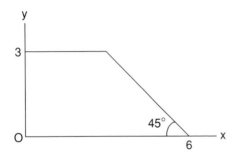

Figure 15.20. A uniform plane lamina.

and 3/4 the way along the beam from A. If $W_1 = 200\,\text{N}$, $W_2 = 200\,\text{N}$, $W_3 = 600\,\text{N}$ and the length $AB = 1\,\text{m}$, find the distance AC, where C is the centre of action of the three weights along the beam. Use this result to find the tensions T_a and T_b in the ropes supporting the beam at A and B, respectively.

PROBLEM 23

Let four non-coplanar parallel forces $F_1 = 5\,\text{N}$, $F_2 = 10\,\text{N}$, $F_3 = 7\,\text{N}$ and $F_4 = 3\,\text{N}$ act at the points $A_1(1, -2, -3)$, $A_2(-4, 1, 3)$, $A_3(2, 2, -2)$ and $A_4(-2, -1, 4)$, respectively, where the unit of length for the Cartesian coordinates of position is $0.1\,\text{m}$. Find the coordinates x_c, y_c and z_c for the centre of action of the parallel forces.

PROBLEM 24

A uniform plane lamina has the shape shown in Figure 15.20, the unit of length being $0.1\,\text{m}$. With coordinate axes as shown, find the coordinates x_g and y_g of the centre of gravity of the lamina.

PROBLEM 25

Figure 15.21 shows a triangular bracket, formed from half of a uniform square plate of side $10\,\text{cm}$, with a triangular hole such that the width of each of the three sides of the bracket is $1\,\text{cm}$. Show that the area of the hole is $21.69\,\text{cm}^2$ and, using the coordinates shown in the diagram, show that the x- and y-coordinates of the centre of gravity of the

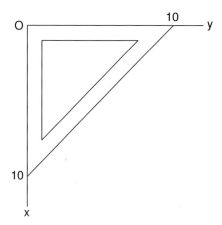

Figure 15.21. A triangular bracket.

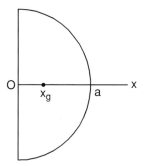

Figure 15.22. A hemispherical shell.

hole are both 3.195 cm. Hence, find the coordinates x_g and y_g for the centre of gravity of the bracket.

PROBLEM 26

Find the position of the centre of gravity of a uniform plane lamina bounded in Cartesian coordinates by the lines: $x = 1$, $x = 3$, $y = 0$ and the curve $xy = 3$.

PROBLEM 27

Find the position of the centre of gravity of a uniform plane lamina bounded in Cartesian coordinates by the two curves: $ay = x^2$ and $(x - a)^2 + y^2 = a^2$, with $y > 0$.

PROBLEM 28

(a) A hemispherical shell has constant weight per unit area and its radius is a. Find the distance x_g (see Figure 15.22) of its centre of gravity from its base. (b) Find the new value for x_g when the base is covered over with a sheet of material with the same weight per unit area.

PROBLEM 29

A uniform solid has the shape of a half ellipsoid with circular base of radius 0.1 m and height 0.2 m. Find the height of the centre of gravity.

PROBLEM 30

Two vertical poles are set 60 m apart. The length of each pole is 12 m and the bottom 2 m is embedded in concrete. A light cable is stretched between the tops of the poles. A weight of 200 N is attached to the mid-point of the cable which causes it to sag by 1 m below the tops of the poles. Assuming it to be linear, sketch the transverse load diagram for the embedding concrete of one of the poles and determine the maximum load intensity.

PROBLEM 31

A uniform solid is in the shape of a cube of side 1m weighing 10 kN. The cube is mounted on the vertical side of a wall, the attachment being made along an upper edge of the cube which is horizontal and in contact with the wall. Suppose that the whole of one face of the cube is in contact with the wall. Assume that the pressure from the cube onto the wall increases linearly from zero with distance down from the top of the cube. Sketch the corresponding load diagram and find its maximum intensity. Find the value of the horizontal component of the reaction from the wall where the cube is attached.

PROBLEM 32

A 1 m square hole in a vertical partition between two water tanks is closed by a flap hinged along the top A and resting against a stop along the bottom B, as shown in Figure 15.23. If the levels of water on either side of the partition are as shown in the diagram, find the total reaction exerted by the stop B, given that the weight of water per unit volume is 9.81 kN/m³.

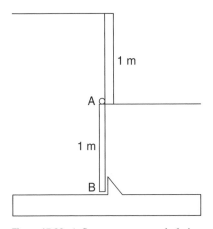

Figure 15.23. A flap over a square hole in a vertical partition between two water tanks.

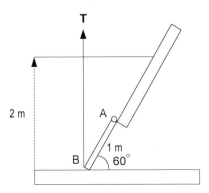

Figure 15.24. A flap over a square hole in an inclined face of a water tank.

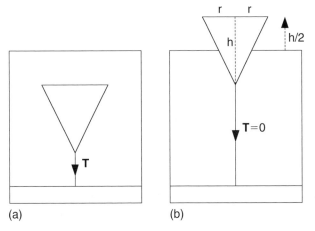

(a)   (b)

Figure 15.25. A solid cone (a) immersed and (b) floating in a liquid.

PROBLEM 33

The end of a water tank is inclined away from the vertical by $30°$, as shown in Figure 15.24. AB represents a flap covering a 1 m square hole. If the flap is hinged along the top at A, find the vertical pull T which must be exerted to open the flap against the water pressure, assuming that the weight of water per unit volume is 9.81 kN/m^3.

PROBLEM 34

A uniform solid cone is held completely immersed in a liquid by an anchor rope of tension T attached to the apex of the cone, as shown in Figure 15.25a. Find T, if V is the volume of the cone and the weight/unit volume of liquid and solid are w and w_s, respectively, with $w > w_s$. Then, find the ratio w_s/w if the cone floats $(T = 0)$ with half its height above the surface of the liquid, as shown in Figure 15.25b. (Volume of cone $= \pi r^2 h/3$.)

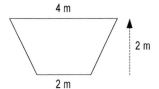

Figure 15.26. Vertical end of a tank filled with water.

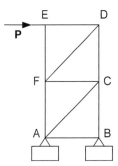

Figure 15.27. A light vertical truss with a horizontal force exerted along the top.

PROBLEM 35

A uniform solid right circular cylinder of length 30 cm and radius 5 cm floats in water with its axis vertical and 3 cm protruding above the surface of the water when a lump of lead is attached to the base. Find the volume of lead if the cylinder, water and lead have weights per unit volume of $0.5w$, w and $11.4w$, respectively.

PROBLEM 36

Figure 15.26 represents the vertical end of a tank which is filled to the top with water (weight per unit volume, $w = 9.81\,\text{kN/m}^3$). Calculate the total thrust exerted on this surface and also the depth of the centre of pressure.

PROBLEM 37

A cubic tank, with 2 m side, has top and bottom horizontal and is half filled with water. Then, the top half is filled with oil of specific gravity 0.8. (*Specific gravity* is the ratio of its density to that of water.) With the water weight/unit volume $w = 9.81\,\text{kN/m}^3$, find the total thrust on one vertical side and the depth of the centre of pressure.

PROBLEM 38

A horizontal force P is applied at joint E of the light vertical truss ABCDEF, shown in Figure 15.27, in which all vertical and horizontal struts have equal length. Use the method of sections to find the forces in the struts AF, AC and BC.

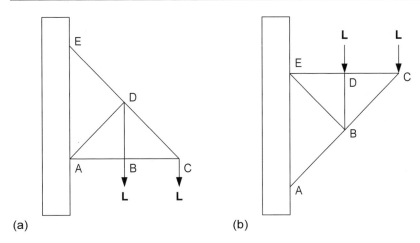

Figure 15.28. Light trusses supporting loads at two joints.

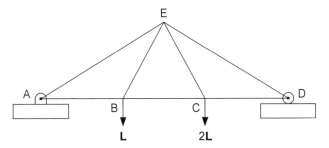

Figure 15.29. A light truss supporting loads L and $2L$.

PROBLEM 39

The light truss shown in Figure 15.28a supports loads L at B and C. $AB = BC = BD$ and $AD = CD = DE$. Use the method of sections to find the forces in struts AB, AD and DE.

PROBLEM 40

The light truss illustrated in Figure 15.28b supports loads L at D and C. Struts ED, DC and DB have equal lengths and so do struts EB, BC and AB. Use the method of joints to find the forces in all of the struts.

PROBLEM 41

Figure 15.29 shows a light truss ABCDE supported at A and D, and loaded with L at B and $2L$ at C. The struts AB, BC, CD, BE and CE have equal lengths. Use the method of joints to find the forces in all of the struts.

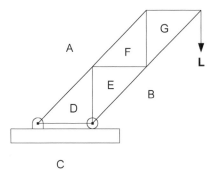

Figure 15.30. A light truss supporting a load L.

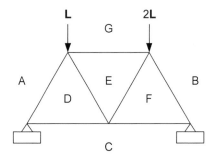

Figure 15.31. A light truss supporting loads L and $2L$.

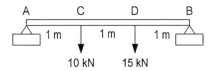

Figure 15.32. A light beam loaded at two points.

PROBLEM 42

Use Bow's notation and a single diagram to find the strut forces in the light truss shown in Figure 15.30, which supports a load L.

PROBLEM 43

Repeat the process used in Problem 42 to find the strut forces in the light truss of seven equal struts, subject to two loads L and $2L$, as shown in Figure 15.31.

PROBLEM 44

A light beam AB rests horizontally on supports at A and B. The length $AB = 3$ m and the beam is subject to loads of $10\,$kN and $15\,$kN at C and D, respectively, as shown in Figure 15.32. Draw the shearing force and bending moment diagrams, and determine the magnitude and location of the maximum value of each.

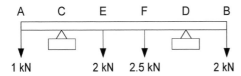

Figure 15.33. A light beam loaded at four points.

Figure 15.34. A heavy beam on two point supports.

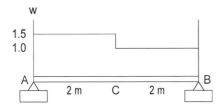

Figure 15.35. Load distribution on a horizontal beam.

PROBLEM 45

A light beam AB rests horizontally on supports at C and D, as shown in Figure 15.33, and is subject to loads of 1 kN, 2 kN, 2.5 kN and 2 kN at A, E, F and B, respectively. The lengths $AC = CE = EF = FD = DB = 1$ m. Draw the shearing force and bending moment diagrams, and determine the magnitude and location of the maximum value of each.

PROBLEM 46

A uniform heavy beam AB, of length $4a$ and weight intensity w per unit length, rests horizontally on supports at A and C, as shown in Figure 15.34, with the length $AC = 3a$. Sketch the shearing force and bending moment diagrams, and find the magnitude and location of the maximum value of each.

PROBLEM 47

A horizontal beam AB of length 4 m is supported at A and B. It has different distributed loads over the two halves as illustrated in Figure 15.35. The load intensities are $w = 1.5$ kN/m from A to C and $w = 1$ kN/m from C to B, where C is the centre point of the beam. Sketch the shearing force and bending moment diagrams, and find the maximum bending moment.

PROBLEM 48

A horizontal beam AB of length 3 m is supported at A and B. The load intensity w increases linearly from 300 N/m at A to 900 N/m at B, as shown in Figure 15.36.

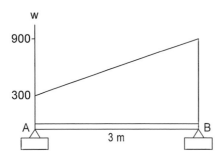

Figure 15.36. Linearly varying load on a horizontal beam.

Figure 15.37. A steel wedge splitting a log.

Denoting the distance along the beam from A by x, find the function of x for the shearing force F and the bending moment M. Hence, find the location and magnitude of the maximum bending moment.

PROBLEM 49

If a steel wedge of angle α is used for splitting a log, as indicated in Figure 15.37, what is the minimum coefficient of friction between the steel and wood for the wedge not to slip out between hammer blows?

PROBLEM 50

A block of wood weighing 20 N rests on a horizontal plank of length 2 m. If one end of the plank is slowly raised, the block will start to move as soon as the height exceeds 0.8 m. What is the coefficient of friction between the block and the plank? If the block is stopped from sliding by being pressed against the plank, what is the minimum such pressure required to prevent sliding when the end of the plank is at a height of 1 m?

PROBLEM 51

Figure 15.38 shows a uniform rectangular block resting on a plane surface inclined at angle α to the horizontal. P is a force applied to the centre of the top edge of the block, that edge being horizontal. Find how small the coefficient of friction must be in terms of a, b and α for the block to eventually slide rather than topple as P is increased.

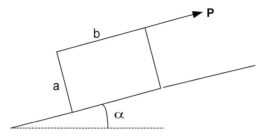

Figure 15.38. Condition on coefficient of friction for block to slide rather than topple as force P is increased.

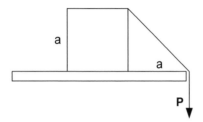

Figure 15.39. A cube being pulled by a string attached to its upper edge.

PROBLEM 52

A uniform solid cube of side a rests on a horizontal table at a distance a from the edge. A smooth string is attached to the top of the cube and then passed over the edge of the table to where a downward force P is applied, as shown in Figure 15.39. Find how large the coefficient of friction between the cube and the table must be for the block to topple rather than slide as P is increased. (The string is attached to the centre of the upper edge of the cube, that edge being parallel to the edge of the table.)

PROBLEM 53

A uniform rectangular block of weight W is prevented from sliding down a steep inclined plane by a force P as indicated in Figure 15.40. Given that the angle of friction is $\lambda < \alpha$, the angle of inclination of the plane to the horizontal, find the angle θ to minimize the necessary P and also find the corresponding magnitude of P.

PROBLEM 54

A ladder is leant up against a wall at an angle of $70°$ to the horizontal, as shown in Figure 15.41. The wall is rough with a coefficient of friction $\mu_a = 1$ but the floor is relatively smooth with a coefficient of friction $\mu_b = 0.2$. The force W in the diagram represents the resultant of the weight of the ladder and that of a climber on the ladder. If l is the length of the ladder, find the maximum distance a that W may be up the ladder before the ladder slips.

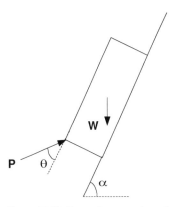

Figure 15.40. Force P preventing a block sliding down a plane.

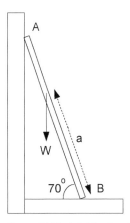

Figure 15.41. A ladder leaning against a wall.

PROBLEM 55

A ladder AB is leant against a wall at an angle θ to the horizontal, which is just sufficiently large for the ladder not to slip. This is illustrated in Figure 15.42 along with the reaction forces R_a and R_b from wall and floor which must be concurrent with the line of action of the resultant weight acting through the centre of the ladder. Given that the coefficient of friction is the same at both A and B, find the angle of friction λ in terms of θ. (Hint: lines joining the ends of a diameter to a point on the circumference of a circle are perpendicular.)

PROBLEM 56

A clutch plate has a flat driven plate gripped between a driving plate and a pressure plate giving two driving surfaces each with an inner radius of 5 cm and outer radius of 10 cm. Springs supply an axial force $P = 4\,\text{kN}$. Find the maximum torque which may be transmitted if the coefficient of friction $\mu = 0.4$ when (a) the pressure intensity is constant over the surface and when (b) it is a constant wear clutch.

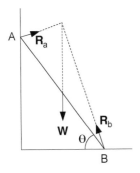

Figure 15.42. A ladder with angle θ just large enough to prevent slipping.

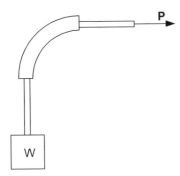

Figure 15.43. A weight suspended from a rope which passes through a bent pipe.

PROBLEM 57

A weight W is suspended from a rope which passes through a pipe bent through a quarter circle before emerging to be pulled by a horizontal force P, as shown in Figure 15.43. If the weight W exerts a downward force of 400 N and the coefficient of friction between the rope and the pipe is $\mu = 0.25$, find the range of the values of P for which the weight will remain where it is.

PROBLEM 58

Let a force $\mathbf{F} = 3\mathbf{i} - 4\mathbf{j} + 5\mathbf{k}$ N act on a rigid body at a point P with coordinates $(-1, 2, -1)$ cm together with a couple of moment $\mathbf{C} = -7\mathbf{i} - 9\mathbf{j} - 3\mathbf{k}$ N cm. The effect on the body is equivalent to the force $\mathbf{F}$ acting by itself at a point Q. Find the coordinates of Q given that its z-coordinate is 1 cm.

PROBLEM 59

Figure 15.44 shows three rigidly connected arms lying in the same plane. Each arm ends in a circular disc: A of radius 15 cm, B of radius 12 cm and C of radius 10 cm. The arms connected to A and B are at right angles to each other. Equal and opposite forces are applied to opposite sides of the discs as shown. If $F_a = 50$ N and $F_b = 100$ N, find

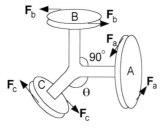

Figure 15.44. Discs on rigidly connected arms lying in the same plane.

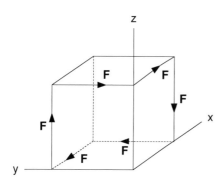

Figure 15.45. Three couples formed by equal forces acting along the edges of a cube.

the value of F_c and the angle θ such that the couple applied to disc C balances the couples applied to discs A and B.

PROBLEM 60

Forces F act along six of the edges of a cube, as shown in Figure 15.45. If the length of each edge is a, use the Cartesian axes in the diagram to specify the moment vector of the resultant couple in terms of the unit vectors $\mathbf{i}$, $\mathbf{j}$ and $\mathbf{k}$ in the x-, y- and z-directions.

PROBLEM 61

A negative wrench applied through a point P with position vector $(\mathbf{i} + \mathbf{j} - \mathbf{k})$ m has a couple of moment 20 N m and force $\mathbf{F} = (20\mathbf{i} - 30\mathbf{j} + 40\mathbf{k})$ N. Find the vector moment of the wrench about the point Q with position vector $(2\mathbf{i} + 2\mathbf{j} + \mathbf{k})$ m.

PROBLEM 62

Assuming appropriate units, let a force $\mathbf{F} = 10\mathbf{i} + 20\mathbf{j} - 20\mathbf{k}$ act through a point P with position vector $\mathbf{r}_p = 2\mathbf{i} + 3\mathbf{j} + \mathbf{k}$. If this is combined with a couple with vector moment $\mathbf{C} = 5\mathbf{i} - 5\mathbf{j} + 10\mathbf{k}$, find the equivalent wrench and the position vector $\mathbf{r}_q$ of its intercept with the xy plane.

PROBLEM 63

Find the force $\mathbf{F}$ acting at the origin and the moment $\mathbf{C}$ of the couple, which together are equivalent to the system of forces shown in Figure 15.46.

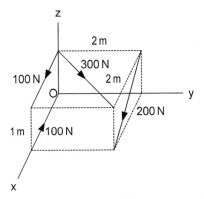

Figure 15.46. A system of forces to be replaced by the equivalent couple and force at the origin.

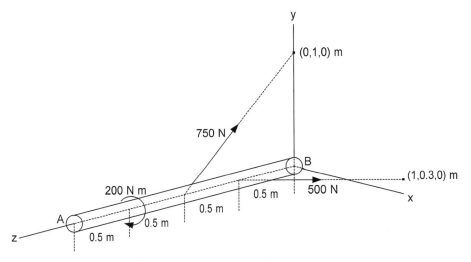

Figure 15.47. Two forces and a couple acting on a bar AB.

PROBLEM 64

Figure 15.47 shows two forces and a couple acting on a bar of length 2 m. What is the equivalent resultant force **F** and couple of moment **C** acting at A?

PROBLEM 65

A uniform straight rod AB of length $\sqrt{2}a$ is freely pivoted with a ball joint at the end A, which is at a point of distance a from a vertical wall. Setting up Cartesian axes as shown in Figure 15.48, with the x-axis corresponding to the horizontal bottom edge of the wall, let θ be the maximum angle between OB and the vertical z-axis for the rod to rest without slipping. Find μ, the coefficient of friction between the rod and the wall.

PROBLEM 66

Referring to Figure 15.49, a uniform straight rod of weight $W = 100\,\text{N}$ and length 2 m is smoothly jointed to a fixed point O, which we take as the origin of Cartesian

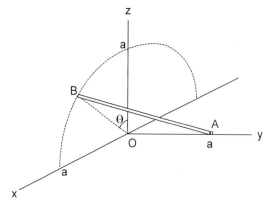

Figure 15.48. Rod AB pivoted at A with end B resting against a vertical wall.

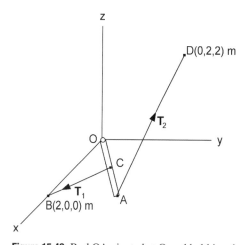

Figure 15.49. Rod OA pivoted at O and held in a horizontal position by strings BC and AD.

coordinates, with the z-axis vertical. The rod is held in a horizontal position at an angle of $30°$ to the x-axis by one string stretched between its other end A and a fixed point $D(0, 2, 2)$ m and another string stretched between its centre C and a fixed point $B(2, 0, 0)$ m. Find the tension T_1 in string BC and the tension T_2 in string AD. Also find R_x, R_y and R_z, the x-, y- and z-components of the reaction at the joint O.

PROBLEM 67

A block weighing 200 N slides a distance of 2 m down a slope of $30°$ to the horizontal against a frictional force of 50 N. Find W_g, the work done by gravity and W_f, the work done by the frictional force.

PROBLEM 68

Find the work done by a force $\mathbf{F} = (200\mathbf{i} - 100\mathbf{j} + 100\mathbf{k})$ N when its point of application moves from $A(-1, 1, 2)$ m to $B(2, 2, 0)$ m.

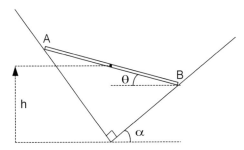

Figure 15.50. A rod resting between two smooth planes.

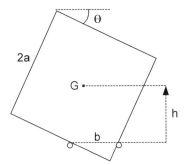

Figure 15.51. A cube resting on two smooth poles.

PROBLEM 69

What is the work done by the torque of moment 3 N m of an electric motor when it turns its load through an angle of 30°?

PROBLEM 70

Find the work done by a couple with moment vector $(10\mathbf{i} + 20\mathbf{j} + 40\mathbf{k})$ Nm when the body on which it acts turns through 10° about an axis with direction $(\mathbf{i} - \mathbf{j} + \mathbf{k})$.

PROBLEM 71

Two smooth perpendicular planes meet in a horizontal line. One plane is at an angle $\alpha < 45°$ to the horizontal. A uniform straight rod AB rests between the planes as indicated in Figure 15.50. Assuming that the rod is in a vertical plane perpendicular to the smooth surfaces, find the angle θ between AB and the horizontal for equilibrium. (Hint: taking W as the weight of the rod, the virtual work for a small change $\delta\theta$ in θ is: $-W\delta h = -W\frac{dh}{d\theta}\delta\theta$.)

PROBLEM 72

Referring to Figure 15.51, a uniform solid cube of side $2a$ rests on two smooth horizontal poles, which are parallel, distance b apart and at the same height. Show that a non-symmetrical position of equilibrium exists if $1/\sqrt{2} < b/a < 1$. Find the corresponding

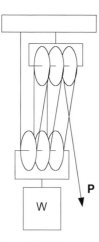

Figure 15.52. A rope and pulley system.

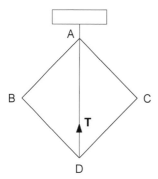

Figure 15.53. A square framework of struts held in place by a light cable AD.

equilibrium value for the angle θ between the lower side and the horizontal when $b = 0.75a$.

PROBLEM 73

Figure 15.52 represents a rope and pulley system with one set of three pulleys and one end of the rope attached to a roof support. The other set of three pulleys, together with a load, has weight W and is suspended in equilibrium with a pull force P applied to the other end of the rope. Neglecting the weight of the rope and any friction in the pulleys, use virtual work to find P in terms of W.

PROBLEM 74

Figure 15.53 represents four equal uniform struts smoothly jointed to form a square which is prevented from collapsing by a light cable AD. Use virtual work to find the tension T in the cable if the weight of each strut is w.

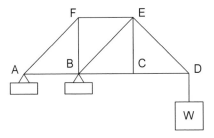

Figure 15.54. A framework of nine struts supporting a load W.

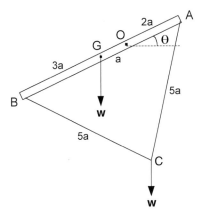

Figure 15.55. A freely pivoted rod supporting a weight on a string attached to the ends of the rod.

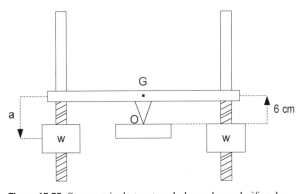

Figure 15.56. Symmetrical structure balanced on a knife-edge.

PROBLEM 75

The framework shown in Figure 15.54 consists of freely jointed uniform struts which are constrained to stay in the same vertical plane. Each strut is of weight w or $\sqrt{2}w$ according to length. The framework is fastened to two fixed points A and B, where AB is horizontal, and a weight W is suspended from D. Use the method of virtual work to find the tension or compression in strut BE.

PROBLEM 76

A uniform straight rod AB of length $6a$ is freely pivoted about O at distance $2a$ along the rod from A, as shown in Figure 15.55, so that it can turn in a vertical plane about O. The weight of the rod is W and another weight w is suspended from the mid-point C of a light string attached to A and B with $AC = BC = 5a$. Show that the string will be taut in an equilibrium position similar to that in the diagram if $W < 2w$. In this case, if θ is the angle between AB and the horizontal, find the equilibrium value for θ in terms of W and w and test it for stability.

PROBLEM 77

Figure 15.56 shows a symmetrical structure balanced on a knife-edge at O. The weight of the structure is $W = 20\,\text{N}$ and its centre of gravity is at G with $OG = 6\,\text{cm}$. Two weights, $w = 15\,\text{N}$ each, are screwed onto the structure so that their centres of gravity are the same depth a below G. Find how large a must be for the structure to rest in stable equilibrium.

16 Dynamics

PROBLEM 78

A motor car accelerates from rest with a constant acceleration f and reaches a speed of 24 m/s after 12 s. Evaluate f and also the distance x travelled during that time.

PROBLEM 79

A car travels a distance of 480 m in 30 s by accelerating from rest at 2 m/s^2 until it reaches a certain speed $v = v_c$. The speed $v = v_c$ is then maintained constant for some time until the brakes are applied to bring the car to rest with a constant deceleration of 3 m/s^2. Evaluate the speed v_c.

PROBLEM 80

A point P moves in a straight line with simple harmonic motion about a fixed point O. If the acceleration of P is 0.8 m/s^2 when its distance from O is 0.2 m, find the period T of the oscillation. Also, if the velocity of P is 0.8 m/s as it passes through O, find the amplitude a of the oscillation.

PROBLEM 81

A point P moves in a straight line with simple harmonic motion about a fixed point O with period $T = 2$ s. As P moves away from O it passes another fixed point Q with velocity 0.2 m/s, the distance OQ being 0.1 m. Find the time which elapses before P passes Q again on the way back.

PROBLEM 82

A crank rotates at a constant angular speed $\dot{\theta} = \omega$ (rad/s). Attached to the crank is a block which slides up and down a guide, the latter being attached to a piston, as shown in Figure 16.1. Let x be the position of the piston measured to the right of its centre of oscillation. If we let θ be the function of time $\theta = \omega t$, find the corresponding functions of time for the position, velocity and acceleration of the piston.

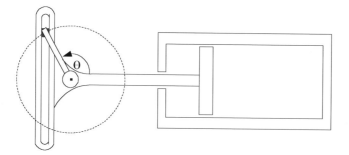

Figure 16.1. Piston and guide connected to a crank.

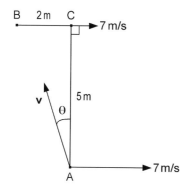

Figure 16.2. A ball passed between two rugby players.

PROBLEM 83

A wheel which is turning initially at 50 rad/s is slowed down at a constant rate of 0.2 rad/s^2. Find how many revolutions the wheel will make before it comes to rest.

PROBLEM 84

Two rugby players A and B are running down the field at 7 m/s along lines 5 m apart and parallel to the touchline. A passes the ball to B so that B can receive it without checking and the flight of the ball is perpendicular to the touchline. If B is 2 m behind A, as shown in Figure 16.2, find the magnitude and direction of the velocity **v** of A's pass (relative to A).

PROBLEM 85

Referring to Figure 16.3, a boat is rowed across a 100 m wide river which is flowing at 1 m/s. If the boat is propelled at 2 m/s relative to the water, what must be its heading (θ) in order to cross to a point 20 m downstream and how long will it take to cross?

PROBLEM 86

Two footpaths cross each other at right angles. When one person A is approaching the crossing from a distance of 50 m at a speed of 4 km/h, another person B is at the crossing

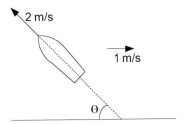

Figure 16.3. A boat being rowed across a river.

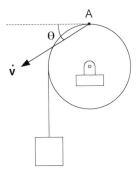

Figure 16.4. Finding the acceleration of the topmost point of the wheel as the string unwinds.

proceeding along the other path at 3 km/h. Find the magnitude v of the velocity of B relative to A, the nearest distance d of approach of A to B and the distance a that A has walked when the point of nearest approach is reached.

PROBLEM 87

A ship A, which is 3 km north of another ship B, is travelling at 12 km/h in the direction of 30° S of E, while B is travelling at 18 km/h in the direction 30° N of E. Find the nearest distance of approach as the ships pass each other.

PROBLEM 88

Referring to Figure 16.4, a wheel of radius 0.1 m has a string wrapped round it which carries a weight as shown. At a given instant the weight is moving downwards with a velocity of 0.5 m/s and acceleration of 2 m/s². Assuming no slipping of the string, find the magnitude and direction of the acceleration of the topmost point A of the wheel at that instant.

PROBLEM 89

A locomotive moves away from rest with constant acceleration along a circular curve of radius $r = 600$ m. If its speed after 60 s is 24 km/h, find what its tangential a_t and normal a_n components of acceleration were at $t = 30$ s after the start.

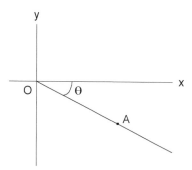

Figure 16.5. A is the point on the smooth slope where a block sliding down the slope reaches a given velocity v.

PROBLEM 90

A mine-cage moves downwards with constant acceleration and travels a distance of 50 m in 10 s after starting from rest. If the total mass of the cage and its passengers is 1500 kg, find the tension T in the suspension cable during this movement.

PROBLEM 91

Continuing on from Problem 90, find the tension T in the suspension cable if the cage is now brought to rest with constant deceleration over the next 40 m.

PROBLEM 92

Let a block slide down a perfectly smooth surface inclined at an angle θ to the horizontal. If A is the point (see Figure 16.5) reached when the block achieves a given velocity v, find the locus of A as the inclination of the surface is set at different values of θ, with the block starting from rest at O in each case.

PROBLEM 93

A block slides down a plane surface inclined at 20° to the horizontal. If the coefficient of kinetic friction is 0.1, find: (a) the speed v of the block after sliding from rest a distance of 1 m and (b) the time taken to reach that position.

PROBLEM 94

A four-wheel drive vehicle with traction control and antilock brakes is driven up a somewhat slippery slope inclined at 6° to the horizontal for a distance of 400 m. It starts and ends at rest using maximum possible acceleration and deceleration, and travels at a constant speed of 80 km/h in between the acceleration and deceleration phases. If the coefficient of static friction between the tyres and the slope is $\mu = 0.2$, find the time taken to complete the journey.

PROBLEM 95

A motor car is travelling along a straight horizontal stretch of road when the driver slams on the brakes. The car skids to a halt in 4 s over a distance of 39.2 m. Find the coefficient of kinetic friction between the tyres and the road.

PROBLEM 96

When a mass of 2 kg is suspended by a particular spring, the spring is stretched by $x_0 = 5$ mm in its equilibrium position. Find the spring stiffness k. If the weight is pulled down a little way and then released, it will oscillate up and down. Derive a formula for the frequency of oscillation in terms of x_0 and then evaluate the frequency in this case.

PROBLEM 97

Find the length a of a simple pendulum if its frequency is 50 cycles/minute.

PROBLEM 98

A cyclist on a horizontal track rounds a bend of radius 25 m at a speed of 36 km/h. What is the least value of the coefficient of static friction between the bicycle tyres and the track for this to be possible?

PROBLEM 99

A rail track on a curve of radius 100 m is banked so that a train travelling at 36 km/h will exert no sideways pressure on the rails. Calculate the height of the outer rail above the inner rail if the distance apart of the rails is 1 m.

PROBLEM 100

If a weight which is suspended by a string of length a (see Figure 16.6) passes through its lowest point ($\theta = 0$) with a speed of $\sqrt{4ga}$, find the angle θ at which the weight will leave its circular path.

PROBLEM 101

A weight of mass 0.5 kg is whirled round in a vertical circle at the end of a string of length 1 m. What is the possible range of speeds v through the highest point if the breaking tension is 36 N?

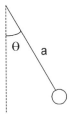

Figure 16.6. A weight swinging on the end of a string.

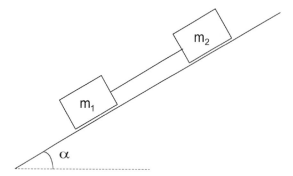

Figure 16.7. Weights connected by a string and sliding down an inclined plane.

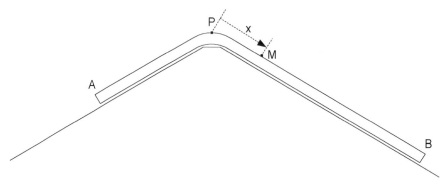

Figure 16.8. A flexible chain straddling two smooth inclined planes.

PROBLEM 102

A projectile is fired over a horizontal range. Find its angle of projection α for its range to be five times its height of trajectory. Find the velocity of projection V and time of flight T if the range is 500 m. (Neglect air resistance.)

PROBLEM 103

The horizontal ranges of a projectile fired at elevations $\alpha = 30°$ and $45°$ are $(R - 20)$ m and $(R + 20)$ m, respectively, where R is the desired range. Find the value of α to achieve the desired range R. Also, find R and the projected velocity V. (Neglect air resistance.)

PROBLEM 104

Two blocks with masses $m_1 = 1$ kg and $m_2 = 1$ kg are connected by a string as shown in Figure 16.7 and slide down a surface inclined at angle $\alpha = 30°$ to the horizontal. If the coefficients of kinetic friction with the surface are $\mu_1 = 0.1$ for m_1 and $\mu_2 = 0.2$ for m_2, find the acceleration $\dot{v}$ of the blocks down the surface and the tension T in the string.

PROBLEM 105

Figure 16.8 is supposed to represent a flexible chain AB, of length $2l$ and with mid-point M. The chain straddles the join of two smooth planes, each with a slope of $30°$

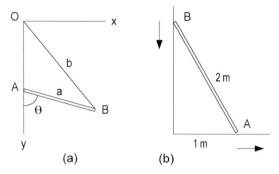

Figure 16.9. (a) Rod AB with A in a vertical track and B suspended from O by a string. (b) Rod AB with A in a horizontal track and B in a vertical track.

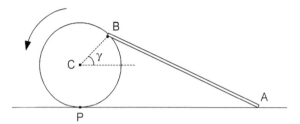

Figure 16.10. A rolling wheel with rod hinged to side of rim.

to the horizontal. If x is the distance of M from the peak P, find the expression for x as a fuction of time t, while A is still to the left of P, given that the chain is released from rest at $t = 0$ with $x = x_0 > 0$.

PROBLEM 106

A rod AB can move in a vertical plane with A in a vertical track and B suspended by a string from a fixed point O vertically above A. Denoting the lengths $AB = a$ and $OB = b$, then $b > a$. Using the coordinate system shown in Figure 16.9a, derive the coordinates of the instantaneous centre of rotation of the rod parametrically in terms of θ as A slides down the track and θ changes from 0 to $\pi/2$.

PROBLEM 107

A straight rod AB of length 2 m is constrained to move in a vertical plane with A moving along a horizontal track while B moves in a vertical track (see Figure 16.9b). Find the angular velocity of the rod at the instant when A is 1 m away from the vertical track and moving with velocity $v_a = 1$ m/s.

PROBLEM 108

Referring to Figure 16.10, a wheel of radius 0.3 m is rolling along a horizontal surface at a speed of 1 m/s. A rod AB is hinged to a point on the side of the rim at B, while the end A drags behind on the horizontal surface. Find the magnitude v_b and the angle β to the horizontal of the velocity $\mathbf{v}_b$ of B when CB makes an angle $\gamma = 30°$ to the horizontal, C being the centre of the wheel. If the length of the rod AB is 1.2 m, find the velocity v_a of A when B is in the same position.

PROBLEM 109

A uniform circular hoop rolls from rest without slipping down a slope of $30°$ to the horizontal. Find its velocity when it has rolled 10 m.

PROBLEM 110

A uniform circular hoop is released from rest on a slope of $45°$ to the horizontal. If the coefficient of static friction μ_s between the hoop and the sloping surface is $\mu_s < 0.5$, the hoop will slide as well as roll down the slope. Given that $\mu_s < 0.5$ and that the coefficient of kinetic friction is $\mu_k = 0.2$, find the velocity of the hoop when it has moved 10 m from the rest position. If the radius of the hoop is 0.2 m, find its angular velocity at the same instant.

PROBLEM 111

The shape of the surface of a uniform solid body is formed by rotating the ellipse $x^2 + 4y^2 = 4$ about the x-axis. The unit of length is the decimetre and the body has mass $M = 10$ kg. Find its moment of inertia about the x-axis.

PROBLEM 112

A uniform triangular lamina of mass M has edges of lengths a and b, which are perpendicular to each other. Find the moment of inertia of the lamina about an axis perpendicular to the plane of the lamina and passing through the right-angled corner.

PROBLEM 113

A uniform plane lamina is bounded by the parabola $x + y^2 = a^2$ with $x \geq 0$ and the y-axis with $-a \leq y \leq a$. If M is the mass of the lamina, find its moment of inertia about an axis perpendiculer to its plane and passing through the origin $(0, 0)$.

PROBLEM 114

Two masses M_1 and M_2 are suspended by a light string which passes over a pulley, as shown in Figure 16.11. The mass of the pulley is M_3 and its moment of inertia is the same as that of a uniform circular disc with the same mass and radius. Let $M_1 = 2$ kg and $M_2 = M_3 = 1$ kg. Neglecting friction in the bearing and assuming that the string does not slip, find the acceleration of the masses and the tensions T_1 and T_2 in the corresponding ends of the string after the masses have been released.

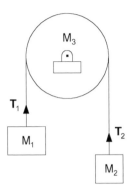

Figure 16.11. Weights with different masses suspended from pulley.

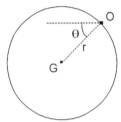

Figure 16.12. A uniform circular disc swinging about a point on its rim.

PROBLEM 115

Referring to Figure 16.12, a uniform circular lamina can turn freely in its own vertical plane about a point O in its rim. It is released from rest with its centre G at the same height as O. If its mass $M = 1$ kg and its radius $r = (2/3)$ dm, find its angular velocity $\dot{\theta}$ as it swings through its lowest position. Measuring θ as the angle of OG from the horizontal, find the angles θ for which: (a) The horizontal component R_h of the hinge reaction is maximum and (b) the vertical component R_v of the hinge reaction is maximum. Evaluate the corresponding components of force in each case.

PROBLEM 116

Analyse the system of weights and pulleys shown in Figure 16.13. Assume that the light rope to which the mass $2M$ is attached does not slip on the pulleys and that the moment of inertia of each pulley corresponds to that of a uniform circular disc of mass M and radius r. Find the acceleration $\ddot{x}$ of the mass $2M$ and the tensions T_1, T_2 and T_3 in the three vertical sections of the rope.

PROBLEM 117

Two uniform circular cylinders have the same mass M and radius r but one is solid and the other is hollow, as indicated in Figure 16.14. Their axles are connected by a light rigid connection AB and they are allowed to roll down a plane inclined at angle α to

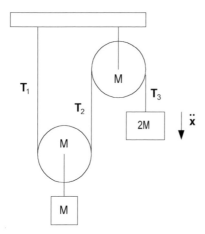

Figure 16.13. Two weights suspended from a rope and pulley system.

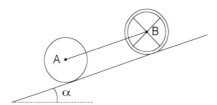

Figure 16.14. Solid and hollow cylinders with a rigid connection.

the horizontal. Assuming no slipping at the points of contact with the inclined plane and neglecting axle friction, find the acceleration down the plane and the force in the connection AB.

PROBLEM 118

A block of soft wood of mass 1 kg is at rest on a smooth horizontal surface. The block is struck by a bullet of mass 30 g travelling horizontally at 650 m/s. The bullet passes through the block and emerges from the other side at a speed of 150 m/s. Calculate the velocity v which is imparted to the block.

PROBLEM 119

Two people, each of mass 70 kg, are standing on a raft of mass 400 kg which is floating at rest on still water. The people run together and dive off the end with a horizontal velocity relative to the raft of 4 m/s. Neglecting the water resistance, find the velocity V_1 which is imparted to the raft. Repeat the calculation to find the velocity V_2 imparted to the raft when the people dive off in succession, the second one starting to run immediately after the first one has dived.

PROBLEM 120

Two particles, A of mass $2m$ travelling with velocity $2v$ and B of mass $3m$ with velocity v, collide and coalesce, as indicated in Figure 16.15. If their original paths are at $60°$ to each other, find the velocity V of the combined particle and the angle α through which the path of A is deflected.

PROBLEM 121

A ball is dropped from a height of 2 m onto a horizontal surface. If the coefficient of restitution when the ball strikes the surface is $e = 0.9$, find the total distance D travelled by the ball before it comes to rest on the surface. Also find the total time T which elapses between the ball being released from rest at the height of 2 m and finally coming to rest on the surface.

PROBLEM 122

Two identical smooth spheres travelling along a horizontal surface with the same speed u collide as indicated in Figure 16.16. Find the magnitudes (v_1 and v_2) and directions (ϕ_1 and ϕ_2) of their velocities after the collision assuming a coefficient of restitution $e = 0.94$.

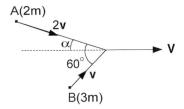

Figure 16.15. Two particles collide and coalesce.

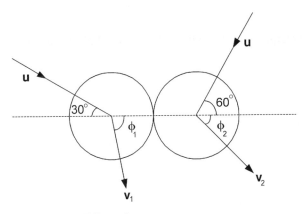

Figure 16.16. Colliding spheres.

PROBLEM 123

Referring to Figure 16.17, find the height h above the table one would need to strike a billiard ball of radius a with a horizontal impulse I in order to avoid any sliding on the table at the point of contact P.

PROBLEM 124

A uniform straight rod AB of length $2a$ and mass M is falling without turning with its length horizontal. When travelling with velocity v, its end A hooks onto a fixed pivot, as shown in Figure 16.18. Find the reactive impulse I supplied by the pivot at A and the angular velocity ω induced in the rod.

PROBLEM 125

Referring to Figure 16.19a, find the position P of the centre of percussion of a uniform circular lamina which is smoothly hinged at a point A on the circumference.

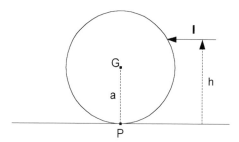

Figure 16.17. Striking a ball so as to avoid any sliding at P.

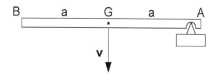

Figure 16.18. Falling rod AB striking a fixed pivot point at A.

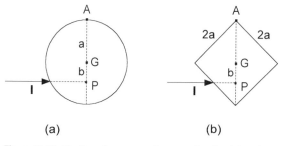

Figure 16.19. Finding the centre of percussion for (a) a circular lamina hinged on the circumference and (b) a square lamina hinged at a corner.

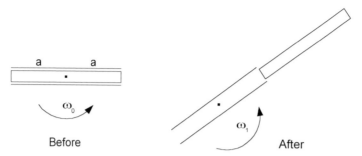

Figure 16.20. A smooth rod sliding out of a rotating tube.

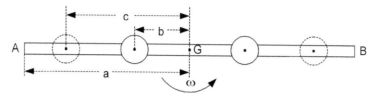

Figure 16.21. A spinning rod carrying sliding weights.

PROBLEM 126

Referring to Figure 16.19b, a uniform square lamina of side $2a$ is smoothly hinged at a corner A. Find the position P of the centre of percussion.

PROBLEM 127

A uniform straight tube of mass M and length $2a$ can rotate freely in a horizontal plane about a vertical axis at its centre. A smooth rod, also of mass M and length $2a$, rests inside the tube as indicated in the left-hand diagram of Figure 16.20. The tube is set spinning with an angular velocity ω_0. The rod slides out of the tube due to the centrifugal effect. Find the angular velocity ω_1 when the rod is about to emerge completely from the tube, as indicated in the right-hand diagram.

PROBLEM 128

Referring to Figure 16.21, a uniform straight rod AB can turn freely in a horizontal plane about a vertical axis through its centre. Two equal weights, each of mass 5 kg, have holes through their centres so that they can slide freely along the rod. The weights are attached to each other by a string and they are placed with the string taut and their centres each at a distance $b = 0.4$ m from the centre G of the rod. The system of rod and weights is set spinning with an angular velocity $\omega_0 = 10$ rad/s. The string is then cut so that the weights slide out to rest against stops with their centres each at distance $c = 0.8$ m from G. If the rod has length $2a = 2$m and mass $M = 4$ kg, find the new angular velocity ω_1 of the system, neglecting the individual moments of inertia of the weights.

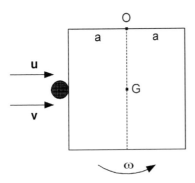

Figure 16.22. A pivoted cube struck by a ball-bearing.

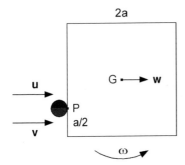

Figure 16.23. A free cube struck by a ball-bearing.

PROBLEM 129

A uniform solid cube is at rest on a smooth horizontal surface. It can turn freely about a vertical axis through the mid-point O of one of its sides (see Figure 16.22). The cube is struck by a ball-bearing of mass $m = 10$ g travelling with velocity $u = 100$ m/s perpendicular to and at the mid-point of an adjacent side. If the cube has mass $M = 1$ kg and side of length $2a = 0.1$ m, find the velocity v of the ball-bearing and the angular velocity ω of the cube immediately after the impact, given that the coefficient of restitution is $e = 0.5$.

PROBLEM 130

Assume the same set-up as in Problem 129 but this time the cube is not constrained to turn about an axis. Also the ball-bearing strikes the cube at point P a distance $a/2 = 0.025$ m from the vertical edge of the cube (see Figure 16.23). Again with $u = 100$ m/s, find the velocity v of the ball-bearing, the velocity w of the centre G of the cube and the angular velocity ω of the cube immediately after the impact.

PROBLEM 131

A ball is projected from a smooth horizontal tube by placing it against a compressed spring and then releasing the spring (see Figure 16.24). Find the velocity of projection

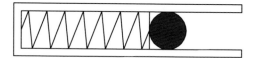

Figure 16.24. A ball projected from a tube by a compressed spring.

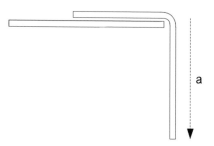

Figure 16.25. A chain about to slide off a smooth table.

if the spring constant is $k = 4\,\text{N/cm}$, its amount of compression is $5\,\text{cm}$ and the mass of the ball is $m = 10\,\text{g}$.

PROBLEM 132

A weight is attached to a fixed point by a string and is released from rest with the string taut and horizontal. If the length of the string is $2.5\,\text{m}$, find the velocity of the weight as it swings through its lowest point.

PROBLEM 133

Figure 16.25 represents a flexible chain hanging over the edge of a smooth horizontal table. If the chain is released from rest with the overhang $a = 16\,\text{cm}$, the length of the chain being $l = 32\,\text{cm}$, find its velocity when it finally leaves the table.

PROBLEM 134

Repeat Problem 131 by equating the kinetic energy of the projected ball to the potential energy in the compressed spring.

PROBLEM 135

An engine with a power output of $400\,\text{kW}$ pulls at constant speed a train of total mass (i.e. including the mass of the engine) of $2 \times 10^5\,\text{kg}$ up a slope of 1 in 80 against a resistance of $10\,\text{kN}$. Find the speed of the train.

PROBLEM 136

A vehicle of mass $M = 1800\,\text{kg}$ is clinbing a constant incline of 1 in 60 at a steady speed of $45\,\text{km/h}$. If the traction power is $6\,\text{kW}$, what is the total resistive force? If subsequently the vehicle is subject to the same resistive force when travelling along the horizontal at a speed of 45 km/h, what will be its acceleration at that instant if its traction power is 10 kW?

PROBLEM 137

A uniform circular lamina of mass M and radius a is turning in its own plane about a point O on its rim, as shown in Figure 16.26a. Find its kinetic energy when its angular velocity is ω. Repeat the process for a uniform square lamina of side $2a$ turning about a corner (see Figure 16.26b).

PROBLEM 138

Let Figure 16.26b represent a uniform square lamina of side $2a = 0.3$ m turning freely in its own vertical plane about a corner O. If it is released from rest with OG horizontal, G being its centre of gravity, find its angular velocity ω through its lowest point.

PROBLEM 139

A uniform solid sphere of radius $r = 7$ cm is allowed to roll across the bottom of a horizontal circular cylinder of radius $R = 2r = 14$ cm (see Figure 16.27). Assuming sufficient friction for it to roll without slipping, find the velocity of the sphere as it passes through the lowest position given that it was released from rest with the line of centres OG at $30°$ down from the horizontal.

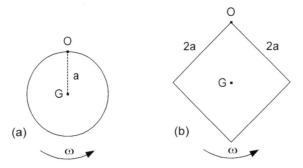

Figure 16.26. (a) Uniform circular lamina turning about a point on its rim. (b) Uniform square lamina turning about a corner.

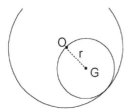

Figure 16.27. Solid sphere rolling through the bottom of a hollow cylinder.

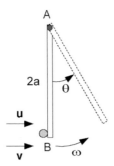

Figure 16.28. Rod AB suspended from A and struck by ball at B.

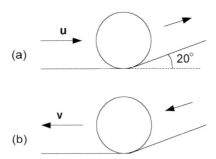

Figure 16.29. Uniform solid sphere rolling (a) up and (b) down a slope.

PROBLEM 140

A uniform straight rod hangs in equilibrium from a frictionless hinge at A. The free end B is struck by a small ball travelling horizontally with velocity u (see Figure 16.28). If $u = 8$ m/s, the ball has mass $m = 0.4$ kg, the rod has length $2a = 1$ m and mass $M = 4$ kg, and the coefficient of restitution between the ball and the rod is $e = 0.92$, find the angle θ to which the rod will rise after the impact.

PROBLEM 141

A uniform solid sphere rolls along a horizontal surface with velocity u. It strikes a slope of $20°$ to the horizontal (Figure 16.29a), rolls up it some way before coming to rest and rolling back down the slope (Figure 16.29b), eventually rolling back along the horizontal surface with velocity v. Assuming no bouncing or slipping, find v in terms of u.

Part IV

Background mathematics

17 Algebra

17.1 Indices

If n is a positive integer, the nth power of a is $a^n = a \times a \times a \times \cdots \times a$ (n terms) and n is called the *index*. If both m and n are positive integers,

$$a^m \times a^n = a \times a \times a \times \cdots \times a \ (m+n \text{ terms}) = a^{m+n}.$$

If $m > n$, $\dfrac{a^m}{a^n} = \dfrac{a \times a \times \cdots \times a \ (m \text{ terms})}{a \times a \times \cdots \times a \ (n \text{ terms})} = a \times a \times \cdots \times a \ (m-n \text{ terms}) = a^{m-n}.$

For any positive integers m and n,

$$(a^m)^n = a^m \times a^m \times \cdots \times a^m \ (n \text{ terms}) = a \times a \times \cdots \times a \ (mn \text{ terms}) = a^{mn}.$$

If we assume that this rule also holds for a rational $m = p/q$ with p and q both positive integers, then:

$$a^{p/q} \times a^{p/q} \times \cdots \times a^{p/q} \ (q \text{ terms}) = (a^{p/q})^q = a^{qp/q} = a^p.$$

Now, if $b \times b \times \cdots \times b$ (n terms) $= b^n = c$, then $b = \sqrt[n]{c}$, the nth root of c. It follows that $a^{p/q} = \sqrt[q]{a^p}$, the qth root of a^p.

a^0: $\quad a^m a^n = a^{m+n}$, so $a^0 a^n = a^n$ and $a^0 = 1$.

a^{-n}: $\quad a^{-n} a^n = a^0 = 1$ and $a^{-n} = 1/a^n$.

a^{m-n}: $a^{m-n} = a^m a^{-n} = a^m/a^n$ (no need for $m > n$ as before).

Finally, the above may be generalized so that a^n exists for any real number n and positive real number a.

17.2 Logarithm

Definition: the log of a number (which is real and positive) to a given base is the power to which the base must be raised to equal the number. Thus, if $x = \log_a N$, then $a^x = N$. It follows that $a^{\log_a N} = N$, which is often referred to as the *anti-log*.

Now $a^0 = 1$, so $\log_a 1 = 0$, which is true for any base a. Also, if $x = 1$, $a = N$ and $\log_a a = 1$.

$$a^{\log_a(MN)} = MN = a^{\log_a M} a^{\log_a N} = a^{(\log_a M + \log_a N)}.$$

On comparing the indices: $\log_a(MN) = \log_a M + \log_a N$, i.e. the log of the product of two numbers equals the sum of their logs. Obviously, the same would be true for more numbers, e.g. $\log_a(MNP) = \log_a M + \log_a N + \log_a P$. Furthermore,

$$a^{\log_a(M/N)} = M/N = \frac{a^{\log_a M}}{a^{\log_a N}} = a^{(\log_a M - \log_a N)}.$$

Again, on comparing indices: $\log_a(M/N) = \log_a M - \log_a N$.

Next, we investigate the log of a number N raised to the power of p:

$$a^{\log_a(N^p)} = N^p = (a^{\log_a N})^p = a^{p \log_a N}.$$

Thus, $\log_a(N^p) = p \log_a N$. Similarly,

$$\sqrt[r]{N} = N^{1/r}, \text{ so } \log_a(\sqrt[r]{N}) = \frac{1}{r} \log_a N.$$

Finally, let $x = \log_b N$, so that $b^x = N$. Then, $\log_a N = \log_a b^x = x \log_a b$ and

$$x = \log_b N = \frac{\log_a N}{\log_a b}.$$

17.3 Polynomials

$P(x)$ is a polynomial in x of degree n if:

$$P(x) = a_0 + a_1 x + a_2 x^2 + \cdots + a_n x^n,$$

where the a terms are constant coefficients.

Remainder theorem

$$P(x) = (x - a)Q(x) + R,$$

where $Q(x)$ is a polynomial of degree $(n - 1)$ and R is the remainder. It can be seen that $R = P(a)$. If $(x - a)$ is a factor of $P(x)$, then $R = P(a) = 0$.

EXAMPLE

To find k if remainder $R = 40$ when the polynomial

$$P(x) = 14 + 3x - 5x^2 + kx^3 + 3x^4$$

is divided by $(x - 2)$.

$$R = P(2) = 14 + 3(2) - 5(2)^2 + k(2)^3 + 3(2)^4 = 8k + 48 = 40.$$

Therefore, $8k = -8$, $k = -1$.

Principle of undetermined coefficients

If two polynomials

$$P(x) = a_0 + a_1 x + \cdots + a_n x^n \quad \text{and} \quad Q(x) = b_0 + b_1 x + \cdots + b_n x^n$$

are equal for all x, their corresponding coefficients must be equal, i.e.
$a_0 = b_0,\ a_1 = b_1, \ldots, a_n = b_n$.

EXAMPLE

Find a, b and c such that:

$$ax(x - 1) + b(x + 1) + c = 3x^2 + 4x + 5.$$

Comparing coefficients of x^2: $a = 3$, of x: $-a + b = -3 + b = 4$, $b = 7$ and of
1: $b + c = 7 + c = 5$, $c = -2$.

Numerical evaluation

To evaluate a polynomial $P(x)$ for a particular value of x, we need to consider the
polynomial in nested form which for degree 4 would be:

$$P(x) = a_4 x^4 + a_3 x^3 + a_2 x^2 + a_1 x + a_0 = \{[(a_4 x + a_3)x + a_2]x + a_1\}x + a_0.$$

Using our calculator, we put the value of x into the memory M and proceed as follows:

$$a_4 \times M + a_3 = P_3, \quad P_3 \times M + a_2 = P_2, \quad P_2 \times M + a_1 = P_1 \quad \text{and}$$
$$P_1 \times M + a_0 = P(M).$$

Notice that P_3, P_2 and P_1 do not need to be written down and, after a little practice, the calculation may be performed without even writing the polynomial in nested
form.

17.4 Partial fractions

If $f(x) = P(x)/Q(x)$, where P and Q are polynomials in x with the degree of P less
than the degree of Q, then Q may be factorized into simple and/or quadratic factors
and $f(x)$ may be split into the sum of partial fractions, e.g.

$$f(x) = \frac{\alpha x^4 + \beta x^3 + \gamma x^2 + \delta x + \epsilon}{(x - a)(x - b)^2(x^2 + cx + d)} = \frac{A}{x - a} + \frac{B_1}{x - b} + \frac{B_2}{(x - b)^2} + \frac{Cx + D}{x^2 + cx + d}$$
$$= N(x)/[(x - a)(x - b)^2(x^2 + cx + d)], \quad \text{where the numerator is:}$$
$$N(x) = A(x - b)^2(x^2 + cx + d) + [B_1(x - b) + B_2](x - a)(x^2 + cx + d)$$
$$+ (Cx + D)(x - a)(x - b)^2.$$

The constants A, B_1, B_2, C and D can be found from the five equations formed by equating the coefficients in the numerators of x^4, x^3, x^2, x and 1.

Note the following.

1. The degree of the numerator must be less than that of the denominator. Otherwise, the denominator must be divided into the numerator until the degree of the remainder is less than that of the denominator.

2. Repeated factors in the denominator lead to a number of partial fractions equal to the power of the factor. For instance, if $(x - b)^3$ had been a factor in the denominator, we would have partial fractions:

$$\frac{B_1}{x - b} + \frac{B_2}{(x - b)^2} + \frac{B_3}{(x - b)^3}.$$

3. Repeated quadratic factors in the denominator require more partial fractions. For instance, if $(x^2 + cx + d)^2$ were a factor, we would have the two partial fractions:

$$\frac{C_1 x + D_1}{x^2 + cx + d} + \frac{C_2 x + D_2}{(x^2 + cx + d)^2}.$$

4. For simple factors, the corresponding constants, i.e. A and B_2 in our example, can be found immediately without comparing coefficients. In our example:

$$(x - a) f(x)|_{x=a} = A = \frac{\alpha a^4 + \beta a^3 + \gamma a^2 + \delta a + \epsilon}{(a - b)^2 (a^2 + ca + d)}$$

and

$$(x - b)^2 f(x)|_{x=b} = B_2 = \frac{\alpha b^4 + \beta b^3 + \gamma b^2 + \delta b + \epsilon}{(b - a)(b^2 + cb + d)}.$$

EXAMPLE

Suppose $f(x) = x^3/(x^2 - x - 2)$. We start by using long division to divide $(x^2 - x - 2)$ into x^3.

$$
\begin{array}{r}
x + 1 \\
x^2 - x - 2 \overline{\smash{\big)}\ x^3 } \\
\underline{x^3 - x^2 - 2x} \\
x^2 + 2x \\
\underline{x^2 - x - 2} \\
3x + 2
\end{array}
$$

This gives $f(x) = x + 1 + g(x)$, where $g(x) = (3x + 2)/(x^2 - x - 2)$. Then, $x^2 - x - 2 = (x - 2)(x + 1)$ and

$$g(x) = \frac{3x + 2}{(x - 2)(x + 1)} = \frac{8}{3(x - 2)} + \frac{1}{3(x + 1)},$$

where the partial fractions have been found using (4) above.

Hence, $f(x) = x + 1 + \dfrac{8}{3(x - 2)} + \dfrac{1}{3(x + 1)}.$

EXAMPLE

$$f(x) = \frac{9}{(x+1)(x-2)^2} = \frac{1}{x+1} + \frac{3}{(x-2)^2} + \frac{A}{x-2}$$

$$= \frac{(x-2)^2 + 3(x+1) + A(x+1)(x-2)}{(x+1)(x-2)^2}.$$

Coefficient of x^2: $0 = 1 + A$, $A = -1$.

Hence, $f(x) = \dfrac{1}{x+1} + \dfrac{3}{(x-2)^2} - \dfrac{1}{x-2}$.

EXAMPLE

$$f(x) = \frac{3x^2}{x^3 - 1} = \frac{3x^2}{(x-1)(x^2+x+1)} = \frac{1}{x-1} + \frac{Ax+B}{x^2+x+1}$$

$$= \frac{x^2 + x + 1 + (Ax+B)(x-1)}{(x-1)(x^2+x+1)}.$$

Coefficient of x^2: $3 = 1 + A$, $A = 2$.
Coefficient of 1: $0 = 1 - B$, $B = 1$.

Hence, $\dfrac{3x^2}{x^3 - 1} = \dfrac{1}{x-1} + \dfrac{2x+1}{x^2+x+1}$.

Note: Instead of comparing coefficients, we could give x particular values. For instance, in the last example, compare the numerators with:

$x = 0$: $0 = 1 - B$, $B = 1$,

$x = -1$: $3 = 1 + 2A - 2B = -1 + 2A$, $A = 2$.

17.5 Sequences and series

Sequence

$$a_1, a_2, a_3, a_4, \ldots, a_n, \ldots.$$

EXAMPLES:

2, 4, 6, 8, ... i.e. $a_n = 2n$.

2, 4, 8, 16, ... i.e. $a_n = 2^n$.

1^2, 2^2, 3^2, 4^2, ... i.e. $a_n = n^2$.

Series

$$a_1 + a_2 + a_3 + \cdots + a_n + \cdots = \sum_n a_n.$$

Arithmetical progression (A.P.)

$a_{n+1} = a_n + d$, $d = $ constant. Thus, 2, 4, 6, 8, ... is an A.P. with $d = 2$; d is called the *common difference*. Starting from a, the A.P. is: a, $a + d$, $a + 2d$, $a + 3d$,.... The nth term in the sequence is: $l = a + (n - 1)d$.

If we now have a *series* of n terms in A.P., the sum is:

$$s_n = a + (a + d) + (a + 2d) + \cdots + (l - 2d) + (l - d) + l$$

$$\text{or } s_n = l + (l - d) + (l - 2d) + \cdots + (a + 2d) + (a + d) + a.$$

Adding corresponding terms of the two series gives:

$$2s_n = (a + l) + (a + l) + (a + l) + \cdots + (a + l) + (a + l) + (a + l) = n(a + l).$$

Hence, $s_n = \dfrac{n}{2}(a + l)$.

EXAMPLE

Find three numbers in A.P. with sum 33 and product 935. In this case, it is convenient to let a be the middle term, so that the three terms become: $a - d$, a, $a + d$. Adding gives: $3a = 33$ and $a = 11$.

Then, the product is: $a(a^2 - d^2) = 11(121 - d^2) = 935$, $121 - d^2 = 85$, $d^2 = 36$, $d = 6$. Hence, the three terms are: 5, 11, 17.

EXAMPLE

Find n if $a_1 = 4$, $a_3 = 14$ and $s_n = 70$. The first term is 4, so $a = 4$ and $a + 2d = 4 + 2d = 14$, so $d = 5$.

$$s_n = \frac{n}{2}(a + l) = \frac{n}{2}[a + a + (n - 1)d] = \frac{n}{2}(3 + 5n) = 70.$$

Hence, $5n^2 + 3n - 140 = 0$ and $n = \dfrac{-3 \pm \sqrt{9 + 2800}}{2 \times 5} = \dfrac{-3 \pm 53}{10}$.

n must be a positive integer, so $n = 5$.

Geometrical progression (G.P.)

Let $a_{n+1} = ra_n$ with $r = $ constant. Then, the geometrical progression is the sequence: $a, ar, ar^2, ar^3, \ldots$ and r is called the *common ratio*. If the nth term is also the last term l in the sequence, then $l = ar^{n-1}$. The sum of the corresponding series is:

$$s_n = a + ar + ar^2 + \cdots + ar^{n-2} + ar^{n-1}$$

$$\text{and } rs_n = ar + ar^2 + \cdots + ar^{n-2} + ar^{n-1} + ar^n.$$

Subtracting gives: $s_n - rs_n = (1 - r)s_n = a - ar^n = a(1 - r^n)$.

Hence, $s_n = \dfrac{a(1 - r^n)}{1 - r}$.

EXAMPLE

Find three numbers in G.P. with sum 21 and product 216. As with the corresponding A.P. problem, it is convenient to let a be the middle term so that the three terms become $a/r, a, ar$. The product then gives $a^3 = 216$ and hence $a = 6$. Then, the sum is:

$$a\left(\frac{1}{r} + 1 + r\right) = 6\left(\frac{1}{r} + 1 + r\right) = 21, \text{ i.e. } r^2 + r + 1 = \frac{21r}{6} = \frac{7}{2}r \text{ or } r^2 - \frac{5}{2}r + 1 = 0.$$

Hence, $r = \dfrac{\frac{5}{2} \pm \sqrt{\frac{25}{4} - 4}}{2} = \dfrac{\frac{5}{2} \pm \sqrt{\frac{9}{4}}}{2} = \dfrac{5 \pm 3}{4} = 2 \text{ or } \dfrac{1}{2}$.

Consequently, $(a/r, a, ar) = (3, 6, 12)$ or $(12, 6, 3)$.

EXAMPLE

Find the sum of the 8 terms: $3, -6, 12, -24, \ldots$. $a = 3$ and $r = -2$, so

$$s_8 = \frac{3[1 - (-2)^8]}{1 - (-2)} = \frac{3(1 - 256)}{3} = -255.$$

Convergence of a geometric series

A geometric series converges if the sum s_n tends to a finite number as the number of terms n tends to infinity. The sum of the first n terms is:

$$s_n = a + ar + ar^2 + \cdots + ar^{n-1} = \frac{a(1 - r^n)}{1 - r} = \frac{a}{1 - r} - \frac{ar^n}{1 - r}.$$

The first of these two terms is constant and the second will tend to zero as n tends to ∞ if the magnitude of r is less than 1, i.e. $|r| < 1$. Thus, the geometric series converges if $|r| < 1$ and in that case s_n converges to the *limit*:

$$\lim_{n \to \infty} s_n = \frac{a}{1 - r}.$$

EXAMPLE

Find the limit as $n \to \infty$ of the geometric series: $16 - 8 + 4 - 2 + \cdots$. $a = 16$ and $r = -1/2$, so:

$$\lim_{n \to \infty} s_n = \frac{a}{1 - r} = \frac{16}{1 - (-1/2)} = \frac{16}{3/2} = \frac{32}{3} = 10\frac{2}{3}.$$

17.6 Binomial theorem

The *factorial* of a positive integer n is defined as the product:

$$n! = n(n-1)(n-2)\cdots 3 \cdot 2 \cdot 1.$$

Then, the symbol:

$$^nC_r = \frac{n!}{r!(n-r)!},$$

where r is a positive integer with $r \leq n$. We define $0! = 1$, so that:

$$^nC_0 = \frac{n!}{0!(n-0)!} = 1 \quad \text{and} \quad ^nC_n = \frac{n!}{n!(n-n)!} = 1.$$

(This symbolism arises from the fact that ^nC_r is the number of combinations of n things r at a time.)

The *binomial theorem* for a positive integer n states that:

$$(a+x)^n = a^n + {}^nC_1 a^{n-1}x + {}^nC_2 a^{n-2}x^2 + \cdots + x^n = \sum_{r=0}^{n} {}^nC_r a^{n-r} x^r.$$

Another way of writing this series is:

$$(a+x)^n = a^n + na^{n-1}x + \frac{n(n-1)}{2}a^{n-2}x^2 + \frac{n(n-1)(n-2)}{3!}a^{n-3}x^3 + \cdots$$

and so on until it terminates with x^n.

In fact the latter is also the form of a series expansion for $(a+x)^n$ when n is either a negative integer or a non-integer. This is then an infinite series which is valid, i.e. it converges, if $|x| < |a|$. The vertical lines in this expression mean *modulus*, i.e. $|x| = x$ if $x > 0$ and $|x| = -x$ if $x < 0$.

EXAMPLE

$$(2+x)^5 = 2^5 + 5 \cdot 2^4 x + \frac{5 \cdot 4}{2}2^3 x^2 + \frac{5 \cdot 4 \cdot 3}{3!}2^2 x^3 + \frac{5 \cdot 4 \cdot 3 \cdot 2}{4!}2x^4 + \frac{5 \cdot 4 \cdot 3 \cdot 2 \cdot 1}{5!}x^5$$

$$= 2^5 + 5 \cdot 2^4 x + \frac{5 \cdot 4}{2}2^3 x^2 + \frac{5 \cdot 4}{2}2^2 x^3 + 5 \cdot 2x^4 + x^5$$

$$= 32 + 80x + 80x^2 + 40x^3 + 10x^4 + x^5.$$

EXAMPLE

$$(2 + x)^{1.5} = 2^{1.5} + 1.5 \times 2^{0.5}x + \frac{1.5 \times 0.5}{2}2^{-0.5}x^2 + \frac{1.5 \times 0.5 \times (-0.5)}{3!}2^{-1.5}x^3$$

$$+ \frac{1.5 \times 0.5 \times (-0.5) \times (-1.5)}{4!}2^{-2.5}x^4 + \cdots$$

$$= 2.83 + 2.12x + 0.265x^2 - 0.0221x^3 + 0.00414x^4 - \cdots.$$

This series will converge if $|x| < 2$, i.e. $-2 < x < 2$.

One particularly simple and useful binomial series is:

$$\frac{1}{1-x} = (1-x)^{-1} = 1 + x + x^2 + x^3 + \cdots = \sum_{n=0}^{\infty} x^n.$$

This series converges if $|x| < 1$.

18 Trigonometry

18.1 Introduction

Consider the triangle ABC with a right-angle (90°) at C, as shown in Figure 18.1. Let the lengths of the adjacent sides BC and CA be x and y, respectively, and the length of the hypotenuse AB be r.

We define the following *trigonometrical ratios*:

$$\sin\theta = \frac{y}{r}, \quad \cos\theta = \frac{x}{r}, \quad \tan\theta = \frac{y}{x} = \frac{\sin\theta}{\cos\theta}.$$

Their reciprocals (cosec(csc), sec and cot) are then given as:

$$\csc\theta = \frac{1}{\sin\theta}, \quad \sec\theta = \frac{1}{\cos\theta}, \quad \cot\theta = \frac{1}{\tan\theta} = \frac{\cos\theta}{\sin\theta}.$$

Pythagoras' theorem for a right-angle triangle states that $x^2 + y^2 = r^2$. This leads to the following *trigonometrical relations*:

$$\sin^2\theta + \cos^2\theta = \frac{y^2 + x^2}{r^2} = 1,$$

$$1 + \tan^2\theta = \frac{\cos^2\theta + \sin^2\theta}{\cos^2\theta} = \sec^2\theta,$$

$$1 + \cot^2\theta = \frac{\sin^2\theta + \cos^2\theta}{\sin^2\theta} = \csc^2\theta.$$

In order to deal with both positive and negative angles and also angles with magnitude greater than 90°, we measure the angle θ about the origin of a Cartesian system of coordinates. The angle θ is measured from the positive x-axis, as shown in Figure 18.2. An anti-clockwise θ as shown is positive, whereas a clockwise θ would be negative. Also, the *quadrants* are numbered anti-clockwise from 1 to 4 as shown.

If θ lies in the first quadrant, then all of the trigonometrical ratios are positive. However, the signs of the ratios vary with the quadrants, since x is negative on the left of the origin and y is negative below the origin. Thus, referring to Figure 18.3a, $\sin\theta = y/r$ is positive in quadrants 1 and 2 but negative in quadrants 3 and 4. Then,

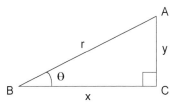

Figure 18.1. A right-angle triangle.

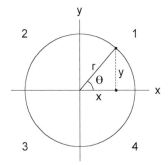

Figure 18.2. Angular measure referred to Cartesian coordinates.

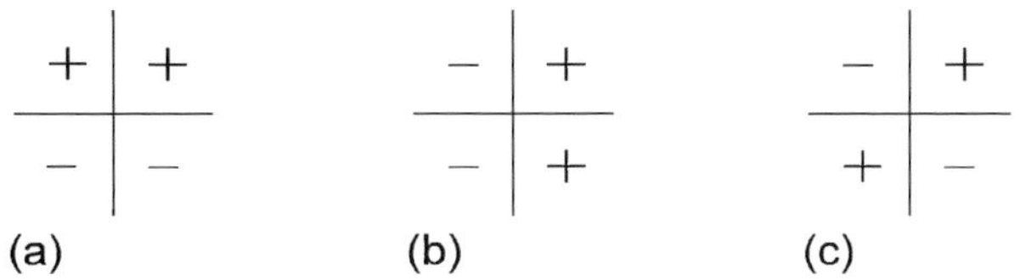

+	+
−	−

−	+
−	+

−	+
+	−

(a)   (b)   (c)

Figure 18.3. Signs of trig. ratios: (a) $\sin\theta$, (b) $\cos\theta$ and (c) $\tan\theta$.

from Figure 18.3b, $\cos\theta = x/r$ is positive in quadrants 1 and 4 but negative in quadrants 2 and 3. From Figure 18.3c, $\tan\theta = y/x$ is positive in quadrants 1 and 3 but negative in quadrants 2 and 4.

We can now deduce the following relations:

$$\sin(-\theta) = -\sin\theta, \ \cos(-\theta) = \cos\theta, \ \tan(-\theta) = -\tan\theta,$$
$$\sin(\theta + 180°) = -\sin\theta, \ \cos(\theta + 180°) = -\cos\theta, \ \tan(\theta + 180°) = \tan\theta.$$

Referring to Figure 18.4, we see that changing θ by $90°$ interchanges the magnitudes of x and y. Hence,

$$\sin(\theta + 90°) = \frac{x}{r} = \cos\theta, \ \cos(\theta + 90°) = \frac{-y}{r} = -\sin\theta,$$
$$\tan(\theta + 90°) = \frac{x}{-y} = -\cot\theta.$$

Similarly, we may deduce that:

$$\sin(\theta - 90°) = -\cos\theta, \ \cos(\theta - 90°) = \sin\theta, \ \tan(\theta - 90°) = -\cot\theta.$$

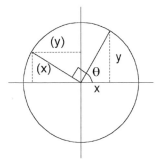

Figure 18.4. Changing the angle by 90°.

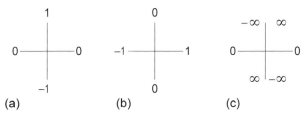

Figure 18.5. Trig. ratios of multiples of 90°: (a) $\sin\theta$, (b) $\cos\theta$ and (c) $\tan\theta$.

Figure 18.6. A right-angle triangle with $\theta = 45°$.

18.2 Trigonometrical ratios to remember

Consider multiples of 90°. Starting from $\theta = 0$, $\sin\theta$ is 0, 1, 0, −1 (see Figure 18.5a) and then repeats. Again, starting from $\theta = 0$, $\cos\theta$ is 1, 0, −1, 0 (see Figure 18.5b) and then repeats. $\tan\theta$ is more complicated because it becomes infinite for odd multiples of 90°. Furthermore, $\tan\theta$ changes sign as θ passes through an odd multiple of 90° (see Figure 18.5c).

The trigonometrical ratios for $\theta = 45°$, 30° and 60° are also worth remembering. If $\theta = 45°$, $x = y$ and consequently, $r = \sqrt{2}x = \sqrt{2}y$ (see Figure 18.6). It follows that:

$$\sin\theta = \frac{y}{r} = \frac{1}{\sqrt{2}} = \frac{x}{r} = \cos\theta \quad \text{and} \quad \tan\theta = \frac{y}{x} = 1,$$

i.e. $\sin 45° = \cos 45° = 1/\sqrt{2}$ and $\tan 45° = 1$.

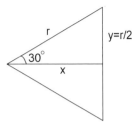

Figure 18.7. A 30° right-angle triangle as half an equilateral triangle.

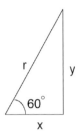

Figure 18.8. A 60° right-angle triangle.

A 30° right-angle triangle is half an equilateral triangle, as shown in Figure 18.7. Hence,

$$y = \frac{r}{2} \quad \text{and} \quad x^2 = r^2 - y^2 = \frac{3}{4}r^2, \text{ i.e. } x = \frac{\sqrt{3}}{2}r.$$

It follows that the sides of the 30° right-angle triangle have the proportions: $y : x : r = 1 : \sqrt{3} : 2$. Thus,

$$\sin 30° = \frac{y}{r} = \frac{1}{2}, \quad \cos 30° = \frac{x}{r} = \frac{\sqrt{3}}{2} \quad \text{and} \quad \tan 30° = \frac{y}{x} = \frac{1}{\sqrt{3}}.$$

Since the third angle in a 30° right-angle triangle is 60°, we see that for $\theta = 60°$, as in Figure 18.8, $y : x : r = \sqrt{3} : 1 : 2$. Consequently,

$$\sin 60° = \frac{\sqrt{3}}{2}, \quad \cos 60° = \frac{1}{2} \quad \text{and} \quad \tan 60° = \sqrt{3}.$$

18.3 Radian measure

So far, we have measured all angles in degrees – 90° for a right-angle and 360° for a complete revolution. It is often more convenient and sometimes essential to use the *radian* as the unit of angular measure. If we draw a circular arc of radius r such that the length of the arc is also equal to r, then the angle subtended by the arc at the centre of the circle is one radian (see Figure 18.9). Defining the number $\pi \approx 3.1416$ as the

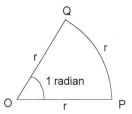

Figure 18.9. Defining the radian as a unit of angular measure.

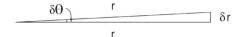

Figure 18.10. Small angle $\delta\theta \approx \delta r/r$.

length of the circumference of a circle divided by the length of its diameter, we see that there are 2π radians in a complete revolution. Similarly,

$$90° = \frac{\pi}{2} \text{ rad, } 180° = \pi \text{ rad, } 60° = \frac{\pi}{3} \text{ rad, } 45° = \frac{\pi}{4} \text{ rad } \text{ and } 30° = \frac{\pi}{6} \text{ rad.}$$

It follows that some of the trig. ratios using radians for angular measure are:

$$\sin\frac{\pi}{2} = 1, \ \cos\frac{\pi}{2} = 0, \ \sin\pi = 0, \ \cos\pi = -1, \ \sin\frac{3\pi}{2} = -1, \ \cos\frac{3\pi}{2} = 0,$$

$$\sin\frac{\pi}{4} = \cos\frac{\pi}{4} = \frac{1}{\sqrt{2}}, \ \sin\frac{\pi}{6} = \frac{1}{2}, \ \cos\frac{\pi}{6} = \frac{\sqrt{3}}{2}, \ \sin\frac{\pi}{3} = \frac{\sqrt{3}}{2} \text{ and } \cos\frac{\pi}{3} = \frac{1}{2}.$$

If $\delta\theta$ is a small angle measured in radians, we see from Figure 18.10 that:

$$\delta\theta \approx \frac{\delta r}{r} \approx \sin\delta\theta \approx \tan\delta\theta \quad \text{and} \quad \cos\delta\theta \approx 1.$$

$\approx$ means 'is approximately equal to'. In fact, the approximation becomes more accurate as $\delta\theta$ is reduced in size.

The periodic nature of trig. ratios leads to the following relations for any integer n:

$$\sin(\theta + 2n\pi) = \sin\theta, \ \cos(\theta + 2n\pi) = \cos\theta \quad \text{and} \quad \tan(\theta + n\pi) = \tan\theta.$$

18.4 Compound angles

Referring to Figure 18.11, let the length of OA be 1 unit. Let us examine the trigonometrical ratios for the compound angle $(\alpha + \beta)$.

$\sin(\alpha + \beta) = AC$, since $OA = 1$,

$= AE + BD = AB\cos\alpha + OB\sin\alpha = \sin\beta\cos\alpha + \cos\beta\sin\alpha$

$= \sin\alpha\cos\beta + \cos\alpha\sin\beta$.

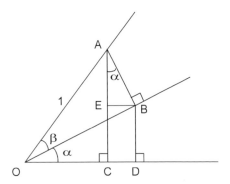

Figure 18.11. Finding the trig. ratios for the compound angle $(\alpha + \beta)$.

$\cos(\alpha + \beta) = OC = OD - EB = \cos \beta \cos \alpha - \sin \beta \sin \alpha$
$= \cos \alpha \cos \beta - \sin \alpha \sin \beta.$

On dividing these two expressions one by the other, we obtain:

$$\tan(\alpha + \beta) = \frac{\sin \alpha \cos \beta + \cos \alpha \sin \beta}{\cos \alpha \cos \beta - \sin \alpha \sin \beta}$$

$$= \frac{\tan \alpha + \tan \beta}{1 - \tan \alpha \tan \beta}.$$

These formulae have been derived for $\alpha > 0$, $\beta > 0$ and $(\alpha + \beta) < \pi/2$ but in fact they hold good for any angles α and β. Hence,

$\sin(\alpha - \beta) = \sin \alpha \cos \beta - \cos \alpha \sin \beta,$
$\cos(\alpha - \beta) = \cos \alpha \cos \beta + \sin \alpha \sin \beta,$
$$\tan(\alpha - \beta) = \frac{\tan \alpha - \tan \beta}{1 + \tan \alpha \tan \beta}.$$

Putting $\beta = \alpha$ in the addition formulae gives:

$\sin 2\alpha = 2 \sin \alpha \cos \alpha,$
$\cos 2\alpha = \cos^2 \alpha - \sin^2 \alpha = 2 \cos^2 \alpha - 1 = 1 - 2 \sin^2 \alpha,$
$$\tan 2\alpha = \frac{2 \tan \alpha}{1 - \tan^2 \alpha}.$$

We may now use the addition and subtraction formulae to obtain *factor formulae* as follows:

$\sin(\alpha + \beta) + \sin(\alpha - \beta) = 2 \sin \alpha \cos \beta,$
$\sin(\alpha + \beta) - \sin(\alpha - \beta) = 2 \cos \alpha \sin \beta,$
$\cos(\alpha + \beta) + \cos(\alpha - \beta) = 2 \cos \alpha \cos \beta,$
$\cos(\alpha + \beta) - \cos(\alpha - \beta) = -2 \sin \alpha \sin \beta.$

Putting $A = \alpha + \beta$, $B = \alpha - \beta$ and consequently, $\alpha = \frac{1}{2}(A + B)$ and $\beta = \frac{1}{2}(A - B)$ leads to the following formulae:

$$\sin A + \sin B = 2 \sin \tfrac{1}{2}(A + B) \cos \tfrac{1}{2}(A - B),$$

$$\sin A - \sin B = 2 \cos \tfrac{1}{2}(A + B) \sin \tfrac{1}{2}(A - B),$$

$$\cos A + \cos B = 2 \cos \tfrac{1}{2}(A + B) \cos \tfrac{1}{2}(A - B),$$

$$\cos A - \cos B = -2 \sin \tfrac{1}{2}(A + B) \sin \tfrac{1}{2}(A - B).$$

18.5 Solution of trigonometrical equations: inverse trigonometrical functions

A sketch of the graph of the trig. function is helpful. To solve the equation $\tan \theta = c$ for a particular value of c, we see from Figure 18.12 that one solution is $\theta = \alpha$. However, we see also that there are other solutions, each separated from the next one by π rad. Thus, having found one solution $\theta = \alpha$, there is an infinite set of solutions $\theta = (\alpha + n\pi)$ rad for all integers n.

The infinite sets of solutions to the equations $\cos \theta = c$ and $\sin \theta = c$, $|c| \leq 1$, are more complicated, but again they may be deduced by examining the graphs of the trig. functions.

Examining our sketch of the graph of the function $\cos \theta$ in Figure 18.13, we see that if $\theta = \alpha$ is one solution, then so is $\theta = -\alpha$. Also, $\theta = 2\pi \pm \alpha$ and $\theta = -2\pi \pm \alpha$ are solutions. Hence, the infinite set of solutions to the equation $\cos \theta = c$ is $\theta = (2n\pi \pm \alpha)$ rad for all integers n.

From our sketch of the graph of $\sin \theta$ in Figure 18.14, we see that if α is a solution to $\sin \theta = c$, then so is $\theta = \pi - \alpha$. Other solutions are $-2\pi + \alpha$, $-\pi - \alpha$ and $2\pi + \alpha$. Taking the solutions in order from left to right of the graph, they are: $-2\pi + \alpha$,

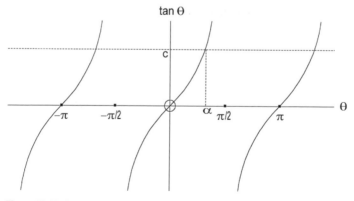

Figure 18.12. Sketch graph of $\tan \theta$.

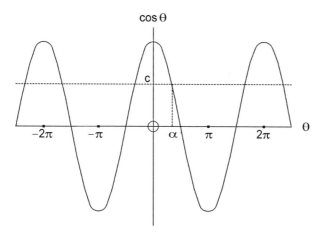

Figure 18.13. Sketch graph of $\cos\theta$.

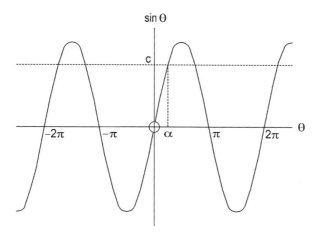

Figure 18.14. Sketch graph of $\sin\theta$.

$-\pi - \alpha$, α, $\pi - \alpha$, $2\pi + \alpha$. From this, we deduce that the infinite set of solutions is: $\theta = [n\pi + (-1)^n\alpha]$ rad, for all integers n.

The solutions to the trigonometrical equations $\tan\theta = c$, $\cos\theta = c$ and $\sin\theta = c$ may be regarded as inverse trigonometrical functions. A clumsy but clear nomenclature is: $\theta = \arctan c$, $\theta = \arccos c$ and $\theta = \arcsin c$. A more concise, but obviously ambiguous, way of writing the inverse functions is simply to use the index -1: $\theta = \tan^{-1} c$, $\theta = \cos^{-1} c$ and $\theta = \sin^{-1} c$.

Since there are infinitely many inverse values, it is convenient to specify *principal values* as follows:

$$-\frac{\pi}{2} < \tan^{-1} c < \frac{\pi}{2}, \quad 0 \le \cos^{-1} c \le \pi, \quad -\frac{\pi}{2} \le \sin^{-1} c \le \frac{\pi}{2}.$$

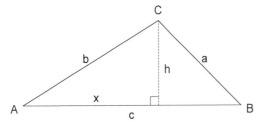

Figure 18.15. Triangle ABC with opposite sides of length a, b, c.

18.6 Sine and cosine rules

Let A, B and C denote the angles of the triangle ABC and a, b and c be the lengths of the opposite sides, as shown in Figure 18.15. The area of the triangle is half base times height, which by alternating the base gives:

area $= \frac{1}{2}cb \sin A = \frac{1}{2}ba \sin C = \frac{1}{2}ac \sin B$.

Divide this by $abc/2$ to give the *sine rule*:

$$\frac{\sin A}{a} = \frac{\sin C}{c} = \frac{\sin B}{b}.$$

Making use of the two extra internal measurements x and h, as shown in Figure 18.15, we can write:

$$b^2 = x^2 + h^2 = x^2 + a^2 - (c - x)^2 = a^2 - c^2 + 2cx$$
$$= a^2 - c^2 + 2cb \cos A.$$

This equation can then be written as the *cosine rule*:

$$a^2 = b^2 + c^2 - 2bc \cos A.$$

19 Calculus

19.1 Differential calculus

Let y be a function of x, i.e. $y = f(x)$, such that the graph of y against x is a smooth curve, as shown in Figure 19.1. The *slope* of the curve at any point (x_1, y_1) along it is defined as the tan of the angle θ which the tangent to the curve at (x_1, y_1) makes with the x-axis. Thus, slope $= \tan \theta$ with upwards positive and downwards negative.

If the same scale is used for both x and y, the *derivative* of $y = f(x)$ at $x = x_1$ is the slope of the curve at the point (x_1, y_1). Let (x_2, y_2) be another point on the curve as indicated in Figure 19.1. Then the slope of the curve at (x_1, y_1) is equal to $(y_2 - y_1)/(x_2 - x_1)$ in the limit as the point (x_2, y_2) slides back along the curve towards the point (x_1, y_1). This leads to the following definition of derivative:

$$\frac{dy}{dx}\bigg|_{x=x_1} = f'(x_1) = \lim_{x_2 \to x_1} \frac{f(x_2) - f(x_1)}{x_2 - x_1}.$$

More generally, if we replace x_1 by x and x_2 by x_1, this becomes:

$$\frac{dy}{dx} = f'(x) = \lim_{x_1 \to x} \frac{f(x_1) - f(x)}{x_1 - x}.$$

$\frac{dy}{dx}$ is the Leibnitz notation for the derivative of y with respect to x. It arises naturally from thinking of the change in x of $(x_1 - x)$ as δx and the corresponding change in y of $(y_1 - y)$ as δy, as shown in Figure 19.2. Then,

$$\frac{dy}{dx} = \lim_{\delta x \to 0} \frac{\delta y}{\delta x} = \lim_{\delta x \to 0} \frac{f(x + \delta x) - f(x)}{\delta x}, \text{ where } y = f(x).$$

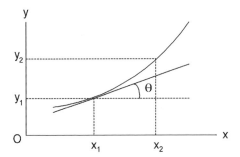

Figure 19.1. Measuring the slope of the graph of x against y.

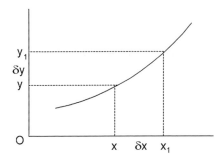

Figure 19.2. Derivation of $\frac{dy}{dx}$ notation for the derivative function.

19.2 Differentiation from first principles

Differentiation is the process of obtaining the derivative and 'from first principles' means that we should find the limit as stated in the definition of derivative.

If $y = 4x^2$, $\dfrac{dy}{dx} = \lim_{\delta x \to 0} \dfrac{4(x + \delta x)^2 - 4x^2}{\delta x} = \lim_{\delta x \to 0} \dfrac{8x\delta x + 4(\delta x)^2}{\delta x}$

$ = \lim_{\delta x \to 0} (8x + 4\delta x) = 8x.$

Next consider $y = x^n$, with n a positive integer. Using the binomial theorem,

$$y + \delta y = (x + \delta x)^n = x^n + nx^{n-1}\delta x + \frac{n(n-1)}{2}x^{n-2}(\delta x)^2 + \cdots .$$

Then $\dfrac{dy}{dx} = \lim_{\delta x \to 0} \dfrac{\delta y}{\delta x} = \lim_{\delta x \to 0} \left(nx^{n-1} + \dfrac{n(n-1)}{2}x^{n-2}\delta x + \cdots \right) = nx^{n-1}.$

Now consider trigonometrical ratios of angles in radians. Then, for small angle θ, $\sin \theta \approx \theta$ and $\cos \theta \approx 1$. If $y = \sin x$,

$$\frac{dy}{dx} = \lim_{\delta x \to 0} \frac{\sin(x + \delta x) - \sin x}{\delta x} = \lim_{\delta x \to 0} \frac{2\cos(x + \delta x/2)\sin(\delta x/2)}{\delta x}$$

(see factor formulae in Section 18.4).

$$= \lim_{\delta x \to 0} \cos(x + \delta x/2) \frac{\sin(\delta x/2)}{(\delta x/2)} = \cos x.$$

Next, if we have $y = \cos x$,

$$\frac{dy}{dx} = \lim_{\delta x \to 0} \frac{\cos(x + \delta x) - \cos x}{\delta x} = \lim_{\delta x \to 0} \frac{-2\sin(x + \delta x/2)\sin(\delta x/2)}{\delta x}$$

$$= \lim_{\delta x \to 0} -\sin(x + \delta x/2) \frac{\sin(\delta x/2)}{(\delta x/2)} = -\sin x.$$

If $y = f(x)$, then y is the *dependent* variable and x is the *independent* variable. In dynamics, the independent variable is t. In that case if $y = t^n$, where n is a positive integer, then $\frac{dy}{dt} = nt^{n-1}$. $\frac{dy}{dt}$ is the *rate of change* of y.

Now consider the sum of two functions, e.g. $y = u + v$, where u and v are each functions of x. Then the derivative of y is:

$$\frac{dy}{dx} = \lim_{\delta x \to 0} \frac{\delta u + \delta v}{\delta x} = \lim_{\delta x \to 0} \frac{\delta u}{\delta x} + \lim_{\delta x \to 0} \frac{\delta v}{\delta x} = \frac{du}{dx} + \frac{dv}{dx}.$$

In other words, the derivative of a sum is the sum of the derivatives.

Unfortunately, this simple rule does *not* carry through to a *product* or a *quotient*. Let $y = uv$ and $\delta y, \delta u, \delta v$ be the changes in y, u, v corresponding to the change δx in x. Then,

$$\delta y = (u + \delta u)(v + \delta v) - uv = \delta u \cdot v + u \cdot \delta v + \delta u \cdot \delta v.$$

Thus, $$\frac{dy}{dx} = \lim_{\delta x \to 0} \frac{\delta y}{\delta x} = \lim_{\delta x \to 0} \frac{\delta u}{\delta x} \cdot v + \lim_{\delta x \to 0} u \cdot \frac{\delta v}{\delta x} + \lim_{\delta x \to 0} \frac{\delta u}{\delta x} \cdot \delta v$$

$$= \frac{du}{dx} v + u \frac{dv}{dx}.$$

In order to find the derivative of the *quotient* of two functions u and v, we rewrite $y = u/v$ as $u = yv$ and use our product formula:

$$\frac{du}{dx} = \frac{dy}{dx} v + y \frac{dv}{dx}.$$

From this, it follows that:

$$\frac{dy}{dx} v = \frac{du}{dx} - \frac{u}{v} \frac{dv}{dx} = \left(\frac{du}{dx} v - u \frac{dv}{dx} \right) / v$$

and $$\frac{dy}{dx} = \left(\frac{du}{dx} v - u \frac{dv}{dx} \right) / v^2.$$

We can now use this formula to find the derivatives of other trigonometrical ratios:

$$\frac{d}{dx}(\tan x) = \frac{d}{dx}\left(\frac{\sin x}{\cos x}\right) = \frac{\cos^2 x + \sin^2 x}{\cos^2 x} = \frac{1}{\cos^2 x} = \sec^2 x,$$

$$\frac{d}{dx}(\cot x) = \frac{d}{dx}\left(\frac{\cos x}{\sin x}\right) = \frac{-\sin^2 x - \cos^2 x}{\sin^2 x} = \frac{-1}{\sin^2 x} = -\csc^2 x,$$

$$\frac{d}{dx}(\csc x) = \frac{d}{dx}\left(\frac{1}{\sin x}\right) = \frac{0 \cdot \sin x - 1 \cdot \cos x}{\sin^2 x} = \frac{-\cos x}{\sin^2 x} = -\csc x \cot x,$$

$$\frac{d}{dx}(\sec x) = \frac{d}{dx}\left(\frac{1}{\cos x}\right) = \frac{\sin x}{\cos^2 x} = \sec x \tan x.$$

Next we consider the derivative of a *function of a function*. For instance, what is $\frac{dy}{dx}$ when $u = u(x)$, i.e. u is a function of x, and $y = y(u)$, i.e. y is a function of u? Let a change δx in x cause a change δu in u and a change δy in y. Then,

$$\frac{dy}{dx} = \lim_{\delta x \to 0} \frac{\delta y}{\delta x} = \lim_{\delta x \to 0} \frac{\delta y}{\delta u}\frac{\delta u}{\delta x}.$$

Assuming that $\delta u \to 0$ and $\delta y \to 0$ as $\delta x \to 0$, then:

$$\frac{dy}{dx} = \lim_{\delta u \to 0} \frac{\delta y}{\delta u} \lim_{\delta x \to 0} \frac{\delta u}{\delta x} = \frac{dy}{du}\frac{du}{dx}.$$

Similarly, if $y = y(v(u(x)))$, then:

$$\frac{dy}{dx} = \frac{dy}{dv}\frac{dv}{du}\frac{du}{dx}.$$

EXAMPLE

If $y = \cos^3(x^2)$, we can regard $y = y(v(u(x)))$ with $u = x^2$, $v = \cos u$ and $y = v^3$. Then,

$$\frac{dy}{dx} = \frac{dy}{dv}\frac{dv}{du}\frac{du}{dx} = 3v^2(-\sin u)2x = 3\cos^2(x^2)[-\sin(x^2)]2x$$

$$= -6x\cos^2(x^2)\sin(x^2).$$

19.3 More derivative formulae

In Section 19.2, we found that $\frac{dy}{dx} = nx^{n-1}$ when $y = x^n$ and n is a positive integer. Consider now the same form for y but with n a negative integer. Thus, if $m = -n$, m is then a positive integer and $y = x^n = x^{-m} = 1/x^m$. We can now apply the formula for the derivative of a quotient as follows:

$$\frac{dy}{dx} = \frac{-mx^{m-1}}{x^{2m}} = -mx^{-m-1} = nx^{n-1}.$$

Hence, the same formula holds when n is a negative integer.

Next, we investigate the derivative of $y = x^n$ with $n = p/q$, where p is any integer and q is any positive integer. Then,

$$y = x^{p/q} = u^p, \quad \text{where } u = x^{1/q} \quad \text{and} \quad x = u^q.$$

Thus, $\dfrac{dy}{du} = pu^{p-1}$ and $\dfrac{dx}{du} = qu^{q-1}$.

Also, $\dfrac{dy}{du} = \dfrac{dy}{dx}\dfrac{dx}{du}$, so $pu^{p-1} = \dfrac{dy}{dx}qu^{q-1}$

and $\dfrac{dy}{dx} = \dfrac{p}{q}u^{p-q} = \dfrac{p}{q}(x^{1/q})^{p-q} = \dfrac{p}{q}x^{\frac{p}{q}-1} = nx^{n-1}$,

which is the same formula as before.

In fact, this implies that if n is any non-zero number and $y = x^n$ then $\frac{dy}{dx} = nx^{n-1}$.

In Section 18.5, we introduced the idea of inverse trig. functions, e.g. if $y = \sin x$, $x = \sin^{-1} y \neq 1/\sin y$. Suppose we now want to find the derivative of an inverse trig. function.

Let us start with any inverse function. Suppose $y = f(x)$ has an inverse function $x = g(y)$. Then, $y = f(g(y))$ and we can differentiate with respect to y using the function of a function rule:

$$1 = \frac{df}{dg}\frac{dg}{dy} = \frac{df}{dx}\frac{dg}{dy}.$$

It follows that:

$$\frac{dg}{dy} = \frac{1}{df/dx}, \quad \text{i.e.} \quad \frac{dx}{dy} = \frac{1}{dy/dx}.$$

We shall use this relation to find the derivatives of the inverse trig. functions.

$$y = \sin^{-1} x, \quad x = \sin y, \quad \frac{dx}{dy} = \cos y = \sqrt{1 - \sin^2 y} = \sqrt{1 - x^2}.$$

Hence, $\dfrac{dy}{dx} = \dfrac{1}{\sqrt{1 - x^2}}.$

$$y = \cos^{-1} x, \quad x = \cos y, \quad \frac{dx}{dy} = -\sin y = -\sqrt{1 - \cos^2 y} = -\sqrt{1 - x^2}.$$

Hence, $\dfrac{dy}{dx} = \dfrac{-1}{\sqrt{1 - x^2}}.$

$$y = \tan^{-1} x, \quad x = \tan y, \quad \frac{dx}{dy} = \sec^2 y = 1 + \tan^2 y = 1 + x^2.$$

Hence, $\dfrac{dy}{dx} = \dfrac{1}{1 + x^2}.$

$$y = \csc^{-1} x, \quad x = \csc y, \quad \frac{dx}{dy} = -\csc y \cot y = -\csc y\sqrt{\csc^2 y - 1} = -x\sqrt{x^2 - 1}.$$

Hence, $\dfrac{dy}{dx} = \dfrac{-1}{x\sqrt{x^2 - 1}}$.

$y = \sec^{-1} x$, $x = \sec y$, $\dfrac{dx}{dy} = \sec y \tan y = \sec y \sqrt{\sec^2 y - 1} = x\sqrt{x^2 - 1}$.

Hence, $\dfrac{dy}{dx} = \dfrac{1}{x\sqrt{x^2 - 1}}$.

$y = \cot^{-1} x$, $x = \cot y$, $\dfrac{dx}{dy} = -\csc^2 y = -(1 + \cot^2 y) = -(1 + x^2)$.

Hence, $\dfrac{dy}{dx} = \dfrac{-1}{1 + x^2}$.

Sometimes the relation between x and y cannot be expressed explicitly by an equation $y = f(x)$ but only *implicitly* by an equation $g(x, y) = 0$, where g is a function of both x and y. It may still be possible to find an expression for $\frac{dy}{dx}$ in terms of x and y.

Consider the relation $xy + \sin y + \cos x = 0$. Now differentiate with respect to x, regarding y as a function of x:

$$y + x\frac{dy}{dx} + \cos y \frac{dy}{dx} - \sin x = 0.$$

Hence, $\dfrac{dy}{dx} = \dfrac{\sin x - y}{x + \cos y}$.

To complete our definitions of derivatives, we must mention *higher derivatives*. If we regard $\frac{dy}{dx}$ as the first derivative of y with respect to x, the corresponding second derivative is:

$$\frac{d^2 y}{dx^2} = \frac{d}{dx}\left(\frac{dy}{dx}\right).$$

Similarly, the third derivative is:

$$\frac{d^3 y}{dx^3} = \frac{d}{dx}\left(\frac{d^2 y}{dx^2}\right),$$

and so on for still higher derivatives.

EXAMPLE

Let $y = x(1 + x)^2$. We could treat this as the product of x with a function of a function of x but it is easier to expand the function as a polynomial in x and differentiate term by term. Hence,

$$y = x + 2x^2 + x^3, \quad \frac{dy}{dx} = 1 + 4x + 3x^2, \quad \frac{d^2 y}{dx^2} = 4 + 6x, \quad \frac{d^3 y}{dx^3} = 6.$$

Since the latter derivative is constant, any higher derivatives are zero.

As an example of the application of both first and second derivatives, let us examine a function $y = f(x)$ for maximum and minimum points. If the graph of y against x is

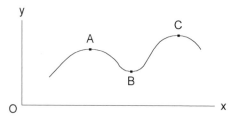

Figure 19.3. A smooth curve with local maxima (A and C) and minimum (B).

a smooth curve as shown in Figure 19.3, we see immediately that the points A and C are relative or local maxima and the point B is a relative or local minimum. Of course, the points may not be maximum or minimum in the absolute sense; for instance, in this case, the value of y at C is greater than that at A and the value of y at the left-hand end of the curve is less than that at B.

Concentrating on local maxima and minima, we can see from the diagram that, provided the curve is smooth, they will occur where the slope of the curve is zero, i.e. where $\frac{dy}{dx} = 0$. Such points are called *stationary points*.

Next, we notice that the slope of the curve goes from positive to negative as x increases through the values corresponding to A and C. This implies that:

$$\frac{d}{dx}\left(\frac{dy}{dx}\right) = \frac{d^2y}{dx^2} < 0$$

at local maximum points. However, as x is increased through the value corresponding to B, we see that the slope changes from negative to positive. Hence, it follows that:

$$\frac{d}{dx}\left(\frac{dy}{dx}\right) = \frac{d^2y}{dx^2} > 0$$

at local minimum points.

As an example, let us examine the function $y = 8x^2 - x^4$. Then,

$$\frac{dy}{dx} = 16x - 4x^3 = 4x(4 - x^2) = 4x(2 - x)(2 + x) = 0$$

at $x = 0, 2, -2$. Hence, the points with coordinates $(0, 0), (2, 16), (-2, 16)$ are stationary points. Next,

$$\frac{d^2y}{dx^2} = 16 - 12x^2 = 16, -32, -32 \text{ at } x = 0, 2, -2,$$

respectively. Hence, we have a local minimum when $x = 0$ and local maxima when $x = 2$ and $x = -2$. This is confirmed by the corresponding y values which are 0 at $x = 0$ and 16 at $x = \pm 2$.

19.4 Complex numbers

Equations such as $x^2 = -a^2$ or $ax^2 + bx + c = 0$, with $b^2 < 4ac$, where a, b and c are real numbers, have no solution in the realm of real numbers. However, it is often useful to give them a solution and to do this, we have to introduce the concept of complex numbers.

Complex numbers incorporate an 'imaginary' number designated i (or j) which is defined by the equation $i^2 = -1$. Then a complex number z has a real part x and an imaginary part y, and it can be written as $z = x + yi$, where x and y are real numbers. Then, x is the real part of z, written alternatively as $x = \Re(z) = \mathrm{Re}(z)$, and y is the imaginary part of z, $y = \Im(z) = \mathrm{Im}(z)$.

Since complex numbers have two components, the real part x and the imaginary part y, they can be represented by points in a Cartesian coordinate system. This is then called the *Argand diagram* in which the x-axis is the real axis and the y-axis is the imaginary axis. Referring to Figure 19.4, the point Z_1 corresponds to the complex number $z_1 = x_1 + y_1 i$ and Z_2 to $z_2 = x_2 + y_2 i$.

A point Z in the Argand diagram may be located by polar coordinates (r, θ), as shown in Figure 19.5.

$$x = r\cos\theta, \ y = r\sin\theta, \ r = +\sqrt{x^2 + y^2} \ (r \geq 0), \ \cos\theta = \frac{x}{r}, \ \sin\theta = \frac{y}{r}.$$

We refer to r as the modulus of z, written as $r = |z| = |x + yi|$. θ is referred to as the *argument* of z, written as $\theta = \arg z = \arg(x + yi)$. The *principal value* of θ is such that $-\pi < \theta \leq \pi$. Notice that *both* equations $\cos\theta = x/r$ and $\sin\theta = y/r$ are required to specify θ since $\tan\theta = y/x$ would give two possible values of θ within the principal value range.

Complex numbers which differ only in the signs of their imaginary parts are *conjugate*. Hence, the conjugate of $z = x + yi$ is $\bar{z} = x - yi$. The bar over the z indicates

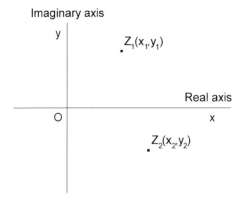

Figure 19.4. The Argand diagram.

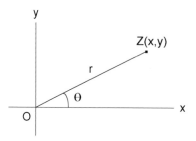

Figure 19.5. The Argand diagram using polar coordinates (r, θ).

conjugate of z. Notice that:

$$z + \bar{z} = 2x, \quad z - \bar{z} = 2yi \quad \text{and} \quad z\bar{z} = x^2 + y^2 = r^2 = |z|^2.$$

We can use the latter in finding the real and imaginary parts of:

$$z = x + yi = \frac{a + bi}{c + di} = \frac{(a + bi)(c - di)}{(c + di)(c - di)} = \frac{(ac + bd) + (bc - ad)i}{c^2 + d^2},$$

where a, b, c, d are all real. Hence,

$$x = \frac{ac + bd}{c^2 + d^2} \quad \text{and} \quad y = \frac{bc - ad}{c^2 + d^2}.$$

Particularly when more than two complex numbers are involved, products and quotients are more easily calculated using the polar form of complex numbers. For two complex numbers z_1 and z_2, let:

$$r_1 = |z_1|, \quad \theta_1 = \arg z_1, \quad r_2 = |z_2| \quad \text{and} \quad \theta_2 = \arg z_2.$$

Then, $z_1 z_2 = r_1(\cos\theta_1 + i\sin\theta_1)r_2(\cos\theta_2 + i\sin\theta_2)$

$$= r_1 r_2[(\cos\theta_1\cos\theta_2 - \sin\theta_1\sin\theta_2) + i(\sin\theta_1\cos\theta_2 + \cos\theta_1\sin\theta_2)]$$

$$= r_1 r_2[\cos(\theta_1 + \theta_2) + i\sin(\theta_1 + \theta_2)].$$

Hence, $|z_1 z_2| = |z_1||z_2| \quad \text{and} \quad \arg(z_1 z_2) = \arg z_1 + \arg z_2.$

The latter may not be the principal value of $\arg(z_1 z_2)$ but, if not, it can easily be converted by addition or subtraction of 2π rad.

Now, $\dfrac{z_1}{z_2} = \dfrac{r_1(\cos\theta_1 + i\sin\theta_1)}{r_2(\cos\theta_2 + i\sin\theta_2)} = \dfrac{r_1(\cos\theta_1 + i\sin\theta_1)(\cos\theta_2 - i\sin\theta_2)}{r_2(\cos\theta_2 + i\sin\theta_2)(\cos\theta_2 - i\sin\theta_2)}$

$$= \frac{r_1}{r_2}\left(\frac{(\cos\theta_1\cos\theta_2 + \sin\theta_1\sin\theta_2) + i(\sin\theta_1\cos\theta_2 - \cos\theta_1\sin\theta_2)}{\cos^2\theta_2 + \sin^2\theta_2}\right)$$

$$= \frac{r_1}{r_2}[\cos(\theta_1 - \theta_2) + i\sin(\theta_1 - \theta_2)].$$

Hence, $\left|\dfrac{z_1}{z_2}\right| = \dfrac{|z_1|}{|z_2|} \quad \text{and} \quad \arg\left(\dfrac{z_1}{z_2}\right) = \arg z_1 - \arg z_2.$

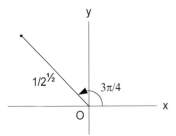

Figure 19.6. The position of z in the Argand diagram.

EXAMPLE

Find $|z|$ and the principal value of arg z for:

$$z = \frac{(1 - 2i)^2}{(2 + i)(1 + 3i)}.$$

Let $z = \dfrac{z_1^2}{z_2 z_3}$, where $z_1 = 1 - 2i$, $z_2 = 2 + i$ and $z_3 = 1 + 3i$.

Then, $|z_1| = \sqrt{1 + 2^2} = \sqrt{5}$, $|z_2| = \sqrt{5}$, $|z_3| = \sqrt{10}$.

Thus, $|z| = \dfrac{|z_1|^2}{|z_2||z_3|} = \dfrac{5}{\sqrt{5}\sqrt{2}\sqrt{5}} = \dfrac{1}{\sqrt{2}} = 0.707$.

Now, $\arg z_1 = \tan^{-1}\left(\dfrac{-2}{1}\right) = -1.107\,\text{rad}$, $\arg z_2 = \tan^{-1}\left(\dfrac{1}{2}\right) = 0.464\,\text{rad}$

and $\arg z_3 = \tan^{-1}\left(\dfrac{3}{1}\right) = 1.249\,\text{rad}$.

Thus, $\arg z = \arg z_1 + \arg z_1 - \arg z_2 - \arg z_3 = -3.927\,\text{rad}$.

and the principal value of arg z is:

$$2\pi - 3.927 = 2.356\,\text{rad} = 0.75\pi\,\text{rad} = 135°.$$

If we wish to write z as $x + yi$, referring to Figure 19.6, we see that:

$$x = \frac{1}{\sqrt{2}} \cos \frac{3\pi}{4} = -0.5 \quad\text{and}\quad y = \frac{1}{\sqrt{2}} \sin \frac{3\pi}{4} = 0.5.$$

Hence, $z = -0.5 + 0.5i = 0.5(-1 + i)$.

19.5 Integral calculus

In differential calculus we obtained the derivative of a function by the process of differentiation. We now look at the reverse of this process, which is referred to as *integration*. In other words, we wish to find y when $\frac{dy}{dx} = f(x)$ is given. For instance, suppose the velocity of a body is given as $\frac{ds}{dt} = u + at$, where u and a are constants, s and t are distance and time. Then, what is the distance s travelled in time t? We must find a function of t which when differentiated with respect to t gives us $u + at$. Such a function would be $s = ut + \frac{1}{2}at^2$. However, since a constant differentiates to zero, the general answer would be $s = ut + \frac{1}{2}at^2 + C$, where $C = $ constant. This is then the general *integral* of $\frac{ds}{dt} = u + at$.

If $\frac{dy}{dx} = f(x)$, we write $y = \int f(x)\,dx$, where y is the *indefinite integral* of $f(x)$ with respect to x. Similarly, $\int (u + at)\,dt = ut + \frac{1}{2}at^2 + C$ is the indefinite integral of $(u + at)$ with respect to t.

Since integration is the inverse operation from differentation, a table of integrals may be formed from a table of derivatives as follows:

$$\frac{d}{dx}(x^n) = nx^{n-1}, \quad \int x^n\,dx = \frac{x^{n+1}}{n+1} + C, \ n \neq -1.$$

$$\frac{d}{dx}(\sin x) = \cos x, \quad \int \cos x\,dx = \sin x + C.$$

$$\frac{d}{dx}(\cos x) = -\sin x, \quad \int \sin x\,dx = -\cos x + C.$$

$$\frac{d}{dx}(\tan x) = \sec^2 x, \quad \int \sec^2 x\,dx = \tan x + C.$$

$$\frac{d}{dx}(\cot x) = -\csc^2 x, \quad \int \csc^2 x\,dx = -\cot x + C.$$

$$\frac{d}{dx}(\sin^{-1} x) = \frac{1}{\sqrt{1 - x^2}}, \quad \int \frac{dx}{\sqrt{1 - x^2}} = \sin^{-1} x + C.$$

$$\frac{d}{dx}(\tan^{-1} x) = \frac{1}{1 + x^2}, \quad \int \frac{dx}{1 + x^2} = \tan^{-1} x + C.$$

The function that is integrated, e.g. x^n in $\int x^n\,dx$, is called the *integrand*.

Since the derivative of the sum of two functions is the sum of the separate derivatives, the converse applies, i.e. $\int [f(x) + g(x)]\,dx = \int f(x)\,dx + \int g(x)\,dx$. Also, if a is constant, $\int af(x)\,dx = a\int f(x)\,dx$.

19.6 The definite integral

Assume for the time being that $y = f(x) > 0$ for $a \leq x \leq b$. Then, the *definite integral* of $f(x)$ from $x = a$ to $x = b$ is the area F_a^b between the graph of $y = f(x)$ and the x axis for $a \leq x \leq b$ (see Figure 19.7a).

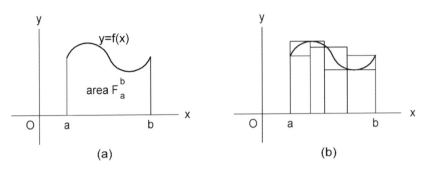

Figure 19.7. Integral as the area under the curve $y = f(x)$.

Suppose we now sub-divide the interval $[a, b]$ into n sub-intervals (not necessarily of equal width) and draw vertical rectangular strips as shown in Figure 19.7b. Each strip has two tops, one at the lowest point and one at the highest point of the curve $y = f(x)$ in the sub-interval. Let:

F_{nl} = the sum of the areas of the lower rectangles and
F_{nh} = the sum of the areas of the higher rectangles.
Obviously, $F_{nl} \leq F_a^b \leq F_{nh}$ and furthermore,

$$\lim_{n \to \infty} F_{nl} = F_a^b = \lim_{n \to \infty} F_{nh},$$

provided that the width of the widest sub-interval $\to 0$ as $n \to \infty$.

Let us now use this process to obtain a rigorous definition for the definite integral. Again, assume that $f(x)$ is positive for $a \leq x \leq b$ and also that $f(x)$ is continuous (finite and with no step changes) for $a \leq x \leq b$. Divide the interval $[a, b]$ into n sub-intervals (not necessarily of equal width) at points $x = x_1, x_2, \ldots, x_{n-1}$ with $x_0 = a$ and $x_n = b$. In each sub-interval, choose an arbitrary point ξ_i, $x_{i-1} \leq \xi_i \leq x_i$, $i = 1, 2, \ldots, n$.

Consider now the discontinuous step function which is constant at $y = f(\xi_i)$ in each sub-interval (see Figure 19.8). The area under this step function is:

$$F_n = \sum_{r=1}^{n} (x_r - x_{r-1}) f(\xi_r), \quad \text{where} \quad \sum_{r=1}^{n}$$

means 'the sum of the following terms as r goes from 1 to n'. If we write Δx_r for the width $(x_r - x_{r-1})$ of the rth sub-interval, F_n becomes:

$$F_n = \sum_{r=1}^{n} f(\xi_r) \Delta x_r.$$

Finally, we let $n \to \infty$ and the width of the widest sub-interval $\to 0$. $F_n \to$ a limiting value which is independent of the manner of sub-division and the way in which the sub-interval points are chosen. This limit is the definite integral of $f(x)$ between a and b. Hence,

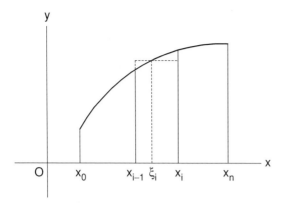

Figure 19.8. A representative sub-interval.

$$\int_a^b f(x)\,dx = \lim_{\substack{n\to\infty \\ \text{largest}\Delta x_r \to 0}} \sum_{r=1}^n f(\xi_r)\Delta x_r.$$

So far, we have assumed that $f(x) > 0$ and $a < b$. Both of these conditions can now be relaxed. If $f(x)$ is negative, the corresponding $f(\xi_r)$ terms are negative, so any area underneath the x axis is counted negatively. Also if $a > b$, the corresponding integration is in the negative x direction and the corresponding Δx_r values are negative. Thus,

$$\int_a^b f(x)\,dx = -\int_b^a f(x)\,dx.$$

The summation definition for the definite integral immediately implies that:

$$\int_a^b cf(x)\,dx = c\int_a^b f(x)\,dx \quad \text{and} \quad \int_a^b f(x)\,dx + \int_b^c f(x)\,dx = \int_a^c f(x)\,dx,$$

c being a constant. The latter relation for the integration over two adjacent intervals means that the integrand may have finite step discontinuities. In the example cited, $f(x)$ could have a finite step discontinuity at $x = b$.

Another relation which emerges from the summation definition is that, if $f(x) = \phi(x) + \psi(x)$, then:

$$\int_a^b f(x)\,dx = \int_a^b \phi(x)\,dx + \int_a^b \psi(x)\,dx.$$

It is useful to bear in mind the summation definition when performing definite integration. However, the process of taking the limit can be difficult. Hence, once the integral has been formulated, it is better to use a relation between definite and indefinite integrals which will now be derived.

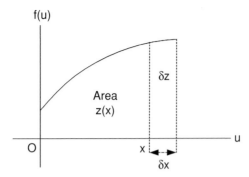

Figure 19.9. Deriving the relation between definite and indefinite integrals.

The indefinite integral y of $f(x)$ is defined by $\frac{dy}{dx} = f(x)$. Referring to Figure 19.9, let:

$$z(x) = \int_0^x f(u)\, du.$$

Then, $\dfrac{dz}{dx} = \lim_{\delta x \to 0} \dfrac{z(x + \delta x) - z(x)}{\delta x} = \lim_{\delta x \to 0} \dfrac{\delta z}{\delta x} = \lim_{\delta x \to 0} \dfrac{f(x)\delta x}{\delta x} = f(x),$

if $f(x)$ is continuous. Consequently, $z(x) + C$ is the indefinite integral of $f(x)$.

Now, the definite integral of $f(x)$ from a to b is:

$$\int_a^b f(x)\, dx = \int_a^b f(u)\, du = \int_0^b f(u)\, du - \int_0^a f(u)\, du = z(b) - z(a).$$

Hence, the procedure for finding the definite integral is to find the indefinite integral (ignoring C) and subtract the value at a from the value at b. This subtraction is usually denoted:

$$[z(x)]_a^b = z(b) - z(a).$$

EXAMPLE

Find the area A between the curve $y = (x^2 + 2x - 3)/2$ and the x-axis for $1 \le x \le 2$ (see Figure 19.10).

$$A = \int_1^2 \left(\frac{x^2}{2} + x - \frac{3}{2} \right) dx = \left[\frac{x^3}{6} + \frac{x^2}{2} - \frac{3x}{2} \right]_1^2$$

$$= \left(\frac{2^3}{6} + \frac{2^2}{2} - \frac{3 \times 2}{2} \right) - \left(\frac{1}{6} + \frac{1}{2} - \frac{3}{2} \right) = \frac{7}{6}.$$

EXAMPLE

Find the volume of a right circular cone of height h and base radius r.

The equation for the top edge of the cone, as drawn in Figure 19.11, is $y = rx/h$.

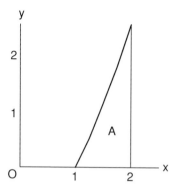

Figure 19.10. Finding the area A.

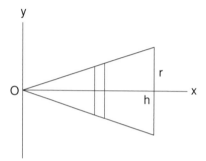

Figure 19.11. Side view of the cone with its axis coinciding with the x-axis.

Imagine a slice through the cone at distance x from the y-axis and of thickness δx. The volume of the slice is approximately:

$$\delta V \approx \pi y^2 \delta x = \frac{\pi r^2}{h^2} x^2 \delta x.$$

Then the total volume (summing over all such slices) is:

$$V = \lim_{\substack{n \to \infty \\ \delta x \to 0}} \sum \pi y^2 \delta x = \frac{\pi r^2}{h^2} \int_0^h x^2 \, dx = \frac{\pi r^2}{h^2} \left[\frac{x^3}{3} \right]_0^h = \frac{\pi r^2 h}{3}.$$

19.7 Methods of integration

Since analytical methods of integration are based on the indefinite integral, we need to explore the different methods for finding the indefinite integral of various functions.

Suppose we know that $f(x) + C$ is the indefinite integral of another function of x, which of course must be $f'(x)$, i.e. the derivative of $f(x)$. For constants a and b, $f(ax + b)$ is the function of a function of x and hence, $\frac{d}{dx} f(ax + b) = a f'(ax + b)$.

Therefore, it follows that:

$$\int f'(ax+b)\,dx = \frac{1}{a}f(ax+b)+C.$$

EXAMPLE

$$\int x^2\,dx = \frac{x^3}{3}+C.$$

Therefore, $\displaystyle\int (3x+2)^2\,dx = \frac{1}{3}\frac{(3x+2)^3}{3}+C = \frac{1}{9}(3x+2)^3 + C.$

This leads to the following list of integrals.

$$\int (ax+b)^n\,dx = \frac{(ax+b)^{n+1}}{a(n+1)}+C,\ n \ne -1,$$

$$\int \cos(ax)\,dx = \frac{1}{a}\sin(ax)+C,\ \int \sin(ax)\,dx = -\frac{1}{a}\cos(ax)+C,$$

$$\int \sec^2(ax)\,dx = \frac{1}{a}\tan(ax)+C,\ \int \csc^2(ax)\,dx = -\frac{1}{a}\cot(ax)+C.$$

Sometimes we need to make *trigonometrical substitutions*, as in the following two cases. For

$$\int \frac{dx}{\sqrt{a^2-x^2}},$$

we substitute $x = a\sin\theta$ and $dx = \frac{dx}{d\theta}d\theta = a\cos\theta\,d\theta$. Then:

$$\int \frac{dx}{\sqrt{a^2-x^2}} = \int \frac{a\cos\theta\,d\theta}{\sqrt{a^2-a^2\sin^2\theta}} = \int \frac{a\cos\theta\,d\theta}{a\cos\theta} = \theta + C$$

$$= \sin^{-1}\left(\frac{x}{a}\right)+C,\ |x| \le |a|.$$

For $\displaystyle\int \frac{dx}{a^2+x^2},$

we substitute $x = a\tan\theta$ and $dx = \frac{dx}{d\theta}d\theta = a\sec^2\theta\,d\theta$. Then:

$$\int \frac{dx}{a^2+x^2} = \int \frac{a\sec^2\theta\,d\theta}{a^2+a^2\tan^2\theta} = \int \frac{\sec^2\theta\,d\theta}{a\sec^2\theta} = \frac{\theta}{a}+C$$

$$= \frac{1}{a}\tan^{-1}\left(\frac{x}{a}\right)+C.$$

EXAMPLE

$$\int_1^5 \frac{dx}{x^2-6x+13} = \int_1^5 \frac{dx}{(x-3)^2+(2)^2}.$$

Put $x - 3 = 2 \tan\theta$, $dx = 2 \sec^2\theta \, d\theta$. Then,

$$\int_1^5 \frac{dx}{x^2 - 6x + 13} = \int \frac{2 \sec^2\theta \, d\theta}{(2)^2(\tan^2\theta + 1)} = \left[\frac{\theta}{2}\right] = \frac{1}{2}\left[\tan^{-1}\left(\frac{x-3}{2}\right)\right]_1^5$$

$$= \frac{1}{2}\left(\frac{\pi}{4} + \frac{\pi}{4}\right) = \frac{\pi}{4}.$$

Factor formulae in trigonometry (Section 18.4) can sometimes be used in the *integration of the products of trig. functions*. We shall use the following formulae:

$$2 \sin\alpha \sin\beta = \cos(\alpha - \beta) - \cos(\alpha + \beta),$$
$$2 \cos\alpha \cos\beta = \cos(\alpha - \beta) + \cos(\alpha + \beta),$$
$$2 \sin\alpha \cos\beta = \sin(\alpha - \beta) + \sin(\alpha + \beta).$$

From these, if $m \neq n$, it follows that:

$$\int \sin mx \sin nx \, dx = \frac{1}{2}\int [\cos(m-n)x - \cos(m+n)x] \, dx$$
$$= \frac{\sin(m-n)x}{2(m-n)} - \frac{\sin(m+n)x}{2(m+n)} + C,$$

$$\int \cos mx \cos nx \, dx = \frac{1}{2}\int [\cos(m-n)x + \cos(m+n)x] \, dx$$
$$= \frac{\sin(m-n)x}{2(m-n)} + \frac{\sin(m+n)x}{2(m+n)} + C,$$

$$\int \sin mx \cos nx \, dx = \frac{1}{2}\int [\sin(m-n)x + \sin(m+n)x] \, dx$$
$$= -\frac{\cos(m-n)x}{2(m-n)} - \frac{\cos(m+n)x}{2(m+n)} + C.$$

If $m = n$, the corresponding factor formulae become:

$$2 \sin^2 mx = 1 - \cos 2mx,$$
$$2 \cos^2 mx = 1 + \cos 2mx,$$
$$2 \sin mx \cos mx = \sin 2mx.$$

Then, $\int \sin^2 mx \, dx = \frac{1}{2}\int(1 - \cos 2mx) \, dx = \frac{x}{2} - \frac{\sin 2mx}{4m} + C,$

$$\int \cos^2 mx \, dx = \frac{1}{2}\int(1 + \cos 2mx) \, dx = \frac{x}{2} + \frac{\sin 2mx}{4m} + C,$$

$$\int \sin mx \cos mx \, dx = \frac{1}{2}\int \sin 2mx \, dx = -\frac{\cos 2mx}{4m} + C.$$

In order to integrate higher even powers of $\sin mx$ or $\cos mx$, it is necessary to repeat the above type of reduction. For instance, suppose we wished to integrate $\sin^4 mx$. Then, we must make the following reduction:

$$\sin^4 mx = [(1 - \cos 2mx)/2]^2 = \tfrac{1}{4}(1 - 2\cos 2mx + \cos^2 2mx)$$

$$= \tfrac{1}{4}(1 - 2\cos 2mx) + \tfrac{1}{8}(1 + \cos 4mx) = \tfrac{1}{8}(3 - 4\cos 2mx + \cos 4mx).$$

Thus, $\displaystyle \int \sin^4 mx \; dx = \frac{3}{8}x - \frac{\sin 2mx}{4m} + \frac{\sin 4mx}{32m} + C.$

Products of integer powers of $\cos x$ and $\sin x$ may be integrated fairly easily if at least one of the powers is odd.

Consider $\int \sin^m x \cos^n x \; dx$ when m is odd. $\sin^{m-1}x$ is even powered and may be changed into the sum of terms involving even powers of $\cos x$ by using the relation $\sin^2 x = 1 - \cos^2 x$. Furthermore, $\sin x \; dx = -\frac{d}{dx}(\cos x) \; dx = -d(\cos x)$, so we are left with the simple task of integrating powers of $\cos x$ with respect to $\cos x$.

EXAMPLE

$$\int \sin^3 x \cos^2 x \; dx = -\int (1 - \cos^2 x) \cos^2 x \; d(\cos x)$$

$$= \int (\cos^4 x - \cos^2 x) \; d(\cos x) = \frac{\cos^5 x}{5} - \frac{\cos^3 x}{3} + C.$$

Returning to $\int \sin^m x \cos^n x \; dx$ but this time with n odd, $\cos^{n-1}x$ is even powered and may be changed into the sum of terms involving even powers of $\sin x$ by using the relation $\cos^2 x = 1 - \sin^2 x$. Also, $\cos x \; dx = \frac{d}{dx}(\sin x) \; dx = d(\sin x)$. We are now left with the task of integrating powers of $\sin x$ with respect to $\sin x$.

EXAMPLE

$$\int \sin^3 x \cos^3 x \; dx = \int \sin^3 x(1 - \sin^2 x) \; d(\sin x)$$

$$= \int (\sin^3 x - \sin^5 x) \; d(\sin x) = \frac{\sin^4 x}{4} - \frac{\sin^6 x}{6} + C.$$

When the function $f(x)$ can be integrated, i.e. when $\int f(x) \; dx$ is known, it is a simple matter to find $\int x^{n-1} f(x^n) \; dx$. Start by differentiating the function of a function $g(x^n)$:

$$\frac{d}{dx} g(x^n) = g'(x^n) n x^{n-1}.$$

It follows from this that:

$$\int x^{n-1} g'(x^n) \; dx = \frac{1}{n} g(x^n) + C.$$

Then, if $f(x) = g'(x)$, $g(x) = \int f(x)\,dx$ and $g(x^n)$ is $\int f(x)\,dx$ with x replaced by x^n.

EXAMPLE

$\int x^4\sqrt{1+x^5}\,dx$. In this case :

$$f(x) = (1+x)^{1/2} \quad \text{and} \quad \int f(x)\,dx = \frac{2}{3}(1+x)^{3/2} + C.$$

Therefore, $\int x^4\sqrt{1+x^5}\,dx = \frac{1}{5} \cdot \frac{2}{3}(1+x^5)^{3/2} + C = \frac{2}{15}(1+x^5)^{3/2} + C.$

EXAMPLE

n need not be an integer, so the same method can be used to find:

$$\int x^{-1/2}(1 + x^{1/2})^2\,dx.$$

$$f(x) = (1+x)^2 \quad \text{and} \quad \int f(x)\,dx = \frac{1}{3}(1+x)^3 + C.$$

Therefore, $\int x^{-1/2}(1 + x^{1/2})^2\,dx = \frac{1}{1/2} \cdot \frac{1}{3}(1+x^{1/2})^3 + C = \frac{2}{3}(1+x^{1/2})^3 + C.$

The following is a very important and useful method of integration called *integration by parts*. If $u(x)$ and $v(x)$ are differentiable functions of x, then:

$$\frac{d}{dx}(uv) = \frac{du}{dx}v + u\frac{dv}{dx}.$$

On taking the definite integral from a to b, this becomes:

$$[uv]_a^b = \int_a^b \left(\frac{du}{dx}v + u\frac{dv}{dx}\right) dx.$$

The useful form of this equation is:

$$\int_a^b u\frac{dv}{dx}dx = [uv]_a^b - \int_a^b \frac{du}{dx}v\,dx.$$

This is often a convenient way to integrate the product of two functions, one corresponding to u and the other to $\frac{dv}{dx}$. For instance, with $\int_0^{\pi/2} x \sin x\,dx$, we let x correspond to u and $\sin x$ to $\frac{dv}{dx}$. Hence,

$$\int_0^{\pi/2} x \sin x\,dx = [-x \cos x]_0^{\pi/2} + \int_0^{\pi/2} \cos x\,dx = [-x\cos x + \sin x]_0^{\pi/2} = 1.$$

The process may be repeated as with:

$$\int_0^{\pi/2} x^2 \sin x \, dx = [-x^2 \cos x]_0^{\pi/2} + \int_0^{\pi/2} 2x \cos x \, dx$$

$$= [-x^2 \cos x + 2x \sin x]_0^{\pi/2} - \int_0^{\pi/2} 2 \sin x \, dx$$

$$= [-x^2 \cos x + 2x \sin x + 2 \cos x]_0^{\pi/2} = \pi - 2.$$

19.8 Numerical integration

Often, it is not possible to find the indefinite integral of a function. However, if the function can be plotted as a graph of y against x, say, the definite integral is still the area between the curve and the x-axis and between the limits of integration. Of course, we must remember to take areas below the x-axis as negative values and also to reverse the sign if we integrate in the negative direction. We shall now look at two ways of approximating the area numerically.

Trapezoidal rule

Divide the area into n strips of equal width $h = (b - a)/n$ when we wish to integrate the function $y = f(x)$ from $x = a$ to $x = b$ (see Figure 19.12). The area of each strip is then approximated by the area of the corresponding trapezium, i.e. by the area $(y_{r-1} + y_r)h/2$ for the rth strip, where $y_r = f(a + rh)$.

Summing over all such areas, the integral from a to b is approximated by:

$$\int_a^b f(x) \, dx \approx \tfrac{1}{2}h(y_0 + y_1) + \tfrac{1}{2}h(y_1 + y_2) + \cdots + \tfrac{1}{2}h(y_{n-1} + y_n)$$

$$= \tfrac{1}{2}h[y_0 + y_n + 2(y_1 + y_2 + \cdots + y_{n-1})].$$

In this case, the curve $y = f(x)$ has been approximated by a series of straight line segments.

Simpson's rule

An improvement in accuracy may be achieved by replacing the straight line segments used in the trapezoidal rule by curved segments described by the quadratic equation $y = Ax^2 + Bx + C$.

Let us start by considering two sub-intervals of width h centered at the origin as shown in Figure 19.13. Write $y_0 = f(-h)$, $y_1 = f(0)$ and $y_2 + f(h)$. Using the quadratic

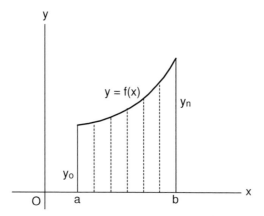

Figure 19.12. Dividing an area into strips for numerical integration.

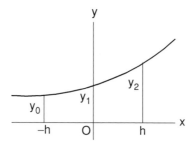

Figure 19.13. Deriving Simpson's rule.

approximation for $y = f(x)$ for $-h \le x \le h$,

$$\int_{-h}^{h} f(x)\,dx \approx \int_{-h}^{h} (Ax^2 + Bx + C)\,dx = \left[\frac{1}{3}Ax^3 + \frac{1}{2}Bx^2 + Cx\right]_{-h}^{h}$$

$$= \tfrac{2}{3}Ah^3 + 2Ch.$$

We must now choose A, B and C to give the true values of y at $x = -h, 0, h$. Hence,

$$y_0 = Ah^2 - Bh + C, \quad y_1 = C \quad \text{and} \quad y_2 = Ah^2 + Bh + C.$$

Thus, $y_0 + y_2 = 2Ah^2 + 2C = 2Ah^2 + 2y_1$.

Therefore, $2Ah^2 = y_0 + y_2 - 2y_1$.

Substituting into the integral approximation gives:

$$\int_{-h}^{h} f(x)\,dx \approx \tfrac{1}{3}h(y_0 + y_2 - 2y_1) + 2hy_1$$

$$= \tfrac{1}{3}h(y_0 + y_2 + 4y_1).$$

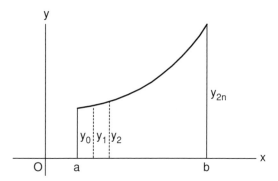

Figure 19.14. An area divided into an even number of strips.

We now proceed to use this formula to approximate the integral over each pair of sub-intervals from a to b. Referring to Figure 19.14, we must have an even number $2n$ of sub-intervals, each of width $h = (b - a)/2n$.

The approximation for the total integral from a to b becomes:

$$\int_a^b f(x)\, dx \approx \tfrac{1}{3}h(y_0 + y_2 + 4y_1) + \tfrac{1}{3}h(y_2 + y_4 + 4y_3) + \cdots$$

$$+ \tfrac{1}{3}h(y_{2n-2} + y_{2n} + 4y_{2n-1})$$

$$= \tfrac{1}{3}h[(y_0 + y_{2n}) + 4(y_1 + y_3 + \cdots + y_{2n-1}) + 2(y_2 + y_4 + \cdots + y_{2n-2})].$$

This is the numerical integration formula called *Simpson's rule*.

Let us take an example which can be integrated analytically for comparison. Assuming the positive square root:

$$\int_0^3 \sqrt{2x}\, dx = \sqrt{2}\left[\frac{2}{3}x^{3/2}\right]_0^3 = 4.899.$$

Now, integrate numerically over six sub-intervals, i.e. with $h = 0.5$

Subscript	x	.	y	.
0	0	0	.	.
1	0.5	.	1	.
2	1	.	.	1.4142
3	1.5	.	1.7321	.
4	2	.	.	2
5	2.5	.	2.2361	.
6	3	2.4495	.	.
Add	:	2.4495	4.9682	3.4142

Applying our numerical formulae, we obtain the following results.

Trapezoidal rule: $\dfrac{0.5}{2}(2.4495 + 2 \times 8.3824) = 4.804.$

Simpson's rule: $\dfrac{0.5}{3}(2.4495 + 4 \times 4.9682 + 2 \times 3.4142) = 4.858.$

In this case, neither numerical result is very good, accuracies being roughly 2% and 1% for the trapezoidal and Simpson's rule, respectively.

19.9 Exponential function e^x and natural logarithm $\ln x$

In the indefinite integral $\int x^n \, dx = x^{n+1}/(n+1) + C$, the index $n = -1$ had to be excluded since this would have made $n + 1 = 0$. However, as we see from Figure 19.15, $\int_a^b x^{-1} \, dx$ does exist for $0 < a < b$. It is, by definition, the area between the curve $y = x^{-1}$ and the x-axis for $a \le x \le b$. To overcome this problem, we define e as the number such that:

$$\frac{d}{dx}\left(e^x\right) = e^x.$$

Then, if we let $y = \log_e x = \ln x$, from the definition of logarithm, $x = e^y$. Furthermore,

$$\frac{d}{dx}(x) = 1 = \frac{d}{dx}(e^y) = \frac{d}{dy}(e^y)\frac{dy}{dx} = e^y \frac{dy}{dx}.$$

Therefore, $\dfrac{dy}{dx} = \dfrac{1}{e^y} = \dfrac{1}{x}$ and $\dfrac{d}{dx}(\ln x) = \dfrac{1}{x}.$

Integrating the latter with respect to x gives:

$$\int \frac{dx}{x} = \ln x + C.$$

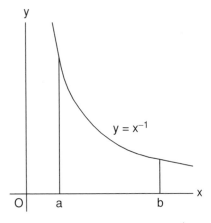

Figure 19.15. Area under curve $y = x^{-1}$.

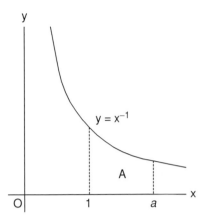

Figure 19.16. $e = a$ when area $A = 1$.

Also, the area under the curve between $x = a$ and $x = b$ in Figure 19.15 is:

$$\int_a^b \frac{dx}{x} = [\ln x]_a^b = \ln b - \ln a = \ln\left(\frac{b}{a}\right).$$

$\ln x$ is defined for positive x only. To allow for negative values, we should write:

$$\int \frac{dx}{x} = \ln |x| + C.$$

EXAMPLE

$$\int_{-2}^{-1} \frac{dx}{x} = [\ln |x|]_{-2}^{-1} = \ln 1 - \ln 2 = -\ln 2.$$

Referring to Figure 19.16, the area A is:

$$A = \int_1^a \frac{dx}{x} = [\ln x]_1^a = \ln a = \log_e a.$$

From this, it follows that $a = e^A = e^1$ if $A = 1$. Hence, the number e is the value of a such that $A = 1$. To four decimal places, $e = 2.7183$.

We can use the ln function to find the indefinite integrals of $\tan x$ and $\cot x$ as follows.

Let $y = \ln(\sec x) = \ln\left(\frac{1}{\cos x}\right) = \ln 1 - \ln(\cos x) = -\ln(\cos x)$.

Then, $\dfrac{dy}{dx} = -\dfrac{1}{\cos x}(-\sin x) = \tan x$.

Hence, $\displaystyle\int \tan x \, dx = y + C = \ln(\sec x) + C$.

Similarly, if $y = \ln(\sin x)$, $\dfrac{dy}{dx} = \dfrac{1}{\sin x} \cos x = \cot x.$

Hence, $\displaystyle\int \cot x \, dx = y + C = \ln(\sin x) + C.$

We defined the number e such that $\frac{d}{dx}(e^x) = e^x$. Correspondingly:

$$\int e^x \, dx = e^x + C.$$

If $a = $ constant, regarding e^{ax} as a function of a function of x:

$$\frac{d}{dx}(e^{ax}) = e^{ax} \cdot a = ae^{ax}.$$

It also follows that:

$$\int e^{ax} \, dx = \frac{1}{a}e^{ax} + C.$$

Now, let $a = $ constant > 0 and $y = a^x$. Then, $\ln y = \ln(a^x) = x \ln a$. Since y is a function of x:

$$\frac{d}{dx}(\ln y) = \frac{d}{dy}(\ln y)\frac{dy}{dx} = \frac{1}{y}\frac{dy}{dx} = \frac{1}{a^x}\frac{dy}{dx} = \ln a.$$

Therefore, $\dfrac{dy}{dx} = \dfrac{d}{dx}(a^x) = a^x \ln a.$

Correspondingly, $\displaystyle\int a^x \, dx = \frac{a^x}{\ln a} + C.$

Suppose we now wish to differentiate $y = x^x$. We start as with a^x by taking the natural logarithm: $\ln y = \ln(x^x) = x \ln x$. Now, differentiate with respect to x ($\ln y$ being a function of a function of x):

$$\frac{1}{y}\frac{dy}{dx} = \ln x + x\frac{1}{x} = \ln x + 1.$$

Therefore, $\dfrac{dy}{dx} = y(1 + \ln x)$, i.e. $\dfrac{d}{dx}(x^x) = x^x(1 + \ln x).$

Finally, we generalize the use of the ln function in integration. Let $f(x)$ be a differentiable function of x. Then $\ln f(x)$ is a function of a function of x, so:

$$\frac{d}{dx}\ln f(x) = \frac{1}{f(x)}f'(x) \quad \text{and} \quad \int \frac{f'(x)}{f(x)}dx = \ln f(x) + C.$$

Consequently, the ratio of two functions of x is integrable if the numerator is the derivative of the denominator.

We can use this property to integrate $\tan x$ and $\cot x$ directly:

$$\int \tan x \, dx = \int \frac{\sin x}{\cos x} dx = -\int \frac{\frac{d}{dx}(\cos x)}{\cos x} dx = -\ln(\cos x) + C$$

$$= \ln(\cos x)^{-1} + C = \ln(\sec x) + C,$$

$$\int \cot x \, dx = \int \frac{\cos x}{\sin x} dx = \int \frac{\frac{d}{dx}(\sin x)}{\sin x} dx = \ln(\sin x) + C.$$

Furthermore, we can use the same property to find the indefinite integral of $\csc x$ and $\sec x$:

$$\int \csc x \, dx = \int \frac{dx}{\sin x} dx = \int \frac{dx}{2 \sin \frac{x}{2} \cos \frac{x}{2}} = \int \frac{dx}{2 \tan \frac{x}{2} \cos^2 \frac{x}{2}}$$

$$\int \frac{\frac{1}{2} \sec^2 \frac{x}{2}}{\tan \frac{x}{2}} dx = \int \frac{\frac{d}{dx}(\tan \frac{x}{2})}{\tan \frac{x}{2}} dx = \ln \tan \frac{x}{2} + C.$$

Then, since $\cos x = \sin(x + \pi/2)$,

$$\int \sec x \, dx = \int \csc \left(x + \frac{\pi}{2} \right) dx = \ln \tan \left(\frac{x}{2} + \frac{\pi}{4} \right) + C.$$

To use the same property for integrating rational functions of x, a certain amount of algebraic manipulation may be required. For example, consider:

$$\int \frac{2x^2 + 1}{x^2 + 2x + 2} dx.$$

Since the derivative of $(x^2 + 2x + 2)$ is of degree one in x, we must start by dividing the denominator of the integrand into the numerator[1] to obtain:

$$\int \left[2 - \frac{4x + 3}{x^2 + 2x + 2} \right] dx = \int \left[2 - \frac{2 \frac{d}{dx}(x^2 + 2x + 2)}{x^2 + 2x + 2} + \frac{1}{1 + (x + 1)^2} \right] dx$$

$$= 2x - 2\ln(x^2 + 2x + 2) + \tan^{-1}(x + 1) + C.$$

19.10 Some more integrals using partial fractions and integration by parts

If we can split rational functions into simple *partial fractions*, we should be able to integrate the latter separately.

---

[1] The method for division is as shown in the first example of Section 17.4.

EXAMPLE

$$\int \frac{dx}{x^2 - a^2} = \int \frac{dx}{(x-a)(x+a)} = \int \left[\frac{1}{2a(x-a)} - \frac{1}{2a(x+a)} \right] dx$$

$$= \frac{1}{2a} \ln(x-a) - \frac{1}{2a} \ln(x+a) + C = \frac{1}{2a} \ln \left(\frac{x-a}{x+a} \right) + C.$$

EXAMPLE

$$\int \frac{x^3 dx}{x^2 + 2x - 15}.$$

For partial fractions, the degree of the numerator should be less than the degree of the denominator. Hence, in this case, we must divide until the latter is true (see Section 17.4):

$$\frac{x^3}{x^2 + 2x - 15} = x - 2 + \frac{19x - 30}{x^2 + 2x - 15}.$$

Also, $\dfrac{19x - 30}{x^2 + 2x - 15} = \dfrac{19x - 30}{(x-3)(x+5)} = \dfrac{27}{8(x-3)} + \dfrac{125}{8(x+5)}.$

Therefore, $\displaystyle\int \frac{x^3 \, dx}{x^2 + 2x - 15} = \int \left[x - 2 + \frac{27}{8(x-3)} + \frac{125}{8(x+5)} \right] dx$

$$= \frac{1}{2}x^2 - 2x + \frac{27}{8} \ln(x-3) + \frac{125}{8} \ln(x+5) + C.$$

In order to integrate $e^{-x} \cos x$, let us *integrate by parts* twice, integrating the trig. term each time:

$$\int e^{-x} \cos x \, dx = e^{-x} \sin x + \int e^{-x} \sin x \, dx$$

$$= e^{-x} \sin x - e^{-x} \cos x - \int e^{-x} \cos x \, dx.$$

Moving the last term over to the left of the equation gives us twice the desired integral. Hence,

$$\int e^{-x} \cos x \, dx = \tfrac{1}{2} e^{-x} (\sin x - \cos x) + C.$$

19.11 Taylor, Maclaurin and exponential series

We shall start by deriving the most general of these series, which is the Taylor series.

Let $f(x)$ be a function of x which has finite derivatives of any order. Assume that we can write $f(x)$, for x near to x_0, as an infinite series of the form:

$$f(x) = a_0 + a_1(x - x_0) + a_2(x - x_0)^2 + \cdots + a_n(x - x_0)^n + \cdots.$$

Differentiating successively with respect to x before putting $x = x_0$, we find that:

$$f(x_0) = a_0, \ f'(x_0) = a_1, \ f''(x_0) = 2a_2, \ f'''(x_0) = 3!a_3, \ldots, \ f^{(n)}(x_0) = n!a_n, \ldots,$$

where $n!$ is n factorial defined as: $n! = n(n-1)(n-2)\ldots \cdot 3 \cdot 2 \cdot 1$.

Substituting for the a coefficients in the series, we obtain the Taylor series:

$$f(x) = f(x_0) + f'(x_0)(x - x_0) + \frac{1}{2}f''(x_0)(x - x_0)^2 + \frac{1}{3!}f'''(x_0)(x - x_0)^3 + \cdots$$

$$\cdots + \frac{1}{n!}f^{(n)}(x_0)(x - x_0)^n + \cdots.$$

This may be referred to as the 'Taylor series expansion for $f(x)$ about $x = x_0$'.

Another form for the Taylor series is obtained by writing $h = x - x_0$. Then:

$$f(x_0 + h) = f(x_0) + f'(x_0)h + \frac{1}{2}f''(x_0)h^2 + \cdots + \frac{1}{n!}f^{(n)}(x_0)h^n + \cdots.$$

The Maclaurin series is the Taylor series with $x_0 = 0$, i.e.

$$f(x) = f(0) + f'(0)x + \frac{1}{2}f''(0)x^2 + \frac{1}{3!}f'''(0)x^3 + \cdots + \frac{1}{n!}f^{(n)}(0)x^n + \cdots.$$

To obtain the exponential series, we let $f(x) = e^x$ in the Maclaurin series. All of the derivatives of e^x are e^x and, when evaluated at $x = 0$, they are all unity. Hence, the exponential series is:

$$e^x = 1 + x + \frac{1}{2}x^2 + \frac{1}{3!}x^3 + \cdots + \frac{1}{n!}x^n + \cdots.$$

20 Coordinate geometry

20.1 Introduction

We have already made use of Cartesian coordinates (x, y) in which the x- and y-axes intersect at right-angles at the origin O. Referring to Figure 20.1, if the point P has coordinates (x, y), then x is the distance from the y-axis and y is the distance from the x-axis. Displacements to the right and above of O are positive but those to the left and down are negative.

We can also use polar coordinates (r, θ) for the position of P, in which case, r is its distance from O and θ is the angle which OP makes with the positive x-axis. θ is measured positively in the anti-clockwise direction and negatively in the clockwise direction.

x- and y-coordinates are sometimes referred to as *abscissa* and *ordinate*, respectively.

The distance between two points $P_1(x_1, y_1)$ and $P_2(x_2, y_2)$ may be expressed in terms of their coordinates. Referring to Figure 20.2, the triangle P_1P_2R is right-angled at R. Therefore, $(P_1P_2)^2 = (P_2R)^2 + (P_1R)^2 = (x_1 - x_2)^2 + (y_1 - y_2)^2$ and the distance $P_1P_2 = \sqrt{(x_1 - x_2)^2 + (y_1 - y_2)^2}$.

The area of a triangle can be expressed in terms of the coordinates of its corners. Referring to Figure 20.3, the area of triangle ABC

$$= \text{area(ALMC)} + \text{area(CMNB)} - \text{area(ALNB)}$$
$$= (x_3 - x_1)(y_1 + y_3)/2 + (x_2 - x_3)(y_2 + y_3)/2 - (x_2 - x_1)(y_1 + y_2)/2$$
$$= [x_1(y_2 - y_3) + x_2(y_3 - y_1) + x_3(y_1 - y_2)]/2.$$

If the three points A, B and C lay on a straight line, the area of the triangle would be zero. We can use this property to find the equation of a straight line through $A(x_1, y_1)$ and $B(x_2, y_2)$, say. The coordinates (x, y) of any point on the line must be such that if (x_3, y_3) is replaced by (x, y), the above expression for area is zero. Hence,

$$x_1(y_2 - y) + x_2(y - y_1) + x(y_1 - y_2) = 0,$$
$$\text{i.e. } (y_1 - y_2)x - (x_1 - x_2)y + (x_1y_2 - x_2y_1) = 0.$$

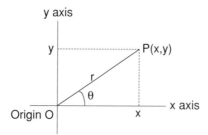

Figure 20.1. Cartesian coordinates.

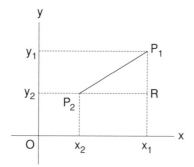

Figure 20.2. Distance between two points.

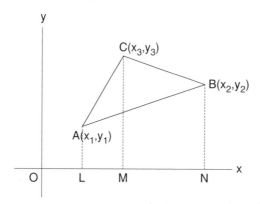

Figure 20.3. Area of a triangle.

EXAMPLE

If the two points are A(3, 2) and B(4, 5), the equation of the straight line through AB is $-3x + y + 7 = 0$.

If a point moves in such a way that its position always satisfies certain conditions, then the path followed is the *locus* of that point. In the straight line example, $-3x + y + 7 = 0$ is the equation of the locus of a point which lies on a straight line through A(3, 2) and B(4, 5).

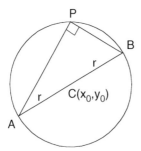

Figure 20.4. Locus of P such that AP ⊥ PB.

To take another example, if A and B are two fixed points, let the locus of P be such that AP is always perpendicular to BP. Then, from the theorem in geometry which states that the angle subtended by the diameter of a circle at any point on the circumference is a right-angle, it follows that the locus of P is a circle on AB as diameter (see Figure 20.4).

The coordinates of the points of intersection of two curves are the (x, y) pairs which simultaneously satisfy the equation of each curve. For example, let the two equations be:

$$x - y + 2 = 0 \quad \text{and} \quad x^2 + y^2 - 4 = 0.$$

From the first equation, $y = x + 2$, and substituting into the second equation gives:

$$x^2 + (x + 2)^2 - 4 = 2x^2 + 4x = 0, \text{ i.e. } x(x + 2) = 0, \quad \text{so } x = 0 \text{ or } -2.$$

From the first equation, if $x = 0$, $y = 2$ and if $x = -2$, $y = 0$. Hence, the points of intersection are $(0, 2)$ and $(-2, 0)$.

20.2 Straight line

In Section 20.1, we found the equation for a straight line passing through two given points. The equation took the form $ax + by + c = 0$, where a, b and c were constants. In fact, this is the general form for the equation of a straight line.

The constants may be chosen to represent certain features of the straight line. With the same scale factors for x and y, the slope of the line is $\tan \theta$ (see Figure 20.5a),where θ is the angle of elevation of the straight line above the x-axis. If P is any point on the straight line, with coordinates given generally as (x, y), and the line cuts the y-axis at $y = c$, then the slope of the line is: $m = \tan \theta = (y - c)/x$. Consequently, the equation for the straight line may be written as: $y = mx + c$, where m is the slope and c is the intercept with the y-axis.

In a similar vein, we may have a straight line with slope $m = \tan \theta$ passing through a point (x_1, y_1), as in Figure 20.5b. Again, taking P as a general point on the line with coordinates (x, y), the slope $m = \tan \theta = (y - y_1)/(x - x_1)$. Hence, the equation for the line is: $y - y_1 = m(x - x_1)$.

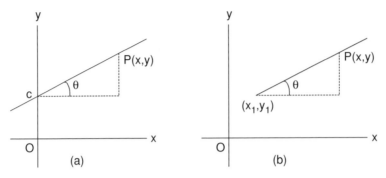

Figure 20.5. Straight line through the point (a) $(0, c)$ and (b) (x_1, y_1).

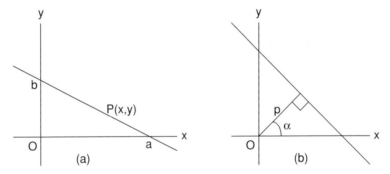

Figure 20.6. Deriving alternative equations for a straight line.

Next, suppose that the straight line cuts the x- and y-axes at $x = a$ and $y = b$, respectively, as shown in Figure 20.6a. Thinking of how we derived the equation $y = mx + c$, in this case $c = b$ and $m = -b/a$. Hence,

$$y = -\frac{b}{a}x + b \text{ or } \frac{x}{a} + \frac{y}{b} = 1.$$

We can use the last form to obtain an equation for the straight line in terms of its perpendicular distance p from the origin and the angle α of elevation of that perpendicular to the x-axis. Referring to Figures 20.6a and b: $a = p \sec \alpha$ and $b = p \csc \alpha$. Hence, the equation for the straight line is:

$$\frac{x}{p \sec \alpha} + \frac{y}{p \csc \alpha} = \frac{x \cos \alpha}{p} + \frac{y \sin \alpha}{p} = 1$$

or $x \cos \alpha + y \sin \alpha = p$.

20.3 Circle

Let the circle have centre $C(a, b)$ and radius r, as shown in Figure 20.7. Then, any point $P(x, y)$ on the circle is distance r from $C(a, b)$ and by Pythagoras' theorem:

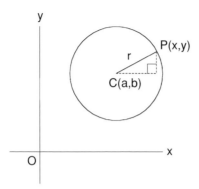

Figure 20.7. Circle with centre C(a, b) and radius r.

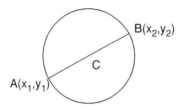

Figure 20.8. Circle with ends of a diameter at A(x_1, y_1) and B(x_2, y_2).

$$(x - a)^2 + (y - b)^2 = r^2,$$

which is the equation for the circle.

The general form for the equation of a circle is:

$$x^2 + y^2 + 2gx + 2fy + c = 0.$$

To compare this with the previous equation, we must complete the squares on x and y to give:

$$(x + g)^2 + (y + f)^2 = g^2 + f^2 - c.$$

Hence, for a circle with the general equation, the centre is at C($-g$, $-f$) and the square of the radius is $r^2 = g^2 + f^2 - c$.

Now, suppose we wish to give the formula for a circle when we know that the ends of a diameter are at the points A(x_1, y_1) and B(x_2, y_2), as in Figure 20.8. Then the square of the radius is:

$$r^2 = \left(\frac{AB}{2}\right)^2 = [(x_2 - x_1)^2 + (y_2 - y_1)^2]/4.$$

Also, the centre C is at:

$$C\left(\frac{x_1 + x_2}{2}, \frac{y_1 + y_2}{2}\right).$$

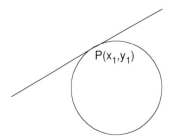

Figure 20.9. Finding the equation for a tangent to a circle.

Hence, the equation for the circle is:

$$\left(x - \frac{x_1 + x_2}{2}\right)^2 + \left(y - \frac{y_1 + y_2}{2}\right)^2 = [(x_2 - x_1)^2 + (y_2 - y_1)^2]/4.$$

On expanding and collecting terms, this becomes:

$$x^2 - (x_1 + x_2)x + x_1 x_2 + y^2 - (y_1 + y_2)y + y_1 y_2$$
$$= (x - x_1)(x - x_2) + (y - y_1)(y - y_2) = 0.$$

We might wish to find the equation for a circle given any three points on the circumference, say (x_1, y_1), (x_2, y_2) and (x_3, y_3). Each of these pairs of (x, y) values must satisfy the equation: $x^2 + y^2 + 2gx + 2fy + c = 0$. Substituting each pair in turn will give three equations in g, f and c, which may be solved in order to complete the equation for the circle.

If we know the equation $x^2 + y^2 + 2gx + 2fy + c = 0$, i.e. we know the values of g, f and c, for a certain circle, then the equation for the tangent to the circle at a given point $P(x_1, y_1)$ on the circumference (see Figure 20.9) may be found in the following way. Regarding y as a function of x, the equation for the circle may be differentiated with respect to x to give:

$$2x + 2y\frac{dy}{dx} + 2g + 2f\frac{dy}{dx} = 0.$$

Thus the slope of the circular curve is: $\frac{dy}{dx} = -(x + g)/(y + f)$. Consequently, the slope of the tangent at $P(x_1, y_1)$ is $-(x_1 + g)/(y_1 + f)$. Hence, the equation for the tangent at P must be:

$$y - y_1 = -\frac{x_1 + g}{y_1 + f}(x - x_1)$$

or $xx_1 + yy_1 + g(x - x_1) + f(y - y_1) = x_1^2 + y_1^2$.

Since (x_1, y_1) lies on the circle:

$$x_1^2 + y_1^2 + 2gx_1 + 2fy_1 + c = 0.$$

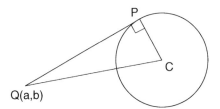

Figure 20.10. Finding the length of the tangent to a circle from a point $Q(a, b)$.

Therefore, on substituting for $x_1^2 + y_1^2$, the equation for the tangent becomes:

$$xx_1 + yy_1 + g(x + x_1) + f(y + y_1) + c = 0.$$

Notice that this is the same as the equation for the circle with x^2 replaced by xx_1, y^2 by yy_1, $2x$ by $x + x_1$ and $2y$ by $y + y_1$.

Referring to Figure 20.10, there is a particularly simple formula for the length PQ of the tangent to a circle from a given point $Q(a, b)$. If the circle is: $x^2 + y^2 + 2gx + 2fy + c = 0$, the centre C will have coordinates $C(-g, -f)$ and the square of the radius is $r^2 = g^2 + f^2 - c$. From the right-angled triangle CPQ:

$$(PQ)^2 = (CQ)^2 - r^2 = (a + g)^2 + (b + f)^2 - g^2 - f^2 + c$$
$$= a^2 + b^2 + 2ga + 2fb + c.$$

This is the left-hand side of the equation for the circle with x replaced by a and y by b.

20.4 Conic sections

Figure 20.11 represents a double-ended circular cone. If the cone is cut by a plane surface, the intersecting curve is called a conic section. The type of conic section depends on the angle at which the plane cuts the cone as indicated in the diagram.

A conic section may also be defined as follows: it is the locus of a point $P(x, y)$ whose distance from a fixed point F(*focus*) is a constant ϵ(*eccentricity*) times the distance of P from a straight line AB(*directrix*), where P, F and AB all lie in the x, y plane. The locus of P is a *parabola* if $\epsilon = 1$, an *ellipse* if $\epsilon < 1$ and a *hyperbola* if $\epsilon > 1$.

20.5 Parabola

Referring to Figure 20.12, let the directrix AB be parallel to the y-axis at $x = -a$ and the focus be the point $F(a, 0)$. Given the directrix and focus, we can always draw Cartesian coordinate axes so that this applies.

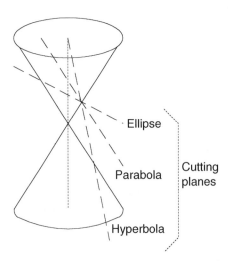

Figure 20.11. Conic sections formed from the intersection of planes with a conical surface.

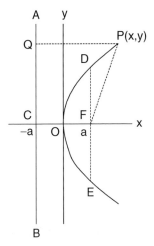

Figure 20.12. Parabola with focus F and directrix AB.

From Figure 20.12, we see that we must have $QP = FP$ for the locus of P to be a parabola. Consequently,

$$(FP)^2 = (x - a)^2 + y^2 = (QP)^2 = (x + a)^2.$$

On cancelling x^2 and a^2, we are left with:

$$y^2 = 4ax,$$

which is the equation for the parabola. The curve is symmetrical about the x-axis, which is therefore the axis of the parabola. Also, the vertex of the parabola is at the origin O(0, 0).

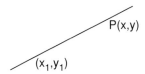

Figure 20.13. Deriving the equation for the tangent at the point (x_1, y_1) on the parabola.

The chord DE of the parabola, which passes through the focus F and is perpendicular to the axis, is called the *latus rectum*. We see from the equation for the parabola that the y-coordinate of D is $2a$ and hence, the length of the latus rectum is $4a$.

To find the equation for the tangent to the parabola at the point (x_1, y_1) on the parabola, we need to find the slope of the parabola at that point. Let the parabola have the equation $y^2 = 4ax$ and differentiate with respect to x: $2y\frac{dy}{dx} = 4a$. Therefore the slope of the parabola at (x_1, y_1) is:

$$\left.\frac{dy}{dx}\right|_{x=x_1} = \frac{2a}{y_1}.$$

Thus, with the help of Figure 20.13, we see that the equation for the tangent is:

$$y - y_1 = \frac{2a}{y_1}(x - x_1) \text{ or } yy_1 - y_1^2 = 2a(x - x_1).$$

Now, the point (x_1, y_1) lies on the parabola and must therefore satisfy the equation $y_1^2 = 4ax_1$. Hence, the equation for the tangent may be written as:

$$yy_1 = 2a(x + x_1).$$

Notice that this corresponds to the equation for the parabola with y^2 replaced by yy_1 and x by $(x + x_1)/2$.

Referring to Figure 20.14, QS is the tangent to the parabola at the point $P(x_1, y_1)$ on the parabola.

Assuming the equation $y^2 = 4ax$ for the parabola, the equation for the tangent is $yy_1 = 2a(x + x_1)$. This crosses the x-axis at Q when $y = 0$, i.e. when $x = -x_1$.

Also, the focus of the parabola is the point $F(a, 0)$. Therefore,

$$(FP)^2 = (x_1 - a)^2 + y_1^2 = (x_1 - a)^2 + 4ax_1 = (x_1 + a)^2 = (QF)^2.$$

Hence, FPQ is an isosceles triangle and $\angle FPQ = \angle FQP$. Furthermore, if PR is drawn parallel to the x-axis, then $\angle SPR = \angle PQF = \angle FPQ$. If we now imagine the parabola rotated round the x-axis to form the surface of a parabolic reflector, any ray striking the surface from a direction parallel to the x-axis will be reflected through the focus F. This property is used to shape the reflector for a search-light and the dish for a satellite television aerial.

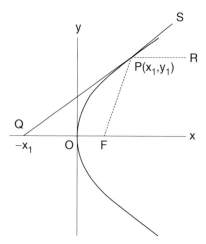

Figure 20.14. Noting where the tangent cuts the axis of the parabola.

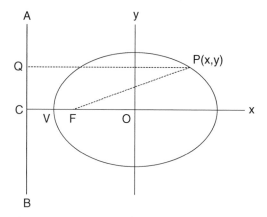

Figure 20.15. An ellipse with focus F and directrix AB.

20.6 Ellipse

In Figure 20.15, we let AB be the directrix drawn parallel to the y-axis and cutting the x-axis at C. We let the focus F be another point on the x-axis nearer to the origin than C. The curve of the ellipse will cut the x-axis at a point V between C and F and such that $(FV)/(VC) = \epsilon < 1$, where ϵ is the eccentricity. Then the ellipse is the locus of the point P(x, y) such that $(FP)/(PQ) = \epsilon$, where Q is on the directrix AB and PQ is perpendicular to AB.

Now, let F and C have coordinates F($-a\epsilon$, 0) and C($-a/\epsilon$, 0). Then, if the coordinates of V are V($-v$, 0),

$$\frac{FV}{VC} = \frac{v - a\epsilon}{a/\epsilon - v} = \epsilon, \text{ i.e. } v - a\epsilon = a - v\epsilon \text{ or } v(1 + \epsilon) = a(1 + \epsilon)$$

and therefore, $v = a$.

The locus of P(x, y) will then be described by:

$$(PF)^2 = \epsilon^2 (PQ)^2, \text{ i.e. } (x + a\epsilon)^2 + y^2 = \epsilon^2 (x + a/\epsilon)^2$$

$$\text{or } (1 - \epsilon^2)x^2 + y^2 = a^2(1 - \epsilon^2), \text{ i.e. } \frac{x^2}{a^2} + \frac{y^2}{a^2(1 - \epsilon^2)} = 1.$$

If we replace $a^2(1 - \epsilon^2)$ by b^2, the equation for the ellipse takes the form:

$$\frac{x^2}{a^2} + \frac{y^2}{b^2} = 1.$$

Given in this form, the eccentricity ϵ for the ellipse is such that: $\epsilon^2 = 1 - (b/a)^2$.

Notice that the ellipse is symmetrical about both the x-axis and the y-axis. When $y = 0$, $x = \pm a$ and when $x = 0$, $y = \pm b$. Furthermore, we could also have the focus at $(a\epsilon, 0)$ and the directrix cutting the x-axis at $(a/\epsilon, 0)$.

As with the parabola, the chord through F and parallel to the y-axis is called the latus rectum. For $x = -a\epsilon$, the point P(x, y) on the ellipse is such that $y = \pm b\sqrt{1 - \epsilon^2} = \pm b^2/a$. Hence, the length of the latus rectum is $2b^2/a$.

Given the equation $x^2/a^2 + y^2/b^2 = 1$ for an ellipse, we can differentiate with respect to x to give:

$$\frac{2x}{a^2} + \frac{2y}{b^2}\frac{dy}{dx} = 0.$$

Thus, at a point P(x, y) on the ellipse, the slope is $dy/dx = -(b^2x)/(a^2y)$. Hence, the equation for the tangent at a point (x_1, y_1) on the ellipse is:

$$y - y_1 = -\frac{b^2 x_1}{a^2 y_1}(x - x_1), \text{ i.e. } b^2 x_1 x + a^2 y_1 y = b^2 x_1^2 + a^2 y_1^2.$$

Dividing through by $a^2 b^2$ gives:

$$\frac{x_1 x}{a^2} + \frac{y_1 y}{b^2} = \frac{x_1^2}{a^2} + \frac{y_1^2}{b^2} = 1,$$

since (x_1, y_1) lies on the ellipse. Notice that this equation for the tangent at (x_1, y_1) on the ellipse corresponds to the equation for the ellipse with x^2 replaced by $x_1 x$ and y^2 by $y_1 y$.

20.7 Hyperbola

For the hyperbola, the eccentricity $\epsilon > 1$. Again, we let the directrix AB cross the x-axis at C$(a/\epsilon, 0)$ and the focus F be at F$(a\epsilon, 0)$ (see Figure 20.16). If V$(v, 0)$ is on the hyperbola, then:

$$\frac{VF}{CV} = \frac{a\epsilon - v}{v - a/\epsilon} = \epsilon \quad \text{and} \quad a\epsilon - v = v\epsilon - a \quad \text{or} \quad a(\epsilon + 1) = v(\epsilon + 1).$$

Consequently, $v = a$.

For a general point $P(x, y)$ on the hyperbola:

$$(PF)^2 = \epsilon^2(QP)^2, \text{ i.e. } (a\epsilon - x)^2 + y^2 = \epsilon^2(x - a/\epsilon)^2$$
$$\text{or } (\epsilon^2 - 1)x^2 - y^2 = a^2(\epsilon^2 - 1).$$

On dividing through by $a^2(\epsilon^2 - 1)$, this becomes:

$$\frac{x^2}{a^2} - \frac{y^2}{a^2(\epsilon^2 - 1)} = 1.$$

Finally, if we write $b^2 = a^2(\epsilon^2 - 1)$, the equation for the hyperbola becomes:

$$\frac{x^2}{a^2} - \frac{y^2}{b^2} = 1.$$

Since this equation describes a curve which is symmetrical about the y-axis as well as the x-axis, the complete hyperbola has another curve which is the reflection of the curve shown in Figure 20.16 about the y-axis. There will also be another focus at $(-a\epsilon, 0)$ and another directrix $x = -a/\epsilon$.

With each curve, there is a latus rectum, which is the chord parallel to the y-axis and passing through the focus F. For $x = a\epsilon$, the point $P(x, y)$ on the hyperbola has $y = \pm b\sqrt{\epsilon^2 - 1} = \pm b^2/a$. Hence, the length of the latus rectum is $2b^2/a$.

We derive the equation for the tangent at a point (x_1, y_1) on the hyperbola in the same way as we did for the ellipse. Differentiating the equation $x^2/a^2 - y^2/b^2 = 1$

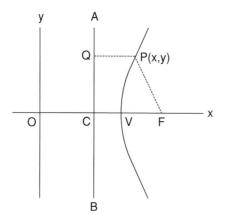

Figure 20.16. Hyperbola with focus F and directrix AB.

with respect to x gives:

$$\frac{2x}{a^2} - \frac{2y}{b^2}\frac{dy}{dx} = 0.$$

At the point P(x, y) on the hyperbola: $dy/dx = (b^2x)/(a^2y)$. Thus, the equation for the tangent at the point (x_1, y_1) on the hyperbola is:

$$y - y_1 = \frac{b^2x_1}{a^2y_1}(x - x_1), \text{ i.e. } b^2x_1x - a^2y_1y = b^2x_1^2 - a^2y_1^2.$$

Dividing through by a^2b^2 gives:

$$\frac{x_1x}{a^2} - \frac{y_1y}{b^2} = \frac{x_1^2}{a^2} - \frac{y_1^2}{b^2} = 1,$$

since (x_1, y_1) lies on the hyperbola. Notice that this equation corresponds to that of the hyperbola with x^2 replaced by x_1x and y^2 by y_1y.

20.8 Three-dimensional coordinate geometry

So far, the coordinate geometry has been restricted to two dimensions, a point P(x_1, y_1) being located by its coordinates $x = x_1$ and $y = y_1$. Moving now to three dimensions, we have the Cartesian system of three mutually orthogonal x-, y- and z-axes. A point P(x_1, y_1, z_1) is now located, as shown in Figure 20.17, by its three coordinates $x = x_1$, $y = y_1$ and $z = z_1$. It is usual to have a right-handed system of axes as shown. If we imagine a *right-hand thread* screw lined up along the z-axis and turn it from positive x direction to positive y direction, the screw will move axially in the positive z direction.

Referring to Figure 20.17, the distance r of point P(x_1, y_1, z_1) from the origin O may be found by using Pythagoras' theorem as follows:

$$r^2 = \rho^2 + z_1^2 = x_1^2 + y_1^2 + z_1^2.$$

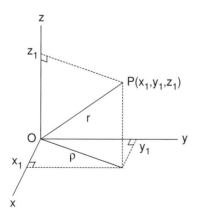

Figure 20.17. Three-dimensional Cartesian coordinates.

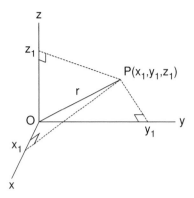

Figure 20.18. Direction of OP relative to the x-, y- and z-axes.

Similarly, if we have two points $P_1(x_1, y_1, z_1)$ and $P_2(x_2, y_2, z_2)$, their distance d apart is given by the equation:

$$d^2 = (x_2 - x_1)^2 + (y_2 - y_1)^2 + (z_2 - z_1)^2.$$

Referring to Figure 20.18, the direction of P from O is specified by the angles between OP and the axes. These are denoted by:

$$\alpha = \angle POx, \quad \beta = \angle POy \quad \text{and} \quad \gamma = \angle POz.$$

In fact, we often use the *direction cosines* which are:

$$\cos\alpha = x_1/r, \quad \cos\beta = y_1/r \quad \text{and} \quad \cos\gamma = z_1/r.$$

Similarly, if we have two points $P_1(x_1, y_1, z_1)$ and $P_2(x_2, y_2, z_2)$ a distance d apart, the straight line from P_1 to P_2 has the direction cosines:

$$\cos\alpha = (x_2 - x_1)/d, \quad \cos\beta = (y_2 - y_1)/d \quad \text{and} \quad \cos\gamma = (z_2 - z_1)/d.$$

Any numbers a, b and c, which are in the same proportion as the direction cosines of the line, are referred to as direction ratios of the line, i.e. of the direction of P_2 from P_1. For instance, the numbers $a = x_2 - x_1$, $b = y_2 - y_1$ and $c = z_2 - z_1$ are direction ratios in this case.

Referring to Figure 20.19, suppose we have two straight lines L_1 and L_2 with direction cosines λ_1, μ_1, ν_1 and λ_2, μ_2, ν_2, respectively, and we wish to find the angle θ between the two lines. Since we are only concerned with directions, we can draw our lines through the origin as shown and pick off points $P_1(x_1, y_1, z_1)$ and $P_2(x_2, y_2, z_2)$ on the respective lines L_1 and L_2. Let d_1 and d_2 be the distances of P_1 and P_2, respectively, from the origin so that the direction cosines of L_1 and L_2 are:

$$(\lambda_1, \mu_1, \nu_1) = \left(\frac{x_1}{d_1}, \frac{y_1}{d_1}, \frac{z_1}{d_1} \right) \quad \text{and} \quad (\lambda_2, \mu_2, \nu_2) = \left(\frac{x_2}{d_2}, \frac{y_2}{d_2}, \frac{z_2}{d_2} \right),$$

respectively. Also, let the distance $P_1P_2 = d$.

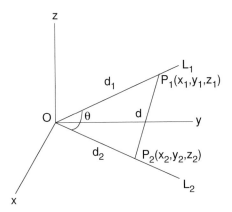

Figure 20.19. Finding the angle θ between two lines.

Referring to our diagram (Figure 20.19), the cosine rule for the triangle P_1OP_2 states that:

$$d^2 = d_1^2 + d_2^2 - 2d_1d_2 \cos\theta, \text{ i.e. } \cos\theta = \frac{d_1^2 + d_2^2 - d^2}{2d_1d_2}.$$

Expanding the numerator terms using the coordinates of P_1 and P_2 gives:

$$\cos\theta = \frac{x_1^2 + y_1^2 + z_1^2 + x_2^2 + y_2^2 + z_2^2 - (x_2 - x_1)^2 - (y_2 - y_1)^2 - (z_2 - z_1)^2}{2d_1d_2}$$

$$= \frac{x_1x_2 + y_1y_2 + z_1z_2}{d_1d_2} = \frac{x_1}{d_1}\frac{x_2}{d_2} + \frac{y_1}{d_1}\frac{y_2}{d_2} + \frac{z_1}{d_1}\frac{z_2}{d_2}$$

$$= \lambda_1\lambda_2 + \mu_1\mu_2 + \nu_1\nu_2.$$

An important result which follows from this is that L_1 and L_2 are perpendicular if and only if: $\lambda_1\lambda_2 + \mu_1\mu_2 + \nu_1\nu_2 = 0$. In fact, the same would be true for their direction ratios a_1, b_1, c_1 and a_2, b_2, c_2, i.e. $a_1a_2 + b_1b_2 + c_1c_2 = 0$ if and only if L_1 and L_2 are perpendicular.

20.9 Equations for a straight line

Let the straight line L pass through two points $P_1(x_1, y_1, z_1)$ and $P_2(x_2, y_2, z_2)$ (see Figure 20.20).

If $P(x, y, z)$ is a general point on the line, direction ratios for the line may be written as $(x - x_1), (y - y_1), (z - z_1)$ or $(x_2 - x_1), (y_2 - y_1), (z_2 - z_1)$. Since they must be in the same proportion, we can use a proportionality factor t and write:

$$(x - x_1) = (x_2 - x_1)t, \quad (y - y_1) = (y_2 - y_1)t \quad \text{and} \quad (z - z_1) = (z_2 - z_1)t$$

$$\text{or} \quad x = x_1 + (x_2 - x_1)t, \quad y = y_1 + (y_2 - y_1)t \quad \text{and} \quad z = z_1 + (z_2 - z_1)t.$$

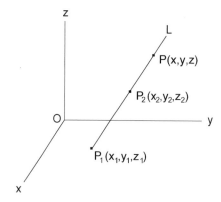

Figure 20.20. A straight line in three dimensional space.

Since $(x_2 - x_1)$, $(y_2 - y_1)$, $(z_2 - z_1)$ are direction ratios, we can say that if the straight line passes through $P_1(x_1, y_1, z_1)$ and has direction ratios a, b, c, the line may be described parametrically by:

$$x = x_1 + at, \quad y = y_1 + bt \quad \text{and} \quad z = z_1 + ct.$$

Eliminating t, we may define the straight line by the equations:

$$\frac{x - x_1}{a} = \frac{y - y_1}{b} = \frac{z - z_1}{c}.$$

This assumes that a, b, c are all non-zero. If one, say c, were zero, we would have the two equations: $(x - x_1)/a = (y - y_1)/b$ and $z = z_1$. In the extreme case of two, say b and c, being zero, we would be left with the two equations: $y = y_1$ and $z = z_1$. In this case, the line would be parallel to the x-axis.

We could also eliminate t for the line through $P_1(x_1, y_1, z_1)$ and $P_2(x_2, y_2, z_2)$ to describe it by the two equations:

$$\frac{x - x_1}{x_2 - x_1} = \frac{y - y_1}{y_2 - y_1} = \frac{z - z_1}{z_2 - z_1}.$$

20.10 The plane

A plane is a flat surface whose position is defined by a point $P_1(x_1, y_1, z_1)$ on the plane together with the direction ratios (a, b, c) of a straight line L which is perpendicular to the plane

Referring to Figure 20.21, let $P(x, y, z)$ be another point in the plane. Then, the straight line P_1P has direction ratios: $(x - x_1)$, $(y - y_1)$ and $(z - z_1)$. Since the line P_1P lies in the plane, it must be perpendicular to the line L. As we discovered at the end of Section 20.8, this implies that:

$$a(x - x_1) + b(y - y_1) + c(z - z_1) = 0.$$

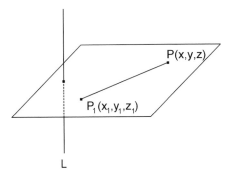

Figure 20.21. Plane perpendicular to a line L.

It follows from this that the general equation for a plane is:

$$ax + by + cz + d = 0,$$

where (a, b, c) are the direction ratios of a straight line perpendicular to the plane and d is another constant.

The position of a plane is completely specified by the positions of any three non-collinear points in the plane.

EXAMPLE

Suppose we wish to find the equation for the plane passing through the three points $P_1(1, 2, 3), P_2(3, 1, 2)$ and $P_3(2, 3, 1)$. Substitute each set of coordinates into the general equation for a plane:

$$a + 2b + 3c + d = 0, \tag{1}$$
$$3a + b + 2c + d = 0, \tag{2}$$
$$2a + 3b + c + d = 0. \tag{3}$$

Eliminating a between (1) and (2) gives: $5b + 7c + 2d = 0$.
Eliminating a between (1) and (3) gives: $b + 5c + d = 0$.
Eliminating b between the latter two gives: $18c + 3d = 0$.
Hence, $c = -d/6$ and substituting back gives: $b = -d/6$ and $a = -d/6$. Finally, putting $d = -6$ and substituting into the general equation specifies the plane as:

$$x + y + z = 6.$$

The angle between two planes is the angle between lines L_1 and L_2 perpendicular to the planes. If one plane is: $a_1 x + b_1 y + c_1 z + d_1 = 0$, its perpendicular L_1 has direction ratios (a_1, b_1, c_1). Similarly, if the other plane is: $a_2 x + b_2 y + c_2 z + d_2 = 0$, its perpendicular L_2 has direction ratios (a_2, b_2, c_2). Now, the direction cosines of L_1 and L_2 are:

$$\frac{(a_1, b_1, c_1)}{\sqrt{a_1^2 + b_1^2 + c_1^2}} \quad \text{and} \quad \frac{(a_2, b_2, c_2)}{\sqrt{a_2^2 + b_2^2 + c_2^2}},$$

respectively, so from Section 20.8, the angle θ between L_1 and L_2 is given by:

$$\cos \theta = \frac{|a_1 a_2 + b_1 b_2 + c_1 c_2|}{\sqrt{a_1^2 + b_1^2 + c_1^2}\sqrt{a_2^2 + b_2^2 + c_2^2}},$$

where we have taken the modulus in the numerator to keep $\theta \leq \pi/2$. Of course, the two planes will be perpendicular if $a_1 a_2 + b_1 b_2 + c_1 c_2 = 0$.

Finally, suppose we wish to find the distance of a point $P_1(x_1, y_1, z_1)$ from a plane: $ax + by + cz + d = 0$. A straight line L_1, through P_1 and perpendicular to the plane, can be described parametrically by:

$$x = x_1 + at, \quad y = y_1 + bt \quad \text{and} \quad z = z_1 + ct.$$

Then, if L_1 intersects the plane at $P_2(x_2, y_2, z_2)$, the distance $d_1 = P_1 P_2$ is given by:

$$d_1^2 = (x_2 - x_1)^2 + (y_2 - y_1)^2 + (z_2 - z_1)^2.$$

Now P_2 is on both the line L_1 and the plane. For some parameter t_2:

$$x_2 = x_1 + at_2, \quad y_2 = y_1 + bt_2 \quad \text{and} \quad z_2 = z_1 + ct_2.$$

Also, $ax_2 + by_2 + cz_2 + d = 0$. Therefore:

$$a(x_1 + at_2) + b(y_1 + bt_2) + c(z_1 + ct_2) + d = 0$$

and $$t_2 = -\frac{ax_1 + by_1 + cz_1 + d}{a^2 + b^2 + c^2}.$$

Also, $$d_1^2 = (a^2 + b^2 + c^2)t_2^2 = \frac{(ax_1 + by_1 + cz_1 + d)^2}{a^2 + b^2 + c^2}.$$

Therefore, $$d_1 = \frac{|ax_1 + by_1 + cz_1 + d|}{\sqrt{a^2 + b^2 + c^2}}.$$

20.11 Cylindrical and spherical coordinates

A circular cylindrical surface with its axis coinciding with the z-axis is defined by the fact that every point on the surface is the same distance ρ from the z-axis. Hence, the equation for the surface is: $x^2 + y^2 = \rho^2 = \text{constant}$.

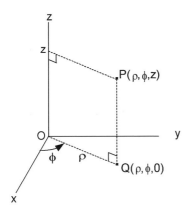

Figure 20.22. Cylindrical polar coordinates (ρ, ϕ, z).

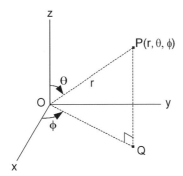

Figure 20.23. Spherical polar coordinates (r, θ, ϕ).

Sometimes, it is more convenient to use *cylindrical* polar coordinates (ρ, ϕ, z), as in Figure 20.22, instead of Cartesian coordinates (x, y, z). Then a point $P(\rho, \phi, z)$ is distance ρ from the z-axis and z is its distance from the x, y plane. The angle ϕ is as shown in the diagram; we drop a perpendicular from P to a point Q in the x, y plane and ϕ is the angle between OQ and the positive x-axis. ϕ is positive in the direction shown in the diagram (right-hand thread rule again). Using this system of cylindrical polar coordinates, the equation for the circular cylindrical surface whose axis is the z-axis is simply $\rho = $ constant.

A spherical surface with its centre at the origin is defined by the fact that every point on it is the same distance r from the origin. Hence, the equation for the surface in Cartesian coordinates is: $x^2 + y^2 + z^2 = r^2 = $ constant.

Sometimes, it is more convenient to use *spherical* polar coordinates (r, θ, ϕ), as in Figure 20.23. ϕ is measured in exactly the same way as for cylindrical polar coordinates but now r is the distance of the point $P(r, \theta, \phi)$ from the origin O. θ is the angle which

OP makes with the positive z-axis. Now the equation for the spherical surface with its centre at the origin is simply $r = $ constant.

The equations which relate cylindrical polar coordinates to Cartesian coordinates are:

$$x = \rho \cos \phi, \ y = \rho \sin \phi, \ z = z.$$

The corresponding relations between spherical polar coordinates and Cartesian coordinates are:

$$x = r \sin \theta \cos \phi, \ y = r \sin \theta \sin \phi, \ z = r \cos \theta.$$

21 Vector algebra

21.1 Vectors

A vector is a quantity which possesses direction as well as magnitude. It is particularly useful in mechanics since force, velocity and acceleration are all vector quantities. By using the axis of rotation and the right-hand thread rule to denote direction, we can also regard torque (or moment of force), angular velocity and angular acceleration as vector quantities.

Bold type is used to indicate that a symbol **a**, say, represents a vector. It may be illustrated diagrammatically by an arrow, the length of which is proportional to the magnitude and the direction by the arrow. Changing the sign to −**a** just reverses the direction of the vector **a**, as in Figure 21.1.

The magnitude of the vector is the modulus |**a**| but written more simply as a in standard type. The symbol **â** is used to indicate a vector which has unit magnitude and the same direction as **a**. Hence, **â** is a *unit vector* and **a** $= a$**â**.

If **a** and **b** are two vectors, their sum **c** $=$ **a** $+$ **b** is obtained by the triangle law of addition, as shown in Figure 21.2. By completing the parallelogram with the dotted lines in the diagram, we see how the triangle law corresponds to the parallelogram law in the parallelogram of forces. Correspondingly, **c** $=$ **a** $+$ **b** may be referred to as the *vector sum* or the *resultant* of **a** and **b**.

Subtraction **a** $-$ **b** is achieved by the triangle law with the direction of **b** reversed, as shown in Figure 21.3. Notice that **a** $-$ **b** corresponds to the other diagonal (not drawn) in the parallelogram of Figure 21.2.

Vectors are *coplanar* if there is a plane which is parallel to all of the vectors. This will always be true for only two vectors but it need not be true for three or more. In that case, the vectors would be *non-coplanar*.

Suppose we wish to find the vector sum **b**$_3$ $=$ **a**$_1$ $+$ **a**$_2$ $+$ **a**$_3$ of three non-coplanar vectors. Two vectors **a**$_1$ and **a**$_2$ are coplanar so we can use the triangle law to find their resultant **b**$_2$ $=$ **a**$_1$ $+$ **a**$_2$. Then **b**$_2$ and **a**$_3$ must be coplanar so that the triangle law gives their resultant **b**$_3$ $=$ **b**$_2$ $+$ **a**$_3$ $=$ **a**$_1$ $+$ **a**$_2$ $+$ **a**$_3$. This is illustrated by Figure 21.4 in which **a$_3$** is meant to be directed out of the plane of the paper. It follows immediately

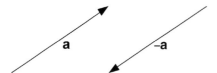

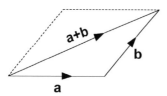

Figure 21.1. Vectors **a** and −**a**.

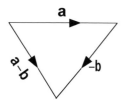

Figure 21.2. Triangle law of addition of two vectors.

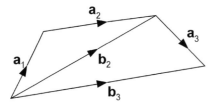

Figure 21.3. Vector subtraction.

Figure 21.4. Vector sum of three non-coplanar forces.

that if we have n vectors $\mathbf{a}_i$, $i = 1, 2, \ldots, n$, either coplanar or non-coplanar, their resultant $\mathbf{b}_n = \sum_{i=1}^{n} \mathbf{a}_i$ may be found by joining the vectors end to end. The resultant $\mathbf{b}_n$ is then the vector joining the starting point to the finishing point. In the case of non-coplanar vectors $\mathbf{a}_i$, this process is easier said than done.

In order to overcome this problem, we introduce the idea of *components of a vector*. Consider the vector **a** in Figure 21.5. It has components in the x-, y- and z-directions of magnitude a_x, a_y and a_z. We can write these components as vectors by introducing unit vectors **i**, **j** and **k** in the x-, y- and z-directions, respectively. Then, we can see from the diagram that, if we apply the triangle law of addition twice:

$$\mathbf{a} = a_x \mathbf{i} + a_y \mathbf{j} + a_z \mathbf{k}.$$

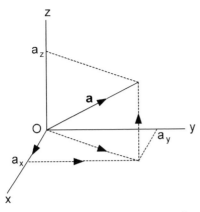

Figure 21.5. x-, y- and z-components of vector **a**.

If we now wish to add several vectors, we just need to add their components separately. For instance, if we have three vectors:

$$\mathbf{a} = a_x\mathbf{i} + a_y\mathbf{j} + a_z\mathbf{k}, \quad \mathbf{b} = b_x\mathbf{i} + b_y\mathbf{j} + b_z\mathbf{k} \quad \text{and} \quad \mathbf{c} = c_x\mathbf{i} + c_y\mathbf{j} + c_z\mathbf{k},$$

then, $\mathbf{a} + \mathbf{b} + \mathbf{c} = (a_x + b_x + c_x)\mathbf{i} + (a_y + b_y + c_y)\mathbf{j} + (a_z + b_z + c_z)\mathbf{k}$.

Splitting a vector **a** into its components $a_x\mathbf{i}$, $a_y\mathbf{j}$ and $a_z\mathbf{k}$ is called *resolving* the vector. Furthermore, if α, β and γ are the angles which **a** makes with the positive x-, y- and z-directions, respectively, then $a_x = a\cos\alpha$, $a_y = a\cos\beta$ and $a_z = a\cos\gamma$. In fact, $\cos\alpha$, $\cos\beta$ and $\cos\gamma$ are the direction cosines of the vector **a**. They correspond exactly to the direction cosines introduced in three-dimensional coordinate geometry in Section 20.8.

The magnitude a of the vector **a** is related to the magnitudes of its x-, y- and z-components by the equation: $a^2 = a_x^2 + a_y^2 + a_z^2$. Hence, the direction cosines are:

$$\cos\alpha = \frac{a_x}{\sqrt{a_x^2 + a_y^2 + a_z^2}}, \quad \cos\beta = \frac{a_y}{\sqrt{a_x^2 + a_y^2 + a_z^2}} \quad \text{and} \quad \cos\gamma = \frac{a_z}{\sqrt{a_x^2 + a_y^2 + a_z^2}}.$$

21.2 Straight line and plane

We now wish to use vectors to define the straight line locus of a point P which passes through a given point A, as shown in Figure 21.6. The position of the point A is determined by a *position vector* **a** from a fixed reference point O. The direction of the straight line will be the same as that of another vector which we shall call **b**. The position of the point P on the straight line will be given by a position vector $t\mathbf{b}$ from A for some value of the parameter t. We can see from the diagram that the position vector of P from O is $\mathbf{r} = \mathbf{a} + t\mathbf{b}$. This is the *vector equation for the straight line* which is the locus of P with variation of the parameter t from negative values to the left of A, through zero at A to positive values to the right of A.

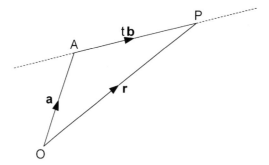

Figure 21.6. Straight line locus of P passing through A.

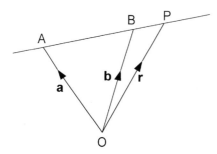

Figure 21.7. Straight line through two given points A and B.

If (x, y, z) and (x_1, y_1, z_1) are the coordinates of P and A, respectively, referred to Cartesian coordinates with origin at O, and the vector $\mathbf{b} = b_1\mathbf{i} + b_2\mathbf{j} + b_3\mathbf{k}$, then the vector equation for the straight line becomes:

$$x\mathbf{i} + y\mathbf{j} + z\mathbf{k} = x_1\mathbf{i} + y_1\mathbf{j} + z_1\mathbf{k} + t(b_1\mathbf{i} + b_2\mathbf{j} + b_3\mathbf{k}).$$

Equating the coefficients of $\mathbf{i}$, $\mathbf{j}$ and $\mathbf{k}$ in turn gives:

$$x = x_1 + b_1 t, \quad y = y_1 + b_2 t \quad \text{and} \quad z = z_1 + b_3 t$$

or $\dfrac{x - x_1}{b_1} = \dfrac{y - y_1}{b_2} = \dfrac{z - z_1}{b_3} = t,$

which correspond to the equations we derived in Section 20.9. Next, suppose that we wish to find the vector equation of a straight line passing through two given points A and B, as shown in Figure 21.7. If $\mathbf{a}$ and $\mathbf{b}$ are the position vectors of A and B, respectively, the vector from A to B is given by $(\mathbf{b} - \mathbf{a})$. Since the latter is directed along the straight line, the position vector $\mathbf{r}$ of another point P on the straight line is given by $\mathbf{r} = \mathbf{a} + t(\mathbf{b} - \mathbf{a})$ for some value of the parameter t. Hence, this is the vector equation for the straight line in this case.

Now, let us find a vector equation for a plane. The orientation of the plane may be described by two vectors $\mathbf{a}$ and $\mathbf{b}$ with different directions but both lying in the plane, as

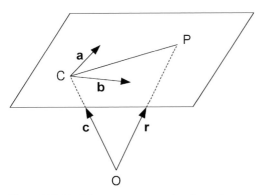

Figure 21.8. Deriving a vector equation for a plane.

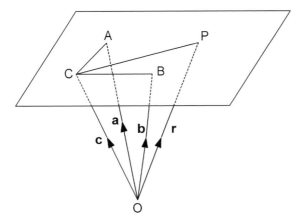

Figure 21.9. Plane through three non-collinear points A, B and C.

shown in Figure 21.8. We also need a fixed point C lying in the plane. In the diagram we have drawn the vectors **a** and **b** from C.

If P is any point in the plane, its position vector $\vec{CP}$ from C can be given in terms of the vectors **a** and **b** using two parameters s and t, i.e. $\vec{CP} = s\mathbf{a} + t\mathbf{b}$. Let the position vectors of C and P from a fixed reference point O be **c** and **r**, respectively. Then:

$$\mathbf{r} = \mathbf{c} + \vec{CP} = \mathbf{c} + s\mathbf{a} + t\mathbf{b},$$

which is a *vector equation for the plane*.

This equation leads us immediately to a vector equation for a plane which passes through three non-collinear points A, B and C, which have position vectors **a**, **b** and **c**, respectively, from the reference point O (see Figure 21.9). The position vector of P from C can be written in terms of the position vectors of A and B from C using parameters s and t:

$$\vec{CP} = s\vec{CA} + t\vec{CB}.$$

Then, $\vec{CP} = \mathbf{r} - \mathbf{c}, \quad \vec{CA} = \mathbf{a} - \mathbf{c} \quad$ and $\quad \vec{CB} = \mathbf{b} - \mathbf{c}.$

Hence, $\mathbf{r} - \mathbf{c} = s(\mathbf{a} - \mathbf{c}) + t(\mathbf{b} - \mathbf{c})$

or, $\mathbf{r} = \mathbf{c} + s(\mathbf{a} - \mathbf{c}) + t(\mathbf{b} - \mathbf{c})$

$= s\mathbf{a} + t\mathbf{b} + (1 - s - t)\mathbf{c}.$

21.3 Scalar product

Since vectors have both direction and magnitude, there is no obvious way of multiplying two vectors together. However, it has been found useful to define two different types of multiplication. One results in a scalar quantity and the other in a vector quantity. We shall deal with the former of the two in this section.

Suppose we have two vectors $\mathbf{a}$ and $\mathbf{b}$ with directions angle θ apart, as in Figure 21.10. Then, the *scalar product* is defined as:

$\mathbf{a}.\mathbf{b} = ab \cos\theta,$

which is the product of their magnitudes multiplied by the cosine of the angle between their directions. Since the scalar product is indicated by a dot between the vectors, it is often referred to as the *dot product*.

The first obvious property of the scalar product is that: $\mathbf{a}.\mathbf{b} = \mathbf{b}.\mathbf{a}$. Also, if the vectors are perpendicular: $\mathbf{a}.\mathbf{b} = 0$, since $\cos\theta = \cos(\pi/2) = 0$. Then, if $\mathbf{a}$ and $\mathbf{b}$ have the same direction: $\cos\theta = \cos 0 = 1$ and $\mathbf{a}.\mathbf{b} = ab$. If they are in opposite directions: $\cos\theta = \cos\pi = -1$ and $\mathbf{a}.\mathbf{b} = -ab$.

The above properties have the following implications:

$$a^2 = \mathbf{a}.\mathbf{a} = a^2, \quad \mathbf{i}^2 = \mathbf{j}^2 = \mathbf{k}^2 = 1 \quad \text{and} \quad \mathbf{i}.\mathbf{j} = \mathbf{j}.\mathbf{k} = \mathbf{k}.\mathbf{i} = 0.$$

We can now use the scalar product in forming an equation for a plane. Let ON be the perpendicular to the plane from the reference point O, N being a point on the plane (see Figure 21.11). If $\hat{\mathbf{n}}$ is the unit vector in the direction of N from O and the distance $ON = p$, then the vector $\vec{ON} = p\hat{\mathbf{n}}$

Let P be any other point on the plane and $\mathbf{r}$ be the position vector of P from O. Then the scalar product $\mathbf{r}.\hat{\mathbf{n}} = r\cos\theta = p$. Similarly, if $\mathbf{n}$ is any vector perpendicular to the plane, then $\mathbf{r}.\mathbf{n} = np = q$, say, and this is a *vector equation for the plane*. If P has Cartesian coordinates (x, y, z) with the origin at O and if $\mathbf{n}$ has Cartesian components

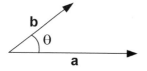

Figure 21.10. Finding the scalar product of two vectors.

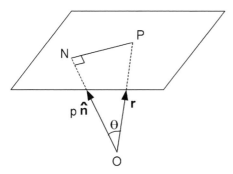

Figure 21.11. Deriving a vector equation for a plane.

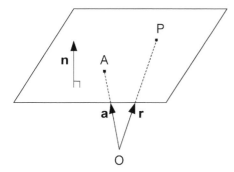

Figure 21.12. Another derivation of a vector equation for a plane.

(n_1, n_2, n_3), then:

$$\mathbf{r.n} = (x\mathbf{i} + y\mathbf{j} + z\mathbf{k}).(n_1\mathbf{i} + n_2\mathbf{j} + n_3\mathbf{k})$$
$$= n_1 x + n_2 y + n_3 z = q,$$

which is the standard equation for a plane in coordinate geometry.

Suppose we know that a vector $\mathbf{n}$ is perpendicular to the plane and that a point A on the plane has position vector $\mathbf{a}$ from O. Then, if P is any other point on the plane (see Figure 21.12) with position vector $\mathbf{r}$, the vector $(\mathbf{r} - \mathbf{a})$ from A to P must be perpendicular to $\mathbf{n}$. Consequently, the scalar product $\mathbf{n.(r - a)} = 0$ and the vector equation for the plane is: $\mathbf{n.r} = \mathbf{n.a}$.

The angle θ between two planes is the angle between vectors $\mathbf{n}$ and $\mathbf{m}$, say, which are perpendicular to the respective planes. Now, $\mathbf{n.m} = nm \cos \theta$ and therefore:

$$\theta = \cos^{-1}\left(\frac{\mathbf{n.m}}{nm}\right).$$

Note the correspondence between this and the equivalent coordinate geometry equation in Section 20.10.

In developing the equation for the plane $\mathbf{r.n} = np = q$, p was the distance of the reference point O from the plane. Hence, if the equation is given as $\mathbf{r.n} = q$, the distance

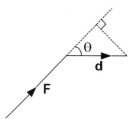

Figure 21.13. Finding the work done by a force.

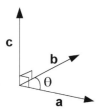

Figure 21.14. Finding the vector product $\mathbf{c} = \mathbf{a} \times \mathbf{b}$.

p of O from the plane is $p = q/n$. Also, if we have another parallel plane containing a point A with position vector $\mathbf{a}$ from O, the distance of O from the second plane must be $(\mathbf{a.n})/n$. It follows that the distance of A from the first plane is:

$$d = \left| \frac{q}{n} - \frac{\mathbf{a.n}}{n} \right| = \frac{|q - \mathbf{a.n}|}{n}.$$

Compare this with the corresponding equation in Section 20.10.

The work done by a force is defined as the magnitude of the force times the distance moved by its point of application resolved in the direction of the force. Hence, if the force is $\mathbf{F}$ and the distance moved is $\mathbf{d}$, the work $W = Fd \cos \theta = \mathbf{F.d}$ (see Figure 21.13). If $\mathbf{F}$ varies and its point of application moves along a path Γ which is not a straight line, we can still apply the scalar product between $\mathbf{F}$ and a corresponding element of displacement $d\mathbf{r}$, so that the *work done* becomes the line integral:

$$W = \int_{\Gamma} \mathbf{F}.d\mathbf{r}.$$

21.4 Vector product

Referring to Figure 21.14, the vector product of two vectors $\mathbf{a}$ and $\mathbf{b}$ is another vector $\mathbf{c}$ written as: $\mathbf{c} = \mathbf{a} \times \mathbf{b}$. By virtue of the $\times$ symbol, the vector product is sometimes called the cross product. The direction of $\mathbf{c}$ is perpendicular to both $\mathbf{a}$ and $\mathbf{b}$. If a right-hand thread screw is placed with its axis perpendicular to both $\mathbf{a}$ and $\mathbf{b}$ and turned from $\mathbf{a}$

round to **b** through the smaller angle θ, the screw will move in the direction of **c**. The magnitude of the product is $c = ab \sin \theta$.

The following relations follow from this definition.

$\mathbf{b} \times \mathbf{a} = -\mathbf{a} \times \mathbf{b}$ so, unlike the scalar product, the order matters.

If **a** and **b** are parallel, $\mathbf{a} \times \mathbf{b} = 0$.

If **a** and **b** are perpendicular, $|\mathbf{a} \times \mathbf{b}| = ab$.

$\mathbf{i} \times \mathbf{i} = \mathbf{j} \times \mathbf{j} = \mathbf{k} \times \mathbf{k} = 0$.

$\mathbf{i} \times \mathbf{j} = \mathbf{k}, \ \mathbf{j} \times \mathbf{k} = \mathbf{i}, \ \mathbf{k} \times \mathbf{i} = \mathbf{j}$.

$\mathbf{j} \times \mathbf{i} = -\mathbf{k}, \ \mathbf{k} \times \mathbf{j} = -\mathbf{i}, \ \mathbf{i} \times \mathbf{k} = -\mathbf{j}$.

$\mathbf{a} \times \mathbf{b} = (a_1 \mathbf{i} + a_2 \mathbf{j} + a_3 \mathbf{k}) \times (b_1 \mathbf{i} + b_2 \mathbf{j} + b_3 \mathbf{k})$

$\qquad = (a_2 b_3 - a_3 b_2)\mathbf{i} + (a_3 b_1 - a_1 b_3)\mathbf{j} + (a_1 b_2 - a_2 b_1)\mathbf{k}$.

A neat way of writing the last expression uses a *determinant*, which we must now define. A determinant is a certain combination of products of elements in a square matrix. Starting with a 2×2 matrix:

$$A = \begin{bmatrix} a_{11} & a_{12} \\ a_{21} & a_{22} \end{bmatrix},$$

the determinant is:

$$\det A = \begin{vmatrix} a_{11} & a_{12} \\ a_{21} & a_{22} \end{vmatrix} = a_{11}a_{22} - a_{21}a_{12}.$$

Moving on to a 3×3 determinant:

$$\begin{vmatrix} a_{11} & a_{12} & a_{13} \\ a_{21} & a_{22} & a_{23} \\ a_{31} & a_{32} & a_{33} \end{vmatrix} = a_{11} \begin{vmatrix} a_{22} & a_{23} \\ a_{32} & a_{33} \end{vmatrix} - a_{12} \begin{vmatrix} a_{21} & a_{23} \\ a_{31} & a_{33} \end{vmatrix} + a_{13} \begin{vmatrix} a_{21} & a_{22} \\ a_{31} & a_{32} \end{vmatrix}.$$

Now use this rule for the 3×3 determinant:

$$\begin{vmatrix} \mathbf{i} & \mathbf{j} & \mathbf{k} \\ a_1 & a_2 & a_3 \\ b_1 & b_2 & b_3 \end{vmatrix} = \mathbf{i}(a_2 b_3 - b_2 a_3) - \mathbf{j}(a_1 b_3 - b_1 a_3) + \mathbf{k}(a_1 b_2 - b_1 a_2)$$

$$= (a_2 b_3 - a_3 b_2)\mathbf{i} + (a_3 b_1 - a_1 b_3)\mathbf{j} + (a_1 b_2 - a_2 b_1)\mathbf{k} = \mathbf{a} \times \mathbf{b}.$$

Hence, $\mathbf{a} \times \mathbf{b} = \begin{vmatrix} \mathbf{i} & \mathbf{j} & \mathbf{k} \\ a_1 & a_2 & a_3 \\ b_1 & b_2 & b_3 \end{vmatrix}$.

The vector product is useful in mechanics for expressing the moment or turning effect of a vector such as force or momentum. For instance, suppose we have a force **F** acting at a point P, which in turn has a position vector **r** from a reference point O (see Figure 21.15). To investigate the moment of **F** about an axis through O perpendicular to

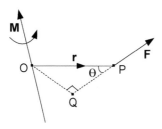

Figure 21.15. Moment $\mathbf{M} = \mathbf{r} \times \mathbf{F}$.

both $\mathbf{r}$ and $\mathbf{F}$, we must imagine our diagram to be in three dimensions. The magnitude of the moment is:

$$F \times (OQ) = F \times (OP)\sin\theta = |\mathbf{r} \times \mathbf{F}|.$$

Then the right-hand thread rule allows us to express the moment as a vector:

$$\mathbf{M} = \mathbf{r} \times \mathbf{F}.$$

Notice that $\mathbf{r}$ must precede $\mathbf{F}$ in this expression.

22 Two more topics

22.1 A simple differential equation

The word 'simple' refers to the method which may sometimes be used to solve a first order differential equation in which the variables are separable. The differential equation takes the form:

$$\frac{dy}{dx} = f(x, y) = F(x)G(y).$$

Then, $\dfrac{1}{G(y)}\dfrac{dy}{dx} = F(x)$ and $\displaystyle\int \frac{dy}{G(y)} = \int F(x)\,dx.$

If these two integrals may be found analytically, then we can derive an algebraic relationship between y and x.

EXAMPLE

For constants g and b, let:

$$\frac{dv}{dt} = g - bv, \text{ then, } \int \frac{dv}{g - bv} = \int dt,$$

$$-\frac{1}{b}\ln(g - bv) = t + C, \ \ln(g - bv) = K - bt,$$

$$g - bv = e^{K-bt} = e^K e^{-bt} = Ae^{-bt}, \ v = \frac{1}{b}(g - Ae^{-bt}),$$

where C, K and A are constants.

EXAMPLE

For constants g and a, let:

$$v\frac{dv}{dx} = g - av^2, \text{ then, } \int \frac{v}{g - av^2}\,dv = \int dx,$$

$$-\frac{1}{2a}\ln(g - av^2) = x + C, \ \ln(g - av^2) = K - 2ax,$$

$$g - av^2 = Ae^{-2ax}, \quad v^2 = \frac{1}{a}(g - Ae^{-2ax}),$$

where C, K and A are constants.

22.2 Hyperbolic sines and cosines

Often exponentials e^a and e^{-a} combine to form an expression: $(e^a \pm e^{-a})/2$. Each of these is given a particular name; with the positive sign, it is called a hyperbolic cosine and with the negative sign, a hyperbolic sine. Hence:

$$\cosh a = \frac{e^a + e^{-a}}{2} \quad \text{and} \quad \sinh a = \frac{e^a - e^{-a}}{2}.$$

EXAMPLE

The differential equation $\ddot{x} - \omega^2 x = 0$ has a solution of the form:

$$x = Ae^{\omega t} + Be^{-\omega t}.$$

If the initial conditions are: $x = x_0$ and $\dot{x} = 0$ when $t = 0$, then $A + B = x_0$ and $\omega(A - B) = 0$. Therefore, $A = B = x_0/2$. Thus, the solution is:

$$x = \frac{e^{\omega t} + e^{-\omega t}}{2} x_0 = x_0 \cosh \omega t.$$

Appendix: answers to problems in Part III

1. 19.1°.
2. 10.9°.
3. $R = 2.457$ N, $\theta = 21.85°$.
4. $\theta = 53.1°$, $\phi = 36.9°$.
5. $W_2 = 14.14$ N, $W_3 = 27.32$ N.
6. $T = W$, $F = \sqrt{2}W$.
7. $R_1 = \sqrt{3}W/2$, $R_2 = W/2$, $R_3 = 2W/\sqrt{3}$, $R_4 = 5W/(2\sqrt{3})$.
8. $T = \sqrt{2}W$, $R = W/\sqrt{2}$.
9. (a) $Ta\cos\alpha$, (b) $2Wa\sin\alpha$.
10. 1.005 kN m.
11. $R = 66.4k$, $\theta = 70.3°$, $a = 3.72$ m.
12. $R = 3.8 \times 10^6$ N, $\phi = 79.4°$, $a = 18.7$ m.
13. $T_1 = 5W/4$, $T_2 = 3W/4$.
14. 100 N vertically downwards and distance $2a$ on the opposite side of A from B, where $AB = 4a$.
15. (a) $P = 4F/5$, $Q = F/5$. (b) $P = 2F$, $Q = -F$. (c) $P = F$, $Q = 3F/2$.
16. $R_c = W$, $R_a = -R_b = aW/b$.
17. $T = 7.07$ kN, $R = 5.10$ kN, $\theta = 101.3°$.
18. $T = 406$ N, $R_a = 1.203$ kN, $R_b = 351$ N.
19. $P = 167$ N, $P = 167$ N.
20. $F_g = 24.5$ kN, $T = 49$ kN.
21. $x_c = 15$ cm, $y_c = -5$ cm.
22. $AC = 0.6$ m, $T_a = 400$ N, $T_b = 600$ N.
23. $x_c = -10.8$ cm, $y_c = 4.4$ cm, $z_c = 5.2$ cm.
24. $x_g = 23.3$ cm, $y_g = 13.3$ cm.
25. $x_g = y_g = 3.44$ cm.
26. $x_g = 1.82$, $y_g = 0.91$.
27. $x_g = 0.447a$, $y_g = 0.516a$.
28. (a) $x_g = a/2$, (b) $x_g = a/3$.
29. 7.5 cm.
30. 17 kN/m.
31. 15 kN/m, 7.5 kN.
32. 4.91 kN.
33. 16.8 kN.
34. $V(w - w_s)$, 1/8.
35. 90.6 cm³.
36. 52.32 kN, 1.25 m.

37. 37.28 kN, 1.37 m.
38. AF, tension P; AC, tension $\sqrt{2}P$; BC, compression $2P$.
39. AB, compression L; AD, compression $L/\sqrt{2}$; DE, tension $3L/\sqrt{2}$.
40. DC, tension L; BC, compression $\sqrt{2}L$; ED, tension L; DB, compression L; EB, tension $L/\sqrt{2}$; AB, compression $3L/\sqrt{2}$.
41. AB, tension $4L/\sqrt{3}$; AE, compression $8L/3$; BC, tension $\sqrt{3}L$; BE, tension $2L/\sqrt{3}$; CD, tension $5L/\sqrt{3}$; CE, tension $4L/\sqrt{3}$; DE, compression $10L/3$.
42. Tensions: $ad = 2\sqrt{2}L$, $af = \sqrt{2}L$, $ag = L$, $ef = L$.
    Compressions: $be = 2\sqrt{2}L$, $bg = \sqrt{2}L$, $cd = 2L$, $de = L$, $fg = L$.
43. Tensions: $cd = 5L/4\sqrt{3}$, $cf = 7L/4\sqrt{3}$, $de = L/2\sqrt{3}$.
    Compressions: $ad = 5L/2\sqrt{3}$, $bf = 7L/2\sqrt{3}$, $ef = L/2\sqrt{3}$, $eg = \sqrt{3}L/2$.
44. $13\frac{1}{3}$ kN between B and D, $13\frac{1}{3}$ kNm at D.
45. $2\frac{2}{3}$ kN between F and D, 2 kN m at D.
46. $5wa/3$ at C, $8wa^2/9$ at distance $4a/3$ from A.
47. 2.52 kN m.
48. $F = 100x^2 + 300x - 750$, $\quad M = -100x^3/3 - 150x^2 + 750x$. Max. $M = 680$ Nm at $x = 1.623$ m.
49. $\tan(\alpha/2)$.
50. 0.436, 5.59 N.
51. $\mu < (b/a - \tan\alpha)/2$.
52. $\mu \geq 1/3$.
53. $\theta = \lambda$ and minimum $P = W\sin(\alpha - \lambda)$.
54. $a = 0.625l$.
55. $\lambda = (\pi - 2\theta)/4$.
56. (a) 249 N m, (b) 240 N m.
57. 270 N $\leq P \leq 592$ N.
58. $(2, -1, 1)$ cm.
59. $F_c = 141.5$ N, $\theta = 122°$.
60. $2Fa(\mathbf{i} + \mathbf{j} + \mathbf{k})$.
61. $(-107.4\mathbf{i} + 11.1\mathbf{j} + 35.1\mathbf{k})$ N m.
62. $\mathbf{F}$ with couple $\mathbf{C}_f = 5(\mathbf{i} + 2\mathbf{j} - 2\mathbf{k})/3$ through Q, $\mathbf{r}_q = 2.08\mathbf{i} + 3.83\mathbf{j}$.
63. $\mathbf{F} = (391\mathbf{i} + 212\mathbf{j} - 89\mathbf{k})$ N, $\mathbf{C} = (-391\mathbf{i} + 491\mathbf{j} - 179\mathbf{k})$ N m.
64. $\mathbf{F} = (432\mathbf{i} + 660\mathbf{j} - 746\mathbf{k})$ N, $\mathbf{C} = (725\mathbf{i} - 648\mathbf{j} - 200\mathbf{k})$ N m.
65. $\mu = \tan\theta$.
66. $T_1 = 107.3$ N, $T_2 = 70.7$ N, $R_x = -54.9$ N, $R_y = 18.3$ N, $R_z = 50.0$ N.
67. $W_g = 200$ J, $W_f = -100$ J.
68. 300 J.
69. $\frac{\pi}{2}$ J.
70. 3.02 J.
71. $\theta = 90° - 2\alpha$.
72. $\theta = 25.5°$.
73. $P = W/6$.
74. $T = 2w$.
75. Compression: $\sqrt{2}W + 3(\sqrt{2} + 1)w$.
76. $\theta = \tan^{-1}[(W + w)/(4w)]$, stable.
77. Stable if $a > 10$ cm.
78. $f = 2$ m/s^2, $x = 144$ m.

79. $v_c = 24$ m/s.

80. $T = \pi$ s, $a = 0.4$ m.

81. 0.36 s.

82. $x = r \cos \omega t$, $\dot{x} = -r\omega \sin \omega t$, $\ddot{x} = -r\omega^2 \cos \omega t$.

83. 995.

84. $v = 18.85$ m/s, $\theta = 21.8°$.

85. $\theta = 72°$, 52.6 s.

86. $v = 5$ km/h, $d = 30$ m, $a = 32$ m.

87. 0.98 km.

88. $|\dot{\mathbf{v}}| = 3.04$ m/s^2, $\theta = 51.34°$.

89. $a_t = 0.111$ m/s^2, $a_n = 0.0185$ m/s^2.

90. $T = 13.2$ kN.

91. 16.6 kN.

92. Horizontal line: $y = -v^2/(2g)$.

93. (a) $v = 2.205$ m/s, (b) $t = 0.907$ s.

94. 33.7 s.

95. $\mu = 0.5$.

96. $k = 3.92$ kN/m, $\sqrt{g/x_0}/2\pi = 7.05$ Hz.

97. $a = 0.36$ m.

98. $\mu \geq 0.408$.

99. 0.102 m.

100. $\theta = 132°$.

101. 3.13 m/s $< v < 4.80$ m/s.

102. $\alpha = 38.66°$, $V = 70.9$ m/s, $T = 9.04$ s.

103. $\alpha = 34.45°$, $R = 279$ m, $V = 54.1$ m/s.

104. $\dot{v} = 3.63$ m/s^2, $T = 0.424 N$.

105. $x = x_0 \cosh(\sqrt{g/2l}\,t)$.

106. $x = (a \sin \theta)[1 - (a \cos \theta)/\sqrt{b^2 - a^2 \sin^2 \theta}]$, $y = \sqrt{b^2 - a^2 \sin^2 \theta} - a \cos \theta$.

107. 0.577 rad/s.

108. $v_b = 1.732$ m/s, $\beta = 30°$, $v_a = 1.85$ m/s.

109. 7 m/s.

110. 10.5 m/s, 13.2 rad/s.

111. 4 kg dm^2.

112. $M(a^2 + b^2)/6$.

113. $M(7 + 8a^2)a^2/35$.

114. 2.8 m/s^2, $T_1 = 14$ N, $T_2 = 12.6$ N.

115. $\dot{\theta} = 14$ rad/s. (a) $\theta = \pi/4$, $R_h = 9.8$ N. (b) $\theta = \pi/2$, $R_v = 22.9$ N.

116. $\ddot{x} = 8g/25$, $T_1 = 28Mg/25$, $T_2 = 30Mg/25$, $T_3 = 34Mg/25$.

117. $\ddot{x} = (4g/7) \sin \alpha$, force is tension, $(Mg/7) \sin \alpha$.

118. $v = 15$ m/s.

119. $V_1 = 1.037$ m/s, $V_2 = 1.114$ m/s.

120. $V = 1.22v$, $\alpha = 25.3°$.

121. $D = 19$ m, $T = 12.1$ s.

122. $v_1 = 0.679u$, $v_2 = 1.2u$, $\phi_1 = 132.6°$, $\phi_2 = 46.4°$.

123. $h = 7a/5$.

124. $I = Mv/4$, $\omega = 3v/(4a)$.

125. $b = a/2$.

126. $b = \sqrt{2}a/3$.
127. $\omega_1 = \omega_0/7$.
128. $\omega_1 = 3.79$ rad/s.
129. $v = -49.1$ m/s, $\omega = 17.9$ rad/s.
130. $v = -48.0$ m/s, $w = 1.48$ m/s, $\omega = 22.2$ rad/s.
131. 10 m/s.
132. 7 m/s.
133. 1.53 m/s.
134. 10 m/s.
135. 41.74 km/h.
136. 186 N, 0.341 m/s^2.
137. $3M(a\omega)^2/4$, $4M(a\omega)^2/3$.
138. $\omega = 8.32$ rad/s.
139. 0.7 m/s.
140. $\theta = 55.1°$.
141. $v = 0.916u$.

Index